# Ecology of Populations

The theme of the book is the distribution and abundance of organisms in space and time. The core of the book lies in how local births and deaths are tied to emigration and immigration processes, and how environmental variability at different scales affects population dynamics with stochastic processes and spatial structure, and it shows how elementary analytical tools can be used to understand population fluctuations, synchrony, processes underlying range distributions, and community structure and species coexistence. The book also shows how spatial population dynamics models can be used to understand life history evolution and aspects of evolutionary game theory. Although primarily based on analytical and numerical analyses of spatial population processes, data from several study systems are also dealt with.

ESA RANTA is Professor in Animal Ecology at the University of Helsinki, Finland.

PER LUNDBERG is Professor in Theoretical Ecology at the Lund University, Sweden.

VEIJO KAITALA is Professor in Population Biology at the University of Helsinki, Finland.

Ecology of Populations

ECOLOGY, BIODIVERSITY, AND CONSERVATION

*Series editors*
Michael Usher    *University of Stirling, and formerly Scottish Natural Heritage*
Denis Saunders    *Formerly CSIRO Division of Sustainable Ecosystems, Canberra*
Robert Peet    *University of North Carolina, Chapel Hill*
Andrew Dobson    *Princeton University*

The world's biological diversity faces unprecedented threats. The urgent challenge facing the concerned biologist is to understand ecological processes well enough to maintain their functioning in the face of the pressures resulting from human population growth. Those concerned with the conservation of biodiversity and with restoration also need to be acquainted with the political, social, historical, economic and legal frameworks within which ecological and conservation practice must be developed. This series will present balanced, comprehensive, up-to-date and critical reviews of selected topics within the sciences of ecology and conservation biology, both botanical and zoological, and both 'pure' and 'applied'. It is aimed at advanced final-year undergraduates, graduate students, researchers, and university teachers, as well as ecologists and conservationists in industry, government and the voluntary sectors. The series encompasses a wide range of approaches and scales (spatial, temporal, and taxonomic), including quantitative, theoretical, population, community, ecosystem, landscape, historical, experimental, behavioural, and evolutionary studies. The emphasis is on science related to the real world of plants and animals, rather than on purely theoretical abstractions and mathematical models. Books in this series will, wherever possible, consider issues from a broad perspective. Some books will challenge existing paradigms and present new ecological concepts, empirical or theoretical models, and testable hypotheses. Other books will explore new approaches and present syntheses on topics of ecological importance.

*Ecology and Control of Introduced Plants*
Judith H. Myers and Dawn R. Bazely

*Invertebrate Conservation and Agricultural Ecosystems*
T. R. New

*Risks and Decisions for Conservation and Environmental Management*
Mark Burgman

*Nonequilibrium Ecology*
Klaus Rohde

# Ecology of Populations

ESA RANTA
*University of Helsinki*

PER LUNDBERG
*Lund University*

VEIJO KAITALA
*University of Helsinki*

CAMBRIDGE
UNIVERSITY PRESS

# CAMBRIDGE
## UNIVERSITY PRESS

University Printing House, Cambridge CB2 8BS, United Kingdom

One Liberty Plaza, 20th Floor, New York, NY 10006, USA

477 Williamstown Road, Port Melbourne, VIC 3207, Australia

314-321, 3rd Floor, Plot 3, Splendor Forum, Jasola District Centre, New Delhi - 110025, India

103 Penang Road, #05-06/07, Visioncrest Commercial, Singapore 238467

Cambridge University Press is part of the University of Cambridge.

It furthers the University's mission by disseminating knowledge in the pursuit of education, learning and research at the highest international levels of excellence.

www.cambridge.org
Information on this title: www.cambridge.org/9780521854351

First published 2006

*A catalogue record for this publication is available from the British Library*

ISBN 978-0-521-85435-1 Hardback
ISBN 978-0-521-67033-3 Paperback

# Contents

# Preface

The ecology of populations is the study of the patterns of distribution and abundance of organisms. It also goes beyond mere description and seeks the evolutionary forces that might produce such patterns, and their ecological constraints. This book is on the ecology of populations but it is not a population ecology textbook. Therefore, there will be much standard textbook material left out. This book does not attempt to establish a new field, or summarize or synthesize an old one. Neither does it provide the student of population biology with all the necessary tools for further exploration, nor does it review the entire discipline, or parts of it. So what does this book do? As with most books, it presents an idiosyncratic world-view. We hope that some well-known problems and phenomena are getting a fresh and novel approach. We also hope that applying basically the same analytical tool to a number of seemingly disparate problems in population biology will be convincing enough to make others do likewise. By using rather simple models of population change to a large number of problems, we hope that conceptual unification will be promoted. Science becomes more and more specialized with the risk of losing track of the bigger picture. Although no bigger picture is presented here in a coherent way, the approach we have taken to address problems in population ecology aims at getting to that more synthetic understanding. We start by investigating simple and well-known models of population change and the extent to which they are reasonable representations of the population phenomena we can observe in natural or laboratory systems. By doing so, we also emphasize the link between evolutionary change, ultimately altering the life history of organisms, and changes in population size in time and space. The spatial dimension in population ecology is such a fundamental one that we have cast many of the problems in one spatial setting or another. In fact, simple assumptions about local births and deaths coupled by the movement of individuals across space (i.e., immigration and emigration) make so much sense, and has indeed the power of explaining a wide range of problems from life history evolution to the management of harvested populations. Including more than one species (or life history strategy), i.e., to move into the field of community ecology, is not a particularly big step. Approaching this wide range of problems with these methods, we believe, is a fruitful way of doing population biology.

Although we have tried to show that the theoretical constructs dealt with here also have a lot to do with actual phenomena in nature, we have not emphasized examples and empirical evidence. Rather, this is more a book on how to approach scientific problems in (evolutionary) population ecology. It does not say anything on how to test models, or how to build models. It emphasizes that it is possible to use rather simple building blocks, the per capita rates of births, deaths, immigration, and emigration, to model and therefore hopefully better understand population processes from an ecological and evolutionary point of view.

Doing theoretical ecology is a hazardous cruise between the formal and rigorous mathematical Scylla, and the overwhelmingly rich empirical Carybdis. Too many equations and too few examples make an empiricist suspicious, uneasy, and sometimes even hostile. Too little rigor and a good deal of careless talking of things we do not *really* understand make the mathematician bored and a disbeliever as well. So, the best pedagogic trick would then be, of course, to show how beautifully the models and theories match the data ecology so painfully has collected over so many years. But our purpose here is not to show that theory works according to some suitable standard. We want to show that if some of the very basic and easily accessible concepts in ecology are taken seriously, and adding some spatial structure, much of what we can observe of the distribution and abundance of organisms can be explained and understood. If we also leave the deterministic world-view and let the processes of concern be the stochastic variables they inevitably are, then even more is gained.

This book should thus be read as a starting point for further explorations rather than a summary of the field. We have probably erred when omitting much of the classic and rather far-reaching knowledge in many of the areas covered here. This is both a consciously biased selection and a flaw in our knowledge and thinking. Hopefully the novelty and in some parts state of the art material compensate for that.

The self-evident, yet joyful fact that science is a collective process does not stop us from acknowledging the contributions from a number of people and organizations to this and other projects of ours. The whole idea of writing this book emerged at a meeting in Møls, Denmark, at a workshop funded by the Nordic Academy for Graduate Teaching

(NorFA) in 1996. We organized a number of workshops and symposia after that, and all the students and colleagues contributing to those meetings have also contributed to this book in various ways. That tradition continued as we launched a network of researchers and graduate students funded by the Finnish Academy (the MaDaMe program). The financial support from NorFA and the Finnish Academy (ER, VK), the Spatial Ecology Program (Ecology and Evolutionary Biology, University of Helsinki), Center of Excellence in Evolutionary Ecology (Department of Biological and Environmental Science, University of Jyväskylä (VK)), the Swedish Research Council (PL), and the Swedish FORMAS Research Council (PL) has been crucial for our joint efforts. Parts of this work were conducted while PL was a Sabbatical Fellow at the National Center for Ecological Analysis and Synthesis, a Center funded by NSF (Grant #DEB-0072909), the University of California, and the Santa Barbara campus. We owe all students and colleagues that have participated in those activities a heartfelt thank-you. People in our immediate neighborhoods have been wonderful friends, colleagues, inspirers and teachers, and hence important throughout the production of this book. Others have influenced our scientific journey more subtly. Many thanks to Sami Aikio, Susanna Alaja, Joel Brown, Katja Enberg, Torbjörn Fagerström, Anna Gårdmark, Mikko Heino, Bob Holt, Peter Hudson, Jouni Laakso, Harto Lindén, Niclas Jonzén, Jörgen Ripa, Nils Christian Stenseth, Stuart Pimm and Peter Turchin. Finnish Game and Fisheries Research Institute has allowed us to use their long-term records of game animals. The Nordic Centre of Excellence Programme by the "Nordiska Ministerrådet" supported the final phase of the manuscript preparation.

The entire manuscript of the book was read and commented upon by Tim Benton and Hannu Pietiäinen. Jordi Bascompte, Mikko Heino, Stuart Humphries, Niclas Jonzén, Hanna Kokko, Jouni Laakso, Elina Lehtinen, Kate Lessells, Jan Lindström, Jari Niemilä, Jörgen Ripa, Graeme Ruxton, Richard Solé, Bill Sutherland, and Erik Svensson have read one or several chapters. We are very grateful to all these people for insightful comments and suggested corrections. We have probably taken their advice too lightly and – of course – all errors or omissions are ours alone.

Finally, we are indebted to our close friends and families for letting us do what we are doing, at times even supporting it!

# 1 · *Introduction*

> ... there is nothing so practical as good theory.
>
> *Richard Feynman*

The scope of this book is almost as wide as it gets. It touches upon a range of topics in ecology and evolution found in many modern textbooks. Instead of going into considerable depth in any one topic, we have chosen to cover quite a few in order to show that the same basic (and well-known) tools are applicable to a wide variety of ecological and evolutionary problems in population biology. However, this is also a narrow-minded book in the sense that it is very "theoretical," i.e., full of mathematical expressions and computer simulation results. We believe ecology becomes a healthier science if it appreciates and acknowledges its strong quantitative and more rigorous nature. It is also narrow-minded in the sense that it reflects our own interests in population ecology without attempting to cover all aspects of the ecology of populations. Yet, the scope remains wide and possibly shallow. We believe that ecology and evolutionary biology have to become far more integrated than the fragmented and disparate impression they give today. We think that this can be done by going back to very simple first principles of births and deaths, immigration, and emigration. From those "simple" entities, we can derive virtually everything that plants and animals do in nature. To do so, however, requires a common thread of theory, the seeds of which at least we believe exist. Extensions of that theory will also be dealt with in this book. The second reason for a wide scope is to show how theory and data can be closely integrated, at least in some (rather important) areas of ecology, and that this integration often is useful for the application of the science of ecology. Although the scope is deliberately wide, there are obvious restrictions and biases involved in our endeavor. For example, plants play a smaller role than animals. In particular, large mammals and birds dominate the case stories and examples. The only excuse for this bias is that we are especially used to those organisms and the literature that covers their biology. We nevertheless think that there is a great deal of generality to be found in our approach. However, one can easily imagine a number of scientific problems with a rather limited selection of organisms as examples and templates for more theoretical considerations. Microorganisms and many plants are, of course, very different from most animals on a number of grounds (e.g., modular structure versus well-defined individuality, reproductive modes and life cycles, and mobility). Even so, and of course depending on exactly in which unit we choose to measure the presence of the organism, the basic and almost trivial relationship is

$$N(t+1) = N(t)(1 + b + i - d - e), \tag{1.1}$$

where $N$ is the abundance of a population at time $t$, $b$ and $d$ are per capita births and deaths, respectively, while $i$ and $e$ are per capita immigration and emigration rates during one time interval. Our task as ecologists and evolutionary biologists is to figure out what determines $b$, $i$, $d$, and $e$, and the dynamical and evolutionary consequences of them.

Our point of departure is hence the simple *renewal function* above that maps the state of the population at one point in time to another. That is, we will – almost without exception – assume that it is possible to read off the state of the population at one point in time and do it again at some other point. The time interval between the observations is in principle arbitrary, but usually matches some natural biological interval, e.g., the sequence of reproductive events. By "state" of the population, we usually understand population size or density. This is of course a very restrictive definition that will be relaxed depending on circumstances. For example, in Chapters 6 and 12, age structure and sex are introduced making "state" more interesting and sophisticated.

Another reason why we have chosen a discrete time approach is that much of the available population and community data come in this form: population size, dispersal, or gene frequencies are measured in discrete time intervals. This does not mean that the time intervals in data are always biologically adequate; for example, many populations are measured once a year for practical rather than biological reasons. This potential mismatch between observation interval and biologically more meaningful sequences is an interesting empirical and applied problem in itself, probably deserving more attention than it usually gets. The format data come in also has implications for the (statistical) analysis of them. Most theory of stochastic processes (e.g., time series analysis, something we will make frequent use of throughout) is based on discrete (random) events. The correspondence between theory and data will consequently be both obvious and close.

Generally, spatial processes will also be viewed as discrete processes. This usually comes less naturally. Although it is almost invariably true that the environment that natural organisms inhabit is heterogeneous, at least at some appropriate scale, the different parts of it are less obviously discrete entities. Much spatial ecology theory nevertheless ignores the fact that landscape or habitat elements really have poorly defined identities and borders. It turns out, however, that this is rarely a problem, both for practical and theoretical purposes. As shall be seen in later chapters (e.g., Chapters 3 and 4), the distinct patches or habitats assumed in the theoretical constructs and models are indeed fair representations of real landscapes when it comes to understanding population and community processes. Partly, this is because many spatial processes are in fact scale invariant in the spatial dimension (see, e.g., Chapter 5).

The way we most often express the renewal functions is hence in discrete time (difference) equations. In Chapter 2 we analyze the discrete time population processes in more detail. As a preamble, consider Box 1.1 for the general discrete time mapping. We also refer to more general textbooks, e.g., Edelstein-Keshet 1988, Roughgarden 1998 and Caswell 2001, for thorough treatments of such processes. Our intention is, however, that most of the material in this book is self-contained. We do not expect the reader to be especially dependent on additional information from other, more technical sources. For obvious reasons, however, many of the topics covered cannot be dealt with in great detail and depth. Therefore, we expect the reader to be familiar with the advanced under-graduate or graduate level of population and community ecology and the associated theoretical and mathematical (and statistical) tools. This does not mean that the somewhat less initiated reader should be left hanging in the air. We have tried to accommodate that by avoiding overly technical jargon, by letting some of the technical problems appear in boxes outside the main text, and by ample references to the literature where more in-depth treatments of technical matters are found (see also the Suggested reading list at the end of this chapter).

## Fitness

The question often arises whether there is really any strong connection between classic population and community dynamics and evolutionary processes. The two branches of evolutionary ecology are often seen as separate which is very unfortunate and misleading. Of course they aren't. Consider the simple renewal processes

**Box 1.1** · *Discrete time mapping*

We are interested in finding out how the number of individuals (or some other adequate entity) is changing from one arbitrary point in time, $t$, to another, $t+1$. Note that the indexing is also arbitrary. Instead of mapping from time $t$ to $t+1$, we can do it from $t-1$ to $t$ (as long as we only deal with one (forward) time step at a time). Let the population size at time $t$ be denoted by $N(t)$ and we then have

$$N(t+1) = f[N(t)],$$

where $f$ is some yet unspecified function. Hence we assume that the population size at future times is dependent on the population size at previous times. Suppose the function $f$ now takes the form

$$f[N(t)] = \frac{\lambda N(t)}{1 + bN(t)}.$$

This is a monotonically increasing function of $N$ determined by two parameters, $\lambda$ and $b$. This is the *renewal function* for the population process (Box 2.1). Suppose now that there is a situation such that $N(t+1) = N(t)$. Denote that population density $N^*$, and we have

$$N^* = \frac{\lambda N^*}{1 + bN^*}.$$

We can now solve this equation for $N^*$ and, after some algebra, we have

$$N^* = \frac{\lambda - 1}{b}.$$

This is the equilibrium population size, i.e., the size at which there is no change from one time to another. That value of $N$ can be illustrated by plotting the function $f$ in the $N(t) - N(t+1)$ plane, and is the point where $f$ is intersecting a straight line with slope 1 in the plane. The slope of $f$ in that point is important because it determines the stability of the equilibrium (see, e.g., Edelstein-Keshet 1988, for a more rigorous and detailed treatment of both general discrete time mapping and its application in population biology).

$$N(t+1) = RN(t), \tag{1.2}$$

where $N$ is some relevant ecological quantity, e.g., population density, and $R$ is some (positive) constant. The rate by which the population density (or size) changes is thus determined by $R$ (cf. eq. 1.1). The exact nature of $R$ is of instrumental importance in population ecology: why is $R$ sometimes large, sometimes small for a given population, why does it vary among populations, and how are, for example, spatial structure and other species in the environment affecting its magnitude and variation? Expressed in this traditional population dynamics ways, $R$ is rather unambiguous – it is (when log-transformed and given the symbol $r$) the finite rate of increase of the population. This is precisely why it appears as a measure of *fitness* (with identical meaning) in the Euler–Lotka equation (here in the discrete time version)

$$1 = \sum l_x m_x e^{-rx}, \tag{1.3}$$

where $l$ and $m$ are age-specific (at age $x$) fecundity and survival, respectively. This is indeed the key equation in all evolutionary ecology. Murray (2001) argues rightly that it has the status of a "law" because it encompasses both evolutionary and ecological change. Evolutionary change in that a trait or strategy (see e.g., Cohen *et al.* 1999 for strategy definitions) that maximizes $r$ in eq. 1.3 is the strategy that will be the evolutionarily most successful one, i.e., by definition having the highest fitness. Should $r$ or $R$ take on values such that eq. 1.3 is no longer satisfied, *ceteris paribus*, then that indicates a population increase or decline. Should we for some reason ignore age (or stage) structure or the entire strategy space (individual variation within the population), then all that is left is population change and we have recovered population dynamics. In the following, we will occasionally slide between the ecological and evolutionary domain, always trying to keep the Euler–Lotka theory in mind.

The heir of Euler–Lotka theory is what is sometimes referred to as adaptive dynamics (Dieckmann and Law 1996; Dieckmann 1997; Abrams 2001). It is a an even more explicit way of incorporating both ecological and evolutionary dynamics within the same theoretical framework. It is a means of characterizing the entire (or at least most of the relevant components) "feedback environment" of an organism (e.g., Heino *et al.* 1998). By doing so, both ecological change (change in abundance and distribution) and evolutionary change (the changes of traits or strategies in time and space) are explicit parts of the analysis. We have refrained from expanding our treatise to include this theoretical approach simply because the focus here is indeed on abundance and distribution, although we will break that rule in later chapters (e.g., in Chapter 11 and 12).

## Ecology of populations

At the very beginning, we stated that this book has a very wide scope, but are now down to a much more narrow and limited one – classic "population dynamics." This is not as restricted as it may seem. It is in fact the study of populations – their distribution and abundance (Andrewartha and Birch 1954) – that is necessarily the core of ecology and evolutionary biology. It is of course true that the individual (or some appropriate similar concept), or even the gene, is the actual scale at which evolution seemingly operates such that those units are the ones that are selected. For both theoretical and practical purposes, however, it is the population level that is the relevant one for our study of the manifestation of evolutionary change; it is at this level that the manifestation of life itself takes place, namely the births and deaths of more than one individual (or gene). This collection of individuals is the population, however we choose to define it more precisely (Berryman 2002). Hence the interesting, measurable, and practical, e.g., in the application of ecology for management or conservation purposes, processes are apparent at this level. Conversely, this approach does not of course preclude the study of biology at all other levels of organization. Anything from molecular biology to ecosystem research will reveal and generate useful biological knowledge. The most obvious and relevant arena for all life is, however, the population – the scale at which molecular processes and vast ecosystems coalesce. Hence the approach taken in this book.

## Theory and data in ecology

Ecology is an empirical science and is therefore ultimately data driven. But it is so only to the extent that we want to explain what is observable rather than there being an unambiguous truth in data. Data, or more correctly, any set of observations of pattern and processes in nature, only get their meaning when interpreted. Theory is what provides conceptual and analytical tools to do that. Strangely, this is rarely an attitude delivered in most undergraduate (or even graduate) teaching in ecology. We strongly advocate the fundamental role of theory not only in the loose and perhaps trivial sense, but as it is formulated in mathematical terms. This is because mathematics provides us with the most powerful analytical tools when things start becoming complicated (as in ecology!) and our intuitive capacity no longer keeps up with that complexity. Also, and perhaps most importantly, a strong theoretical and mathematical foundation of our activities makes them *useful* when we are asked to solve, e.g., conservation or management problems. Qualitative statements or suggestions may be a good start, but can neither replace nor be as operational as quantitative ones, albeit with perhaps disturbingly large confidence limits.

In the following, we are going to transgress the data–theory border as much as possible. That means we are going to be inspired by intriguing patterns we can observe, and by data that have not been satisfactorily explained. Likewise, it means that we are going to analyze old data in new ways, as well as hopefully inspire others to collect the data that theory may indicate are important or interesting. However, this is not a book on applied mathematics or statistics – that is not our intention or within reach of our competence. Instead, we will refer as much as possible to the literature that does a better technical or more rigorous mathematical or statistical job. Much of the data we use, or produce by simulations, are time series of population abundance, density, or some index of it. Such data do, of course, have obvious limitations; they are very "shallow" and are often uninformative. The time series approach is therefore both simplistic and also challenging. Much of the information about population change does indeed come in the form of time series so we do need the tools to analyze them. Time series data are also very inspiring for anyone interested in the demography–environment interaction, i.e., how environmental fluctuations, however generated, affect the mean and the variance of the population or community in question. They also force us to ask what a reasonable population model should look like and to what extent the "environment" should be included in the model or kept aside as "noise." Finally, time series data, and the problems emerging from them, are not confined only to the classic long-term data we recognize from the textbooks (e.g., the Canada lynx – snowshoe hare system in North America). In fact, most ecological research is done over time and whatever phenomenon one is interested in, be it the breeding biology of birds, host plant selection in insects or life history of fish, there is always a statistical problem of model selection and the handling of "noise," or "error" as it is called in the statistical literature. This takes our approach beyond the classic time series domain into all ecology where there is variation across time and space.

The interplay between theory and data becomes perhaps best illustrated in the process of model selection (Hilborn and Mangel 1997; Burnham and Andersson 1998). This goes beyond the standard practice of evaluating null and alternative hypotheses, often from a purely statistical rather than biological point of view. Instead, we may (and should) formulate biologically meaningful models (note the plural) and confront them with the data. This procedure challenges us to keep one foot in each camp at all times, and to think carefully about the biological problem at hand by forcing us to formulate hypotheses as biological models. This approach also becomes particularly intriguing when we are dealing with stochastic processes and when we have to decide what should be regarded as "noise" and what should be included in the biological process. The next chapter takes a closer look at that problem.

## Suggested reading

This is a short list of suggested textbooks and general treatments of theoretical population and community ecology, mathematics, and statistics. Since we are covering neither all relevant theory nor all the analytical tools frequently used at sufficient depth, we refer to the more extensive treatments below. They can be used either as a preamble to the rest of this book, or as references whenever needed.

Burnham, K. P. and Anderson, D. R. 1998. *Model Selection and Inference: A Practical Information-Theoretic Approach*. New York: Springer-Verlag.

Caswell, H. 2001. *Matrix Population Models,* 2nd edn. Sunderland, Mass.: Sinauer.

Chatfield, C. 1999. *The Analysis of Time Series: An Introduction*, 5th edn. Boca Raton, Fla.: Chapman & Hall.

Chiang, A. 1984. *Fundamental Methods of Mathematical Economics*. Singapore: McGraw-Hill.

Edelstein-Keshet, L. 1988. *Mathematical Models in Biology*. New York: Random House.

Hilborn, R. and Mangel, M. 1997. *The Ecological Detective*. Princeton, N.J.: Princeton University Press.

Roughgarden, J. 1998. *Primer of Theoretical Ecology*. New York: Pentice Hall.

Royama, T. 1992. *Analytical Population Dynamics*. New York: Chapman and Hall.

# 2 · *Population renewal*

Population renewal is about how births and deaths of individuals are translated into population level dynamics. Here, we are reviewing some basic concepts and models of population renewal, disregarding both spatial processes (immigration and emigration) as well as interactions with other populations. Those extensions will be addressed in subsequent chapters. We are also briefly reviewing some statistical building blocks necessary for understanding population dynamics as a stochastic process and not only a deterministic route to persistence or extinction. This includes primarily the time series approach to population dynamics. We conclude this chapter by highlighting some very important and disturbing problems when confronting models with data (and the reverse), especially when trying to disentangle the demographic skeleton from "noise."

There is really nothing more to population ecology than births and deaths. If the number of individuals born exceeds the number that dies, the population size increases; should deaths exceed births, the population size decreases. If that simple, how is it so difficult to predict the population size in the future, and to determine what limits – or even regulates – the distribution and abundance of organisms in natural systems? We could argue that it is because the models we inevitably need to perform the above exercises are not good enough. One could also say that the task is difficult because it is not so easy to measure things accurately in nature. It is even problematic to determine what a population really is. One common argument is also that there are so many factors influencing the number of births and deaths, i.e., the problem is so complex, that it will be impossible to solve.

All of the above is probably true, one way or another. In this chapter, we are going to have a closer look at the problem of understanding population renewal. We certainly agree that one of the challenges is to reduce measurement error of, for example, population size estimates. This is true for both the actual counting of individuals (or biomass, or some other relevant measure of population size), and the determination of what constitutes the population in question. The latter involves both relevant time scales over which the population process is measured, and the spatial delimitation. This issue is discussed in an intriguing and important note by Berryman (2002). Our concern in this chapter will, however, primarily be the model formulation problem. If we understand the problem, then we are able to formulate a useful model. This is, of course, not to say that this solves it all, but it would give us considerable mileage. Towards the end of this chapter, we are going to address an important theoretical problem that relates to this issue. How to move from the understanding of individual behaviors and performances to their manifestation at the population level, and how (if ever) we can understand the reverse process. That is, whether we can infer from population level data, e.g., a time series of abundance, what is happening beneath the surface in terms of births and deaths. Before doing that, we are going to prepare ourselves with the basic building blocks and tools for making models of the population renewal process.

## Population growth rate

To begin with, we shall review some of the fundamental population growth processes. There is a rich literature that treats this issue in detail and at depth (e.g., May 1975; Emlen 1984; Edelstein-Keshet 1988; Yodzis 1989; Gotelli 1995; Hastings 1997; Roughgarden 1998). Two very useful accounts are Royama (1992) and Caswell (2001), dealing much with models and data, and structured populations, respectively. What is said in the following sections here is dealt with excellently by those two sources.

### $r$, $R$ and $\lambda$

Recall the basic renewal process outline in Chapter 1 (eq. 1.1)

$$N(t+1) \; = \; N(t)(1 + b + i - d - e), \tag{2.1}$$

where $b$ and $d$ are per capita birth and death rates, respectively, and $i$ and $e$ the per capita immigration and emigration rates, respectively. For illustrative purposes, let us omit immigration and emigration from the population process. We then have

$$N(t+1) \; = \; N(t)(1 + b - d). \tag{2.2}$$

If $b - d$ is equal to 0, i.e., births, $b$, equals deaths, $d$, then $N(t + 1) = N(t)$. That is, the population size does not change. The bracketed term in eq. 2.2 is often referred to as the net growth rate of the population, $R$. A simple reasoning hence says that if $R < 1$, the population declines; if $R = 1$, the population remains the same (is stable); and if $R > 1$ the population grows (Yodzis 1989; Royama 1992; Begon *et al.* 1996; Gurney and Nisbet 1998). We prefer $R$ to be the per capita growth rate for the untransformed densities, and let $R \equiv \exp(r)$ in keeping with the standard notation (e.g., the Ricker equation for population growth; Begon *et al.* 1996; Hastings 1997). In time series analysis, $R$ is often used instead of $r$ (p. 71).

Note that $R$ is the growth rate *per capita*, i.e., the average individual contribution to the population growth. If all individuals in the population are identical, then $R$ is unambiguously related to fitness (it *is* in fact a direct measure of fitness). If there is individual variation in the population, then some individuals contribute with more births and fewer deaths, others do the reverse. At an individual level, there are different net contributions to next generations; hence, there are individual differences in fitness. Yet, the population as a whole may increase ($R > 1$), decrease ($R < 1$), or stay the same ($R = 1$). Even in a decreasing population, some individuals are (possibly) going to do just fine. This is important to remember in the discussion about source-sink dynamics (Chapters 8 and 10). Lucid accounts of the fitness–population growth problem are found in Yodzis (1989) and Roff (1992). In a temporally variable environment, fitness is the *long-term* average growth rate (the geometric mean of the growth rate). This means that, e.g., a population (strategy) that grows faster than another population from one time step to the next does not necessarily beat a temporarily slower growing one.

### Structured populations

The population growth rate definitions given in the preceding section implicitly assume that all individuals in the population can be viewed as one big ensemble of identical individuals. That is, of course, rarely the case. The $N(t + 1) = RN(t)$ process is, however, often a useful proxy for what is actually happening, and is often the starting point when analyzing, e.g., time series data with poor or no resolution of population composition. Leslie (1945, 1948) pioneered the theory development for populations with individuals that differ with respect to fecundity and survival depending on their age. This is the case in most mammals, while in animals with indeterminate growth, such as fish, snakes, and lizards, the fecundity tends to be related to the size of the females (Roff 1992). The corresponding problem with more general stage-structured populations was later dealt with by, e.g., Lefkovitch (1965) and others (see Caswell 2001). An example of a stage-structured population model is given for the flour beetle *Tribolium* in Box 6.2 (p. 145). As a preparation for more detailed extensions of those classic models (Chapter 6), and to complete the understanding of population growth rate, we shall here very briefly recall the simple age-structured model. Caswell (2001) is otherwise an excellent source for a full and clear treatment of the problem.

Assuming that a population can be divided into $x$ age classes, each with their specific net fecundity ($m_x$) and survival ($l_x$, fraction in age class $x$ surviving to age class $x + 1$), and letting the number of individuals in each age class, $n_x$, be a column vector, then we have

$$
\begin{bmatrix} n_1(t+1) \\ n_2(t+1) \\ \vdots \\ n_x(t+1) \end{bmatrix} = \begin{bmatrix} m_1 & m_2 & \cdots & m_x \\ l_1 & 0 & \cdots & 0 \\ 0 & l_2 & \cdots & 0 \\ \vdots & \vdots & \ddots & \vdots \\ 0 & 0 & \cdots & 0 \end{bmatrix} \begin{bmatrix} n_1(t) \\ n_2(t) \\ \vdots \\ n_x(t) \end{bmatrix}. \tag{2.3}
$$

This is the projection of the number of individuals in each age class at time $t$ to time $t + 1$. The $x$ by $x$ matrix with all the age-specific fecundities and survival rates is the transition matrix, $\mathbf{L}$, for the population. Whether the population is going to increase, decrease or remain the same will hence be determined by the properties of $\mathbf{L}$ (i.e., the values of $l_x$ and $m_x$). Particular properties of $\mathbf{L}$ are its *eigenvalues*, i.e., to what extent $\mathbf{L}$ causes elements in the vector $[n_1(t), n_2(t), \ldots, n_x(t)]$ to change. That is, whether the population is going to increase or decrease (and whether the age structure is going to change). An $x$ by $x$ matrix has $x$ eigenvalues, but here it is the dominant eigenvalue, $\lambda_{\text{dom}}$, we are interested in. Should the magnitude of the eigenvalue, $\lambda_{\text{dom}}$, be greater than 1, then the population grows. Should $|\lambda_{\text{dom}}|$ be less than 1, then the population decreases. If the transition matrix is iterated, the population growth will eventually stabilize (to the exponential increase or decrease). $\lambda$ is then the slope of the line of the log population density plotted against time. Hence, precisely like $R$, $\lambda$ tells us the fate of the population. It is therefore sometimes customary to use $R$ and $\lambda$ interchangeably.

# Density dependence or not?

There is clearly only one possibility for a population to persist under the above scenarios, and that is when $R$ is strictly equal to 1. Should $R$ deviate from unity, then the population either eventually explodes to infinity or becomes extinct. The mechanism traditionally invoked to account for the fact that populations tend to grow when small, but decrease when large, is called negative density-dependent feedback. Should such a mechanism operate, then, by definition, the population would be *regulated*. At densities above some critical level, $R < 1$, but below it $R > 1$. Also in a stochastic world (see below), we would then have a population with a tendency to return to some steady state. This problem was intensively debated in the 1950s and 1960s. The debate has had a recent revival (den Boer and Reddingius 1996; Murray 1999; Turchin 1999; White 2001). As is often the case in animated debates, the reason for much of the disagreement lies in the very definition of some key concepts. We shall not scrutinize every detail of that debate here, but we will underscore a few elementary components.

## Per capita versus population growth rate

As we saw in the preceding sections, population growth, $RN(t)$, will necessarily be density-dependent if $R$ is. Let us reformulate eq. 2.2 by making births and deaths explicit functions of population density, $N$. The birth rate is

$$b = b_{max} - \beta N, \qquad (2.4a)$$

the death rate is

$$d = d_{min} + \delta N, \qquad (2.4b)$$

where $b_{max}$ is maximum per capita birth rate, $\beta$ is the rate by which per capita births are decreasing with increasing density, $d_{min}$ is the minimum per capita death rate, and $\delta$ is the rate by which per capita death rate increases with increasing population density. When combined, we have (Emlen 1984)

$$N(t+1) = N(t)\{1 + [b_{max} - \beta N(t)] - [d_{min} + \delta N(t)]\}. \qquad (2.5)$$

At closer inspection, we note that the growth function (2.5) is a second-degree polynomial in $N$. (This is rather easily seen by re-arranging the right-hand side of eq. 2.5; $(1 + b_{max} - \delta_{min})N(t) - (\beta + \delta)N(t)^2$.) That is,

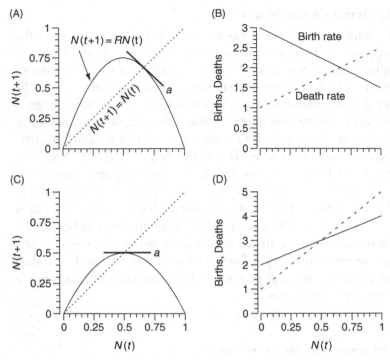

*Fig. 2.1.* (A) Population growth, $N(t+1) = RN(t)$, as a function of population density, $N(t)$. The solid curve shows population growth for a perfectly "symmetric" population, i.e., when the maximum of the population size is reached at half of the equilibrium population density. The equilibrium population size (carrying capacity) is determined by the intersection of the growth curve and the line $N(t+1) = N(t)$, at equilibrium $R = 1$. The steepness of the tangent line *a*, eq. 2.7, defines the stability of the equilibrium. The steeper the negative slope at the equilibrium, the less stable is the population. (B) Per capita birth rate and per capita death rate as a function of population density. The population is at equilibrium when the two lines intersect ($b_{max} = 3$, $d_{min} = 1$, $\beta = \delta = 1.5$)(C), (D) Increasing responses of per capita birth and death rates to density. (C) Population growth, $N(t+1) = RN(t)$, as a function of population density, $N(t)$. (D) Per capita birth and death rates ($b_{max} = 2$, $d_{min} = 1$, $\beta = -2$, $\delta = 4$).

eq. 2.5 is a humped function (fig. 2.1A), whereas the births and deaths are linearly decreasing and increasing functions of $N$, respectively (fig. 2.1B). One can of course imagine any shape of the per capita birth and death rate functions. In fact, both can increase (or decrease) with density. For example, if they both increase, and $b_{max} > d_{min}$, but births do that more slowly than deaths, then there will be a density that provides an equilibrium population size, and we still have the "regulation" scenario (figs. 2.1C,D).

One can easily experiment with various shapes and forms of the responses of per capita births and deaths to changes in population size. It will soon be discovered that many of the standard solutions of equilibrium population density ("carrying capacity") and population densities at which population growth is at maximum will all vary depending on the assumptions made about those relationships. For example, solving eq. 2.5 for the equilibrium population density $N^*$, i.e., when $N(t+1) = N(t)$, results in

$$N^* = \frac{b_{max} - d_{min}}{\beta + \delta}. \qquad (2.6)$$

This leads us to a useful concept in single-population dynamics, namely the renewal function. Sometimes it goes under the name of the "recruitment" function, a concept borrowed from fisheries biology (Ricker 1954; Beverton and Holt 1957; Hilborn and Walters 1992). The renewal function, $F(N)$, is simply a plot of eq. 2.5, or some other appropriate function for population growth. If there is no density dependence, i.e., $R$ is a constant, then we would have a straight line through the origin in the $N(t+1)$ versus $N(t)$ plot. The straight line with slope $= 1$ in this plot is the equilibrium line, $R = 1$ (fig. 2.1A,C). Wherever the renewal function intersects that line, there is an equilibrium $[N(t+1) = N(t) = N^*]$. Typically, the renewal function only intersects at one point, i.e., there is only one equilibrium of the system.

The slope of the renewal function at that intersection determines the stability properties of the equilibrium (Edelstein-Keshet 1988; Yodzis 1989). In more technical terms, the derivative of the renewal function, $F$, with regard to the population size, $N$, at the equilibrium population size

$$a = \left. \frac{\partial F(N)}{\partial N} \right|_{N = N^*} \qquad (2.7)$$

is the slope evaluated at equilibrium (the intersection point). If the slope, $a$, is negative (positive) the dynamics are said to be overcompensatory (undercompensatory). If it is zero, we have perfect compensation (fig. 2.1A,C).

The density-dependent feedback described above is strictly speaking a lagged effect (Moran 1950a). It is the density at time $t$ that affects the population density in time $t+1$. This is the inevitable formulation in discrete time models when we want to represent direct density dependence. However, if we recall the formulation in eq. 2.5, it actually describes the effect of current density on the per capita births and deaths.

It is not unusual that density effects may have even longer lags. For example, several authors (Stenseth et al. 1998a; Stenseth 1999; Turchin et al. 1999; Bjørnstad and Bascompte 2001; Turchin and Batzli 2001) have shown nicely how population interactions, e.g., resource–consumer interactions, introduce longer time lags into the resource population dynamics. This means that population density at time $t + 1$ now becomes a function of the density at time $t$ and time steps further back in time $(t - k; k \geq 1)$. A similar effect is also potentially the result in an age-structured population. For example, the total reproductive output in one year will be contingent on the reproduction and survival of the different age classes many time steps back in time. This is not to say that it is necessarily easy to correctly identify such time lags from data on population density. Both environmental and demographic stochasticity (p. 214) and the demography of the particular population may mask the actual time lags in the system. In Chapter 6, we are going to have a closer look at the dynamics of structured populations.

## Population dynamics: the first step

The mapping of the population density (or size) from one time to another is determined by the renewal function. Hence, we have

$$N(t + 1) = N(t)f[N(t)]. \tag{2.8}$$

As indicated in the previous section, the function $f$ may take various nonlinear forms. Some often used examples of such functions are given in Box. 2.1. We will here exemplify the dynamics predicted by one of them, the Ricker model (Ricker 1954)

$$f[N(t)] = \exp\{r[1 - N(t)/K]\}. \tag{2.9}$$

Two parameters determine the dynamics: $r$ and $K$ (fig. 2.2); $r$ is responsible for the dynamical properties of population fluctuations, while $K$ is a scaling parameter. We recognize $r$ from preceding sections as the per capita population growth rate [recall: $R \equiv \exp(r)$]. This growth rate is discounted, as the population density increases, by the term $1 - N(t)/K$, the strength of the density dependence being inversely related to the parameter $K$. It is easy to see that the function $f$ is equal to 1 when $N(t)$ is equal to $K$. That is, the value of $K$ determines when there is no change in population density from time $t$ to $t + 1$. The parameter $K$ often goes under the name "carrying capacity." This implies that the environment

**Box 2.1** · *Eight nonlinear population models*

The following discrete time population models are frequently used in the literature (Cohen 1995; Kaitala and Ranta 1996). Their dynamics are slightly different, e.g., to what extent they can produce over-compensatory dynamics, but all share the common property of producing logistic growth.

$$N(t+1) = N(t)\exp\{r[1 - N(t)]\} \tag{a}$$

$$N(t+1) = N(t)\{1 + r[1 - N(t)]\} \tag{b}$$

$$N(t+1) = \frac{rN(t)}{1 + \exp\{-a[1 - N(t)]\}} \tag{c}$$

$$N(t+1) = \frac{rN(t)}{[1 + aN(t)]^b} \tag{d}$$

$$N(t+1) = \frac{rN(t)}{1 + aN(t)^b} \tag{e}$$

$$N(t+1) = \begin{cases} rN(t) \text{ if } N(t) \leq C, \\ rN(t)^{1-b} \end{cases} \tag{f}$$

(where $C$ is a constant or threshold)

$$N(t+1) = rN(t)(1 + r\{1 - \exp[-sN(t)]\}[K - N(t)]) \tag{g}$$

$$N(t+1) = rN(t)\{K + L\log[N(t)] - N(t)\} \tag{h}$$

(a) Ricker, (b) Verhulst, (c) Pennycuick, (d) Hassell, (e) Maynard Smith–Slatkin, (f) Varley, (g) Austin-Brewer and (h) Malthus–Condocet–Mill. Note that the parameters $a$, $b$, $r$, etc. may have slightly different interpretations in the different models.

has a certain capacity (related to $K$) to hold an equilibrium density of individuals of the population, which is independent of the growth rate. Despite the formal correctness of that interpretation, it may be misleading. For example, looking at the equilibrium of the population renewal process, we see that, in this model, carrying capacity is not independent of

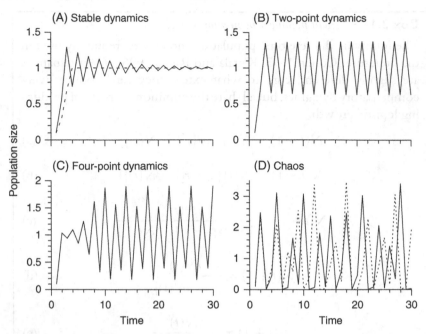

*Fig. 2.2.* (A) Population equilibrium is stable. After a deviation from the equilibrium level the population size returns back to the equilibrium level ($r = 1.2$, dotted line; 1.9, $K = 1$). In an unstable case, the population level may approach (B) a two-point cycle ($r = 2.1$), (C) a four-point cycle ($r = 2.6$), and ultimately (D) a chaotic behavior ($r = 3.5$). Chaos is characterized by the sensitivity to small deviations in the initial population level. The two dynamics in (D) correspond to the $N(1) = 0.1$ (solid line) and $N(1) = 0.11$ (dotted line).

(maximum) per capita population growth rate. $K$ is not a static property of the environment, but a combination of innate fecundity ($b_{max}$), density-independent mortality ($d_{min}$), and how sensitive births and deaths are to changes in density ($\beta$ and $\delta$).

Assume now that the population is at the equilibrium, defined by $N(t + 1) = N(t)$. Then, if we perturb the population away from that point, several things can happen (May 1974, 1975, 1976). Following eq. 2.9, the fate of the perturbation will depend on the value of $r$. If $r$ is small enough, the population will return to the equilibrium. If it is somewhat larger, the population will start oscillating, and when $r$ is big enough, erratic fluctuations will be the result (fig. 2.2). In the former case, the equilibrium point of the dynamics is (asymptotically) stable, whereas in the latter two cases it is unstable. The stability property of a dynamics can be verified analytically by linearizing the renewal function $f$ at the equilibrium point

**Box 2.2** · *Linearization and local stability*

Nonlinear functions can be approximated by linearization. Consider, e.g., the following population model with renewal function $F$

$$N(t+1) = F[N(t)].$$

The function $f$ may be, for example, the Ricker model: $F(N) = N \exp[r(1 - N/K)]$. We are interested in the behavior of the model near equilibrium, which will be the point around which the function is linearized. Call this equilibrium $N^*$. We then have (e.g., Edelstein-Keshet 1988)

$$N(t+1) \approx N^* + \frac{\partial F(N^*)}{\partial N}[N(t) - N^*] + \text{h.o.t.},$$

where "h.o.t." means higher order terms (second and higher order derivatives). Defining $X(t) = N(t) - N^*$ we have, cf. eq. 2.6

$$X(t+1) = \frac{\partial F(N^*)}{\partial N}X(t).$$

If the equilibrium is to be stable, the perturbation away from it should diminish, i.e., $X$ becomes smaller. This happens if the absolute value of the above partial derivative is less than 1. This partial derivative is the slope of the renewal function at the equilibrium point.

Consider next the Ricker equation, eq. 2.9 as an example. Differentiating eq. 2.9 with respect to $N$, and letting $N = K = N^*$, we have

$$\frac{\partial F(N^*)}{\partial N} = 1 - r,$$

where hence $1 - r$ is the growth rate of the perturbation. The only parameter affecting the dynamics (bifurcation parameter) in the Ricker model is $r$. We conclude that the population dynamics is stable when $r < 2$, since for these values $|r - 1| < 1$. We see also that, when $r = 1$, then the slope $= 0$ and we have perfect compensation. If $r > 1$, the dynamics are overcompensatory, when $r < 1$ they are undercompensatory.

(Box 2.2) and recovering eq. 2.7. A way of illustrating the changes in dynamics (local stability) as we change a dynamic parameter is by plotting the population density against that parameter. Thus, we simply simulate, e.g., eq. 2.9 for different values of $r$, allow for a large number of transient time steps, and read off the population density. A figure thus produced is called a bifurcation diagram, and the parameter responsible for the change in dynamics a bifurcation parameter (Box 2.3). The densities where the population ends up after perturbations are called attractors. An attractor may be represented by one point, or it may be a set of points (e.g., two points in a period-two cycle). When the system is chaotic, the attractor is "strange" (Yodzis 1989). Chaos is a complicated form of dynamics and has gained a lot of interest among ecologists. One of its promising attractions has probably been its potential to explain the erratic and irregular dynamics often observed in natural systems. The problem is, of course, that no system in nature is deterministic, but stochastic, and purely deterministic tools therefore have obvious limitations.

---

**Box 2.3** · *Bifurcation diagram*

A bifurcation diagram can be used to summarize a large amount of information about the dynamics of a single-population model in a very compact form. The bifurcation parameter(s) varies among models. As seen in Box 2.2, $r$ is such a parameter in the Ricker model. For each value of the bifurcation parameter, the dynamics of the model are simulated for a few hundred generations to remove the initial transient in order to get a grasp of the final attractor that characterizes the dynamic behavior of the model for this particular value of the bifurcation parameter. Then, a sufficiently large number (usually 50–100, or even more) of the points in the attractor are plotted against the bifurcation parameter. The simulations can be initiated from random or fixed initial values, depending on the purpose of the analysis.

We shall illustrate the bifurcation diagram by using the Ricker model. Equation (2.9) is simulated for a wide range of $r$ values (1.5, 1.51, 1.52, ..., 4.49, 4.5). We used the version of the Ricker model presented in Box. 2.1, i.e., normalized with $K = 1$). The bifurcation diagram confirms the well-known facts that for $r < 2$ the dynamics are represented by one value, $N = 1$. At $r = 2$ a bifurcation occurs, at which the attractor of one point will turn into an attractor of two points. In the range $2 < r < 2.56$, the dynamics are periodic such that the population sizes jump between two points that vary with $r$. Then, new bifurcations occur, and the attractors become four-point, eight-point, etc. cycles. This is called period doubling cascade. Finally, at $r > 2.692$, the population dynamics become chaotic. Note that the chaotic area includes periodic windows (fig. B1).

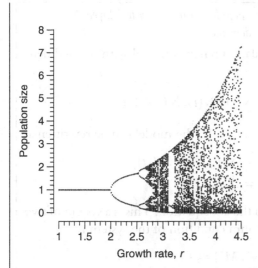

*Fig. B1.* Bifurcation diagram of the Ricker model, eq. (2.8). The points corresponding each value of the growth rate, *r*, represents the attractors of the resulting dynamics.

Bifurcation diagrams can be useful when analyzing simple deterministic models, but useless for stochastic models where different attractors are virtually indistinguishable.

## Population dynamics with delayed density dependence

The preceding section dealt with basically very simple processes – the mapping of $N(t)$ to $N(t+1)$. Such models have obvious limitations. All individuals of the population are lumped into one state variable, and the process has no memory. That is, the manifestation at time $t+1$ is assumed to be unaffected by events and processes at time $t-1$ or further back. A number of factors can, however, have a lagged effect on the current population density, including age structure, biotic interactions, and temporally structured environmental variation (Beckerman *et al.* 2001; Turchin and Batzli 2001; see also below). Thus, instead of the simple population models introduced earlier we may want to use models of the form

$$N(t+1) = N(t)f[N(t), N(t-1)], \qquad (2.10)$$

where the per capita population growth rate is determined not only by the current population density $N(t)$, but also by the population density

---

**Box 2.4** · *Local stability of a population model with delayed density dependence*

Consider the following population renewal model with delayed density dependence

$$N(t+1) = N(t)f[N(t), N(t-1)].$$

Adopting the notation $M(t) = N(t-1)$, the model can be rewritten as

$$N(t+1) = N(t)f[N(t), M(t)]$$
$$M(t+1) = N(t),$$

with the time index shifted in the last equation. This is a system of two linear equations. The equilibrium condition of the system is

$$f(N^*, M^*) = 1$$
$$M^* = N^*.$$

Perturbations away from the equilibrium will increase or decrease depending on the joint change of the two equations around equilibrium. This will be determined by the partial derivatives of the two equations with respect to $N$ and $M$ (evaluated at equilibrium), formulated as the Jacobian matrix (e.g., Edelstein-Keshet 1988)

$$\mathbf{J} = \begin{bmatrix} 1 + N^* \frac{\partial f(N^*, M^*)}{\partial N} & N^* \frac{\partial f(N^*, M^*)}{\partial M} \\ 1 & 0 \end{bmatrix}.$$

If the eigenvalues of matrix $\mathbf{J}$ are both less than 1, then the equilibrium is stable. When the one or both eigenvalues exceed 1, then the system will not return to equilibrium, but instead have cyclic dynamics. The length of the cycle depends on where the largest eigenvalue (which can be a complex number) crosses the unit circle in the complex plane. The length of the cycle is determined by the parameters of the model (Box 2.5).

---

one time step or more steps earlier (Box 2.4). Obviously, the relative weights of these two (or more) densities need to be determined. An example of such a model is

$$f[N(t), N(t-1)] = \exp\{r[1 + a_1 N(t) + a_2 N(t-1)]\}, \qquad (2.11)$$

where $a_1$ and $a_2$ are the weights for the direct and delayed density dependence, respectively. Putting $a_2 = 0$ and defining $a_1 = -1/K$, we have recovered the Ricker model (eq. 2.9). We shall make frequent use of this model and will refer to eq. 2.11 as the delayed Ricker equation.

There are certain advantages when using models with delayed density dependence. Whereas the Ricker model, and other simple models in Box 2.1, yield stable solutions and period-doubling cascades with $2^n$-point cycles, $n = 1, 2, 3, \ldots$, and finally chaotic solutions (May 1974, 1975, 1976), the models with delayed density dependence can provide us with periodic solutions with a continuum of period lengths (Box 2.5). Thus, as we will see, it will be more comfortable to address population dynamics with fluctuation of 3, 6, and 10 years. These period lengths correspond roughly to the population dynamics of small rodents, game birds, and Canada lynx and snowshoe hares, respectively (e.g., Chapters 4, 6, 8).

---

**Box 2.5** · *Periodic solutions under density dependence*

Consider the population dynamics model and its linear approximation presented in Box 2.4. Assume that the eigenvalues are complex as follows

$$\lambda_{1,2} = k \pm li,$$

where $k$ and $l$ are the real and imaginary parts of the eigenvalues. When the magnitude of the eigenvalues is approximately 1, that is $|\lambda_{1,2}| \approx 1$, the cycles in the dynamics are determined by the period length, given as

$$2\pi/\theta,$$

where

$$\theta = \arctan(l/k) \text{ for } k, l > 0 \text{ and}$$

$$\theta = \pi + \arctan(l/k) \text{ for } k < 0,\ l > 0.$$

There are two different types of practical uses of this approach in population modeling. Assume that $|\lambda_{1,2}| < 1$, but close to the unit circle. Then the population dynamics are stable, however, such that the approach to the equilibrium is by damped oscillations. The period length of the dampened oscillations is determined as presented above. Second, assume that $|\lambda_{1,2}| > 1$, but close to the unit circle in the complex plane. Now, the population dynamics are unstable, fluctuating with oscillations characterized above.

Consider next the delayed Ricker model, eq. 2.11, with delayed density dependence, defined as follows

$$N(t+1) = N(t) \exp\{r[1 + a_1 N(t) + a_2 N(t-1)]\},$$

which can be written in the form

$$N(t+1) = N(t) \exp\{r[1 + a_1 N(t) + a_2 M(t)]\},$$
$$M(t+1) = N(t).$$

At a nontrivial (positive) equilibrium, we have

$$\exp[r(1 + a_1 N^* + a_2 M^*)] = 1$$
$$N^* = M^* = -\frac{1}{a_1 + a_2}.$$

Thus, we have the following condition for a positive equilibrium

$$a_1 + a_2 < 0.$$

The Jacobian of the linearized equation becomes

$$\mathbf{J} = \begin{bmatrix} 1 - \frac{ra_1}{a_1 + a_2} & -\frac{ra_2}{a_1 + a_2} \\ 1 & 0 \end{bmatrix},$$

from which we get the eigenvalue equation

$$\lambda \mathbf{I} - \mathbf{J} = \begin{bmatrix} \lambda - (1 - \frac{ra_1}{a_1 + a_2}) & \frac{ra_2}{a_1 + a_2} \\ -1 & \lambda \end{bmatrix} = 0,$$

where $\mathbf{I}$ is the identity matrix, and

$$\lambda^2 - \lambda\left(1 - \frac{ra_1}{a_1 + a_2}\right) + \frac{ra_2}{a_1 + a_2} = 0.$$

Thus, we have two eigenvalues, $\lambda_1$ and $\lambda_2$. We know that

$$\lambda_1 \lambda_2 = \frac{ra_2}{a_1 + a_2}, \text{ and } \lambda_1 + \lambda_2 = \left(1 - \frac{ra_1}{a_1 + a_2}\right).$$

Recalling the notation $\lambda_{1,2} = k \pm li$, we get

$$k = \left(1 - \frac{ra_1}{a_1 + a_2}\right) \Big/ 2.$$

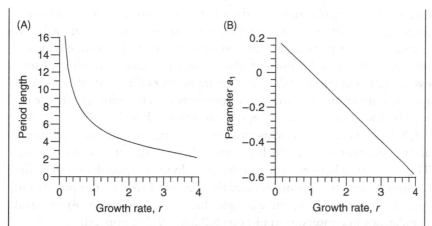

Fig. B2. The length of the cycle period of the delayed Ricker model, eq. 2.11, as a function of $r$ and $a_1$, assuming that $|\lambda_{1,2}| = 1.01$; $a_2 = -0.2$.

Fixing the magnitude $|\lambda_{1,2}|$, we can solve $l$ from equation

$$l = \sqrt{|\lambda|^2 - k^2}.$$

Assume that we aim at study dynamics characterized by $|\lambda_{1,2}| = 1$ and a cycle length of 8 years. Now, we have a condition $8 = 2\pi/\theta$, yielding $\theta = \pi/4$. Here we have $k = l = 1\sqrt{2}$. Recalling that $k$ and $l$ depend on three parameters, we note that the parameter choice is nonunique. The dependence of the length of the cycle period on parameter $r$ and $a_1$ is given in fig. B2. Here, magnitudes of the eigenvalue and parameter $a_2$ are fixed first ($|\lambda_{1,2}| = 1.01$; $a_2 = -0.2$). Then, $a_1$ and the period length will be calculated for different values of the growth rate.

## Population size versus density

The "state" of a population is often expressed as its density or its size. We often use those two measures interchangeably although we really should not. The sometimes confused debate about density dependence arises from the lack of distinction between the two. By its very definition, density dependence is about the number of individuals per unit area (or volume). Area is used here as a proxy for resources available to the organism in question. If we fix the area and the amount of resources contained within it, then, as we increase the number of individuals in that area, every one gets a smaller and smaller share of the resources (assuming scramble competition). As a result, we assume that per capita births decrease and/or per capita deaths increase because of resource deprivation. As a theoretical construct, this is often just fine. Sometimes, however, one has to be careful. Let $N$ be the *number* of individuals within an area $A$, containing the amount of resources $R$. The (biologically meaningful) *density* ($D$) of that population would then be $D = N/R/A = N/(RA)$. Hence, if we decrease the area, keeping the amount of resources and the number of individuals constant, then the density would increase. This distinction between the number of individuals (population size), the amount of resources available, and the area inhabited by the population becomes critical when, for example, habitat loss (Chapter 8) or small populations in conservation biology (Chapter 9) is a concern.

Failure to recognize this potential problem and the distinction between per capita and population rates will continue to inject confusion in the "population regulation" debate. A third, more statistical, problem will be dealt with in the time series section (p. 31) in this chapter.

## Stochasticity

The chapter has so far dealt entirely with deterministic processes. They are the backbone of all ecological and evolutionary theory. The degree to which chance events – stochastic processes – influence the dynamics varies from system to system and from component to component of the population and community processes. The way stochasticity is handled in ecological modeling also varies considerably. For example, one can attempt to make every key process an explicit stochastic variable, e.g., the births and deaths in most models of demographic stochasticity (e.g., Gabriel and Bürger 1992; Burgman *et al.* 1993; Lande 1993), or by combining demographic and environmental stochasticity as in Ripa and Lundberg (2000). No matter how it is modeled, there is more to it than just making a deterministic signal fuzzier or making parameters, variables or equilibria averages of a statistical distribution. Moreover, we cannot understand data unless we understand something about stochastic processes. Covering all relevant theory of stochastic processes is, however, out of the scope of this book and we refer the reader to excellent reviews and textbooks (e.g., Royama 1992; Hilborn and Mangel 1997; Jordan and Smith 1997). Therefore, the remainder of this section only deals with a limited set of problems particularly relevant to the chapters to come.

**Environmental stochasticity**

The environment in which a population lives is never constant. The physical and biotic environment varies from one time to another and from one place to another one. Here, we shall deal with temporal variability and will leave spatial variability to later parts of the book. Traditionally, environmental variability has been treated as "noise," i.e., a disturbance of some kind that masks or disrupts the "signal" we are actually after. The signal–noise terminology reveals the engineering roots of the methods used to analyze the problems. Environmental variability is, however, more than just noise (although we are frequently using that term for convenience).

The variability a population experiences occurs at many temporal scales, from very short-term (e.g., day-to-day) variation to changes over decades or more. It has been argued that in fact all time scales ought to be equally important for the long-term dynamics of a population (Halley 1995). A stochastic variable with such properties is said to have a $1/f$ *spectrum*. Notation $f$ here refers to the frequency of a periodic function. Box 2.6 explains the spectrum in more detail and fig. 2.3 illustrates spectra of some example time series. Of those, one (fig. 2.3(E)) is of special character, because this dynamics also closely corresponds to the $1/f$ process. In more general terms, we let power be a function of $1/f^{\alpha}$, where the exponent $\alpha$ in a ln(power) versus ln(frequency) graph is the slope of the function.

It is often assumed, and sometimes shown, that environmental variability is dominated by low-frequency variation (Steel 1985; Pimm and Redfearn 1988; Halley 1995). Surely there is day-to-day and year-to-year variation in relevant environmental variables (e.g., temperature), but the dominating fluctuations, and also presumably the biologically most important ones occur at longer time scales. It is often practical to let the environmental noise have a temporal structure of its own (Kaitala *et al.* 1997a, b; Ripa *et al.* 1998). One simple, yet incomplete way of implementing such variability is to let the environmental variable be modeled as (Ripa and Lundberg 1996; Ripa and Heino 1999)

$$\omega(t+1) = \kappa\omega(t) + c\varepsilon(t),\qquad(2.12)$$

where $\omega$ is the stochastic variable in question, $\kappa$ is the autocorrelation coefficient and $\varepsilon$ is a series of normal random deviates with zero mean and unit variance, e.g., normally distributed white noise. Equation 2.12 is a first-order autoregressive process, i.e., the value of the variable $\omega$ at

---

**Box 2.6** · *Spectral analysis*

Spectral analysis of a time series is the decomposition of the series into its frequency components. The analogy with visible light makes an intuitive comparison (remembering also that *wavelength* and *frequency* are inversely related). White light is a mixture of all colors, each color representing a particular wavelength. Blue, for example, has a relatively short wavelength (high frequency), whereas red has a long wavelength (low frequency). The environmental variability or a time series from natural populations is usually a mixture of many frequencies. If low frequencies dominate the process, it is said to be "red" and if it is dominated by short wavelengths, it is "blue." If there is a strong cyclic component in the time series, a certain wavelength, corresponding to the cycle period, is dominating.

A discrete time series $x(t)$, $t = 0, 1, 2, \ldots, n-1$ ($n$ is the length of the series), has the periodogram $P_x$, which serves as an estimate of the power spectrum (wavelength composition) of the underlying stochastic process

$$P_x(f) = \frac{1}{n} |X(f)|^2,$$

where $f$ is frequency, $X(f)$ is the discrete Fourier transform of the time series, and $|\ldots|^2$ indicates the squared modulus of the transform. The periodogram, or power spectrum, indicates how much each frequency contributes to the variance of the time series. Note that a periodogram is calculated from a sample time series and a power spectrum is a statistical property of a stochastic process. The power spectrum describes the expected periodogram, but any single periodogram usually deviates substantially from the power spectrum, irrespective of the length of the time series. The noisiness of periodograms is the reason why they are often smoothed in some way to give a better estimate of the underlying power spectrum. In the ecological literature, the terms power spectrum and periodogram are often mixed. See, e.g., Priestly (1981) and Chatfield (1999) for further details.

---

a given time is linearly related to the value of $\omega$ in the previous time step. The parameter $\kappa$ determines how successive values of $\omega$ are related, i.e., sets the "color" of the $\omega$ process. If $\kappa > 0$ then the series is positively autocorrelated (red), if $\kappa < 0$, then it is negatively autocorrelated (blue), and if $\kappa = 0$, then there is no autocorrelation and the $\omega$ process has collapsed back to uncorrelated white noise (fig. 2.3(C)–(E), (c)–(e)). The parameter $c = s\sqrt{1 - \kappa^2}$ determines the amplitude of the $\omega$ process in such a way that the variance of it is independent of the values of $\kappa$. If $\kappa = 0$ (i.e., $c = s$) then $s$ is the standard deviation of the white noise process. That the color of the noise has profound consequences on noise-influenced population renewal has been discussed, e.g., by Kaitala *et al.* (1997a,b), Ripa *et al.* (1998), Heino *et al.* (2000), and will be taken up in various parts of this book.

### How to implement stochasticity

One of the steps in building stochastic population models is, as we have seen, to decide on the properties of the stochastic process. Another is to let the stochastic process be a part of the population system in question. The problem at hand is to define exactly how this is done. Here, a few examples are given. One way of incorporating environmental stochasticity is to assume that a critical demographic parameter in a particular population model is a random deviate. For example, in the Ricker equation, we can let the equilibrium population density $K$ be such a deviate

$$N(t + 1) = N(t)\exp\{r[1 - N(t)/K(t)]\}, \tag{2.13}$$

with $N$ and $r$ as defined earlier. The parameter $K(t)$ is a time-dependent stochastic variable. Using the Ricker equation again, we can also let stochasticity enter as additive noise

$$N(t + 1) = N(t)(\exp\{r[1 - N(t)/K]\} + \sigma\varepsilon(t + 1)), \tag{2.14}$$

where $\varepsilon$ is a random deviate with zero mean and variance $\sigma$.

There are numerous ways of implementing stochasticity into population dynamics (e.g., Gabriel and Bürger 1992; Ludwig 1996; Halley and Iwasa 1998; Ripa and Lundberg 2000). Often the problem at hand, and the model used, will determine how that is done. Throughout this book we are mostly using the following general form

$$X(t + 1) = F[X(t), X(t - 1), X(t - 2), \ldots]\mu(t). \tag{2.15}$$

Here $F$ is the population renewal function, as discussed earlier in this chapter, possibly including one or more lagged terms of density dependence. For the stochastic processes influencing the renewal we have the external noise, $\mu(t)$, which can be white or colored noise within known

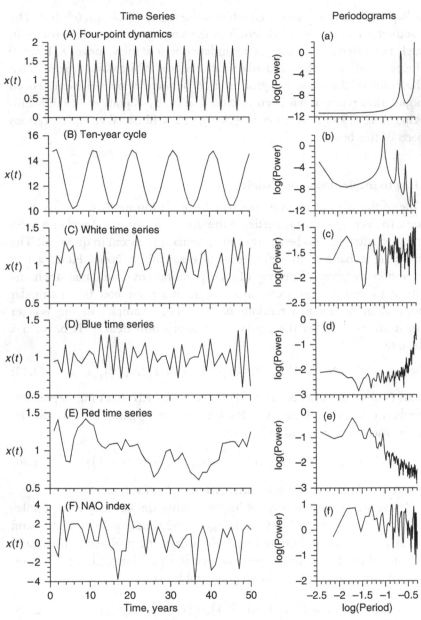

*Fig. 2.3.* Periodograms (power spectra; calculated for time series of length $2^{10}$) of some example stochastic processes. To the left of each spectrum is the actual time series (only 50 time steps displayed). (A)–(a) Period-four cycle created by the Ricker model, eq. 2.9. $r = 2.6$, $K = 1$. (B)–(b) Period-ten dynamics created by the delayed Ricker model eq. 2.11. $r = 0.4$, $a_1 = 0.12$, $a_2 = -0.2$ (Box 2.5).

boundaries (often so that it is drawn from the interval $[1-w, 1+w]$, where $0 < w < 1$). The impact of stochastic processes, like eq. 2.15, on dynamics of populations will be discussed further in Chapter 4.

## Time series

After having discussed deterministic and stochastic models, we now briefly turn our attention to a more data-driven analysis of population renewal. A lot of data on population dynamics come in the form of time series of counts or derived indices of population abundance. The resolution of those data, the precision and accuracy of the data, and the extent to which the data are accompanied by additional information about the biology of the organism in question naturally vary considerably. Here, we are not going to dwell on all the practical problems of gathering population data. Instead, this and the following section are devoted to the ubiquitous and important problem of relating pattern to processes (and the reverse). The aim of much time series analysis in ecology is not only to describe patterns of population dynamics, but also to understand which, primarily biological, processes are responsible for the observed patterns. Apart from that, a mere *description* of data might be relevant for generating hypotheses about processes. One classic example is the regular changes in abundance detected in the fur trade data from the Hudson Bay Company in Canada (Elton 1924). The very existence of apparent regularity often causes ecologists to be on red alert: regular fluctuations (cycles!) must require very special explanations; or do they?

The time series we have to hand in ecology are inevitably stochastic due to the influence of stochastic environmental factors and to measurement error. Disentangling those two sources of variation is not a trivial task (e.g., Quinn and Deriso 1999). Here, we will just acknowledge the fact that they both exist and that the problem itself is of great importance

Caption for fig. 2.3. (*cont.*)
An autoregressive process, eq. 2.12, was used to produce the white time series in (C)–(c), $\kappa = 0$; blue time series (D)–(d), $\kappa = -0.8$; and red time series (E)–(e), $\kappa = 0.8$. A white time series has an entirely flat spectrum whereas a red-shifted one is left-skewed (towards low frequencies) and a blue-shifted is right-skewed (towards high frequencies). A cyclic series has a peak indicating the dominating frequency of the cycle (a), (b). A series can also have several peaks, albeit weak, as illustrated by the time series of the North Atlantic Oscillation (NAO) index, a proxy for primarily winter weather in Northern Europe (F)–(f). In (F) only 1901–1950 is displayed, but (f) is calculated for the period 1866–1999.

for understanding the underlying processes. There is a very rich literature on time series analysis that is also relevant to ecology and evolutionary biology (e.g., Box *et al.* 1994; Powell and Steele 1995; Burnham and Anderson 1998; Chatfield 1999). We strongly encourage the reader to consult that literature for full derivations and details. What follows here is hopefully a more self-contained account for the remainder of this book.

### Characterizing time series

We are by and large going to make use of three basic descriptions of ecological time series. The first is spectral analysis mentioned earlier (Box 2.6). Spectral analysis is done in the frequency domain (Priestly 1981; Chatfield 1999), i.e., the focus is on the wavelength composition of the time series. In the following, focus will be on inference from the *autocorrelation* and *partial autocorrelation* functions, i.e., analysis in the time domain.

The data to be analyzed may come in many different forms. The time interval between measurements may or may not match the relevant time scale of the organism or the problem at hand. The data may also have missing observations. The presence of a trend in time series (Box 4.2, p. 71) is another classic worry and to date there is no foolproof way of handling trends, especially not in the often short time series that ecology offers. Trends are part of a more general problem, referred to as *stationarity*. Virtually all time series techniques (at least the ones usually applicable to ecological time series) hinge on the fact that the series is stationary. That is, the mean and the variance of the series do not change over time [and a couple more details, see Chatfield (1999) for an accessible and detailed account on this]. The stationarity of a series is also influenced by the time interval of sampling relative to the process in question. No matter what the data may look like, they should always be graphed. A plot of the time series data always helps when deciding how to proceed. As a next step, it is often advisable to transform the series, e.g., by using log-transformed data. Transforming stabilizes the variance and tends to make the data normally distributed. A special type of *filtering* (but far from the only one) is differencing as a means of removing trends. The original series $\{x_1, x_2, \ldots, x_k\}$ is transformed to $\{y_1, y_2, \ldots, y_{k-1}\}$ by $y_t = x_{t+1} - x_t$. A common way of detrending data series is to fit linear or nonlinear functions to data and use the residuals as the data. We do caution, however, against a too liberal use of arbitrary function fitting and use of residuals. The problem is that we can rarely be sure that the putative trend is not, in fact, a component of the dynamics we would actually like to retain (cf. fig. 9.1).

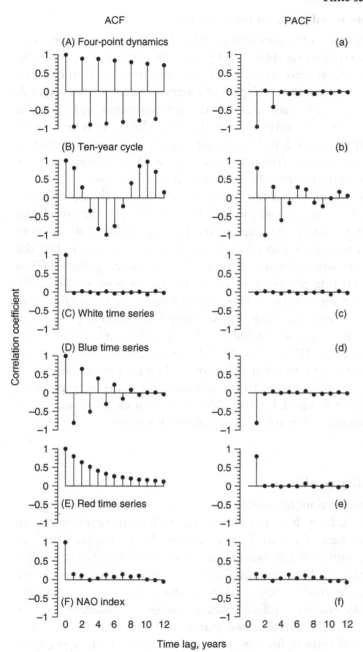

*Fig. 2.4.* The autocorrelation function (ACF, left column) and partial autocorrelation functions (PACF, right column) for sample time series presented in fig. 2.5.

### Autocorrelation and partial autocorrelation

The autocorrelation of a times series is related to the ordinary correlation between $n$ pairs of two variables, $x$ and $y$. The autocorrelation measures the correlation between successive observations with time lag $k$ of the series $\{x_1, x_2, \ldots, x_n\}$ such that $n - k$ pairs of observations are formed, $(x_1, x_2)$, $(x_2, x_3), \ldots, (x_{n-k}, x_k)$, and regarding the first observation in each pair as one variable, and the second pair as the other one. This can now be done for pairs $k$ steps apart. For each such lag the correlation coefficient is calculated. Now, a correlogram (autocorrelation function, ACF) can be constructed by plotting those correlation coefficients for each lag $k$ (fig. 2.4).

The autocorrelation function thus gives us a hint of the lag structure of the series, e.g., if it is strongly positively (or negatively) autocorrelated, or if there is a cyclic component in the time series (fig. 2.4). A more detailed picture of the lag structure of the time series is achieved by the partial autocorrelation function (PACF). Strictly speaking, the PACF is a way of estimating the order of an autoregressive process (Box 2.7), i.e., a model estimation procedure. The details of such procedures are beyond the scope of this book, but we note that estimating the partial autocorrelation is basically as follows. When fitting an autoregressive process of order $p$, AR$(p)$ (Box 2.7), the last coefficient, $\alpha_p$, measures the correlation at lag $p$ which is not accounted for by a model with $p - 1$ coefficients. Partial autocorrelation functions for some example time series are shown in fig. 2.4. Chatfield (1999) is an excellent source for further information on partial autocorrelation functions.

## The visibility problem

The preceding sections have dealt with deterministic models for population renewal, and they have also provided a first look at the interpretation of time series data. They are in a sense very different approaches, yet necessarily intertwined if we want to take both theory and data seriously – which we have to. Data are meaningless unless we have a model, and models are useless if they cannot tell us anything about our observations. The "visibility problem" refers to the challenge of understanding the underlying process from observed patterns (Ranta et al. 2000a). The patterns we will refer to here are primarily the patterns emerging from time series data of population abundance. Note, however, that the problem is entirely general and that there is no inherent restriction to time series data, it is just that they are particularly illustrative.

**Box 2.7** · *Autoregressive processes*

A process (in practice a time series) $\{X(t)\}$ is an autoregressive process of order $p$ if

$$X(t) = \alpha_1 X(t-1) + \alpha_2 X(t-2) + \cdots + \alpha_p X(t-p) + \varepsilon(t),$$

where $\varepsilon(t)$ is an independent random process with zero mean and variance $\sigma^2$. Hence, $X$ is regressed on itself $p$ time steps back in time. An autoregressive process of order $p$ is denoted AR($p$). An AR(1) process is therefore

$$X(t) = \alpha X(t-1) + \varepsilon(t).$$

An AR(1) process is an example of a Markov process. An AR($p$) process is a linear model of a time series $\{X(t)\}$. It is generally (but not always!) assumed that an ecological time series can be approximated by a linear model, at least near the stationary value (the mean) of the series. One can therefore use the linear approximations above to estimate the values of $\alpha_p$, which then give a hint of the lag structure of the time series (the population in focus). This is akin to determining the density-dependence structure of the population, although there are several pitfalls involved in this (Berryman and Turchin 2001).

Here we also show a simple example of how the AR-modeling approach can be applied. A commonly used nonlinear population model is

$$N(t+1) = N(t)\exp[b_0 + b_1 x(t) + b_2 x(t-1) + \varepsilon(t)],$$

where $N(t)$ is the population density at time $t$, $x(t)$ is the natural logarithm of $N(t)$, and $\varepsilon(t)$ is a noise term. This is a generalization of the Ricker model given by eq. 2.9. The parameters $b_1$ and $b_2$ are of interest because they measure the strength of direct and delayed density dependence, respectively, and their values determine the behavior of the deterministic skeleton. $b_0$, however, has no dynamic effect and is only a location parameter often close to zero due to detrending procedures (Stenseth 1999). Rearranging and taking the natural logarithm of both sides, we get

$$x(t+1) = b_0 + (1+b_1)x(t) + b_2 x(t-1) + \varepsilon(t).$$

This is a log–linear AR(2)-model, having well-studied properties (Box *et al.* 1994; Chatfield 1999; Royama 1992).

The visibility problem comes in two different flavors. The first problem deals with the wish to read off some environmental variation (temperature or habitat change, say) from the dynamics of a population. That is, is the environmental variability visible in the time series of some population abundance? This is a correlation problem. Ranta *et al.* (2000a) and Laakso *et al.* (2001) have addressed the problem by letting a deterministic population model be affected by environmental noise of various kinds and they have sought the correlation between that pure noise process and how it is manifested in the population time series. Such a correlation is not always to be expected, however (Ranta *et al.* 2000a). Depending on the intrinsic dynamics (if the deterministic model is stable or not) and on the exact nature of the noise signal (e.g., its color), the putative correlations are often masked. Hence, the time series of population change does not in itself carry any information about the environmental variability affecting the population, unless complemented by additional data about the environment. The visibility problem has since been extended to age-structured populations (Kaitala and Ranta 2001) and to a Ricker dynamics with births and deaths modeled separately (Lundberg *et al.* 2001). Throughout, the basic message is that the fingerprint of the external signal is very difficult to recognize in the noise-modulated dynamics of the focal population, but see Scott and Grant (2004).

The studies by Laakso *et al.* (2001) and Greenman and Benton (2001) illustrate an additional problem, namely that the situation gets even worse if the quite plausible assumption about nonlinear responses to environmental change is made. An environmental variable, temperature say, rarely affects the population renewal in a linear manner. Rather, the effect is often dome-shaped or perhaps sigmoid (or perhaps also a discrete all-or-none response). Should that be the case, the expected correlation between the environmental variable and the population dynamics we may want to use as a probe becomes even more compromised.

The related, and perhaps even more challenging, problem is illustrated by the study by Jonzén *et al.* (2002a). They modeled the population dynamics as a linear process with two lags, i.e., a simple AR(2) process, modified by noise of different color and amplitude. The task was then to recover the underlying model structure given the data thus generated. The question is whether we can learn anything about the demographic (density-dependent) and environmental processes by estimating model parameters from a noisy time series. The answer was that only occasionally could one recollect the skeleton dynamics (fig. 2.5). This illustrates the "inverse" problem (Wood 1997), which means that it may be

PARAMETER ESTIMATES          ORDER ESTIMATES

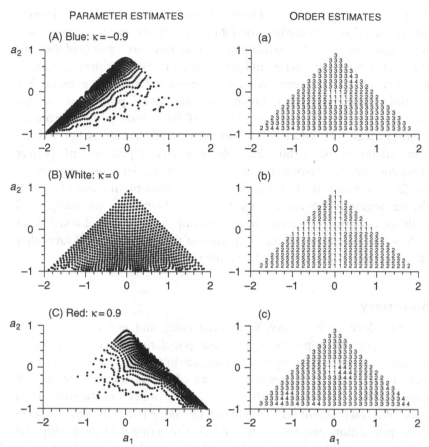

Fig. 2.5. AR(2) parameters, (A), (B), (C), and the process order, (a), (b), (c), estimated from time series with AR(2) skeleton modulated by noise. The gradation of the parameter space ($a_2 + a_1 < 1$, $a_2 - a_1 < 1$, and $-1 < a_2 < 1$) as shown in (a)−(c) is used to generate time series data in three environments ($\kappa = -0.9$, $\kappa = 0$, and $\kappa = 0.9$), where $\kappa$ is the autocorrelation parameter of the colored noise, eq. 2.12. Panels (A)−(C) indicate the values obtained by fitting an AR(2) model to the noise-modulated data. The order estimates [numbers inserted in panels (a)−(c)] were achieved by using the partial autocorrelation technique (Chatfield 1999). Modified after Jonzén et al. (2002a).

impossible to infer from pattern (the time series) the underlying process (the interaction between a revealed demography and a correctly identified environmental variability). Given a time series without prior knowledge about the exact mechanisms behind it, all we can do is some sensible model estimation. The question is what should be in the model and what should be regarded as noise. The purely statistical approach is to devise a model such that, after having fitted it to the data, the residuals are as small as possible, are normally distributed, and are not correlated (and they are stationary). Such a model would generally be regarded as the best model. It is far from clear, however, what we have learned from that exercise. For example, Royama (1981) showed that an AR(1) population process modulated by AR(1) noise, eq. 2.12, will be indistinguishable from an AR(2) process with white noise. Not only in statistical terms, but they are identical processes! A third example is an AR(1) process with perfect compensation (corresponding to the Ricker model with $r = 1$) in an AR(2) environment. This highlights the irreducible uncertainty about the demography–environment interaction. Therefore, we badly need both solid a priori stochastic models of population renewal and at least well-informed guesses about the nature of the environmental variability affecting the demography of the population in question.

## Summary

We introduce the basic tools for understanding and analyzing simple but dynamic population processes. The basic population processes, on which the whole ecological theory will be built, are births and deaths. As we have emphasized, these processes are the core of population fluctuations. Of course, although focusing on the births and deaths helps us also to focus on the ecological problem we are studying, this alone is not enough to understand population processes fully. When furthering and deepening the insight, one needs to specify the ecological problems and framework under consideration. Are we dealing with small populations, where demographic stochasticity will become important, or with larger populations where intrinsic population regulation mechanisms (density dependence), population structures, trophic interactions, or interactions with the environment may become crucial factors? We do not always know which factors are important. Nevertheless, equipped with the basic tools, such as population renewal models, and tools for data analysis, we are ready to proceed to take into account not only births and deaths but also movement between locations, immigration and emigration.

# 3 · Population dynamics in space — the first step

The population renewal processes discussed in the previous chapter assumed spatially homogenous environments. For natural systems this mostly fails to be true. Here, we extend the population renewal in space and will hence come back to the problem of including emigration and immigration in population dynamics. We will do so by arbitrarily delimiting the landscape into well-defined habitat patches connected by redistribution of individuals. This simplified representation of spatial structure and population dynamics allows us to analyze and interpret a wide range of single-species phenomena. Spatial structure can alter the population dynamics significantly and produce emergent phenomena such as synchrony and complex dynamics. This chapter sets the theoretical and conceptual stage for such problems dealt with in more detail in the coming chapters.

In the previous chapter, we deliberately overlooked the important and natural aspect of the import and export of individuals to and from a given focal population. For some populations, ignoring dispersal may be a fair approximation. Most extreme examples of this might be experimental populations of fruitflies in a single container or small aquatic microcosms of protozoans. In such cases, it would be natural to assume that a complete mixing occurs in the whole population. Most populations, however, are spatially structured and the exchange of individuals between landscape elements is an integral part of the dynamics (Hastings 1990; Kareiva 1990; Bascompte and Solé 1997; Tilman and Kareiva 1997). The redistribution of individuals among habitat patches, however defined, is affected at least by two factors: the distance between the local units and local population size. The longer the distance between the localities, the less likely it is that they exchange individuals via dispersal. The larger a population, the better chance it has to send dispersing individuals, while only a few individuals can emigrate from small populations. The reception of dispersing individuals may also be affected by local population size.

## Setting the stage

In Chapter 2, we studied the effects of births and deaths. Here we shall push the pure demographic processes to the background and concentrate on the effects of dispersal on the temporal dynamics of a single, or an isolated population. We simply ask how dispersal may affect population dynamics. Net dispersal is a result of the departure of individuals from their natal patch, and of the arrival of individuals from surrounding subunits to the focal unit. Thus, one is tempted to think that only net dispersal is important. It is convenient, however, to decompose the net dispersal into two components, immigration and emigration.

It will be shown that, as compared with single patch dynamics, new phenomena will emerge when the spatial dimension of the population dynamics is included when addressing the density-dependent feedback. Emigration and immigration may alter the qualitative properties of local population dynamics even in a system with only two habitable patches. In some cases, the effect of the spatial interaction is so crucial that it may be difficult, if not impossible, to predict the dynamics of a single population if the interactions with other populations are neglected. In combining the temporal dynamics with the spatial dimension, we begin to understand, aside from population dynamics aspects, that spatial population dynamics provide a new and extremely interesting approach for structuring and analyzing population patterns. The population sizes may become organized both in time and space and we will attempt here to find out how this comes about.

Classical population ecology assumes that immigration and emigration are negligible compared to the births and deaths. In contrast, metapopulation ecology (Levins 1969; Gilpin and Hanski 1991; Hanski 1999a) focuses entirely on the immigration and extinction processes by ignoring births and deaths and reducing local populations to being either present or absent in well-defined habitat patches. Proper understanding of the ecology of populations requires that all four terms in the process mapping current population size to the future need to be addressed.

### Immigration and emigration and single-population dynamics

There is no more simple setting for studying the consequences of spatial linkages in population dynamics than a one-patch model with net immigration or emigration. Thus, we may reconsider the unstructured single-patch model and modify it by adding constant immigration ($I$) as follows:

$$N(t + 1) = RN(t) + I. \tag{3.1}$$

This simple change of the renewal equation may have greater than expected consequences. Using the normalized Ricker model, we have

$$R(t) = \exp\{r[1 - N(t)]\}. \tag{3.2}$$

We now combine eqs. 3.1 and 3.2, and use the bifurcation diagram as a diagnostic under a slight amount of constant immigration. A careful study of the bifurcation diagrams shows that constant immigration may have drastic effects on the population dynamics (fig. 3.1; McCallum 1992). A comparison of the bifurcation diagrams shows that complex chaotic population behavior, observed under no immigration ($I = 0$, fig. 3.1(A)), may become simpler under a slight amount of immigration ($I = 0.001$, fig. 3.1(B)). Periodic windows become larger around $r = 3.2$ and 3.6, and for a large area of chaotic dynamics ($r > 4$) the population dynamics are maintained as period-four or period-two cycles. This indicates that (constant) immigration tends to stabilize the dynamics. However, the opposite may be true as well. This can be seen by increasing the immigration to $I = 0.01$. In this case, we see that the periodic window around $r = 3.2$, observed under no immigration ($I = 0$, fig. 3.1(A)), is turned to chaotic dynamics ($I = 0.01$, fig. 3.1(C)). That is, constant immigration may also destabilize the population dynamics. The effect may be contradictory, as at the same time as immigration destabilizes population dynamics for $r \approx 3.2$, it will stabilize the chaotic range between $r = 3.3$ and $r = 4.05$ into a periodic window.

Constant emigration is a little bit more complicated to study. We know that for the Ricker model, increasing the growth rate $r$ decreases minimum population size (Box 2.1). Thus, we modify our study such that the population is doomed to become extinct if the population size becomes sufficiently small. Letting individuals emigrate from the population makes it easier to pass that threshold. In fig. 3.1(D), we have assumed that $E = 0.01$. The visible periodic window around $r = 3.2$ is now moved close to $r = 3.1$. Moreover, the periodic window around $r = 3.2$ has become an area of chaotic dynamics. Thus, no unique answer can be given about the stabilizing effect of constant emigration. More thorough investigations of the effect of constant immigration or emigration on single-patch dynamics are presented elsewhere (Holt 1983a,b, 2002; McCallum 1992; Ruxton 1994, 1995a; Doebeli 1995; Rohani and Miramontes 1995a; Sinha and Parthasarathy 1996; Paradis 1997; Gonzalez-Andujar 1998; Ruxton and Rohani 1998; Stone and Hart 1999).

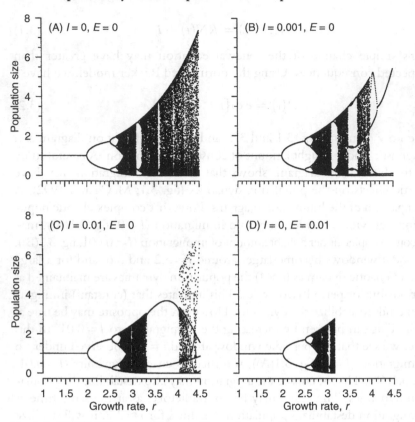

*Fig. 3.1.* (A)–(C) Bifurcation diagrams illustrating the effects of constant immigration on population dynamics after the Ricker equation. (A) $I = 0$; (B) $I = 0.001$; (C) $I = 0.01$. (D) Bifurcation diagram illustrating the effect of constant emigration. When the emigration rate exceeds the lowest population value, the population is doomed to become extinct. $E = 0.01$. For each $r$ value calculated, the simulations were run for 250 generations, and the last 50 points in each time series were used in the bifurcation diagram.

## A two-patch system

The one-patch model with immigration or emigration leaves many questions unanswered. Occasionally, immigration and emigration are constant rates but are varying over time and with local population density. Studying that requires that we explicitly include more than the focal habitat patch in the analysis. By doing so, we also move one level up and have to pay attention to both local and global (including all habitat patches) dynamics.

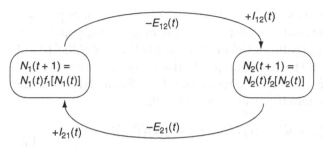

*Fig. 3.2.* Schematic illustration of the two-patch system. The subpopulations reproduce seasonally. After reproduction a fraction of the occupants leaves their natal patch to reproduce in the other one.

A two-patch population system is the simplest landscape for the purposes of showing that spatial structures may matter when studying qualitative properties of population dynamics (Hastings 1993; Lloyd 1995; Kendall and Fox 1998; Solé and Gamarra 1998; Jansen 2001). Interaction between two or more subpopulations will usually maintain the basic patterns observed in the single-patch population dynamics. However, the spatial extension may also result in qualitative changes. Let $N_1$ and $N_2$ denote the population sizes of patches 1 and 2, respectively. The dynamics of this spatially linked system are given as

$$N_1(t+1) = N_1(t)g_1[N_1(t), N_2(t)]$$
$$N_2(t+1) = N_2(t)g_2[N_1(t), N_2(t)], \quad (3.3)$$

where $g_1$ and $g_2$ are functions specifying the effect of both populations on local population growth rate. In each time step, both renewal and dispersal are taking place and the temporal order of those events becomes important. We will assume here that population renewal occurs before individuals move between patches (fig. 3.2). After reproduction, the population size in habitat $i$ is $N_i(t) f_i[N_i(t)]$. Now, eq. 3.3 becomes

$$N_1(t+1) = N_1(t)f_1[N_1(t)] - E_1 + I_1$$
$$N_2(t+1) = N_2(t)f_2[N_2(t)] - E_2 + I_2, \quad (3.4)$$

where $E_1$, $E_2$, $I_1$ and $I_2$ are the net emigration from and net immigration to patches 1 and 2, respectively. In a closed system without mortality while in transit, we may assume that $I_1 = E_2$, and $I_2 = E_1$ yielding

$$N_1(t+1) = N_1(t)f_1[N_1(t)] - E_1 + E_2$$
$$N_2(t+1) = N_2(t)f_2[N_2(t)] - E_2 + E_1. \quad (3.5)$$

Assume next that a fixed and equal fraction $m$ of each subpopulation leaves the natal population and disperses to the neighboring population. Thus, $E_i(t) = mN_i(t)f_i[N_i(t)]$. It follows that a total of $(1-m)N_i(t)f_i[N_i(t)]$ stays in the natal patch $i$ to the next reproductive season, and a total of $E_i(t)$ arrives in the neighboring patch to reproduce there. Put together, the full description of the system becomes (Hastings 1993)

$$N_1(t+1) = (1-m)N_1(t)f_1[N_1(t)] + mN_2(t)f_2[N_2(t)]$$
$$N_2(t+1) = (1-m)N_2(t)f_2[N_2(t)] + mN_1(t)f_1[N_1(t)]. \tag{3.6}$$

Following Hastings (1993), we let the per capita population growth rate be

$$f_i[N_i(t)] = r[1 - N_i(t)]. \tag{3.7}$$

With equal growth rates, $r$, and with equal fractions ($m_i = 0.1$) of individuals dispersing, the bifurcation diagram is identical to the one–patch case if the system is initiated with identical population sizes. This is due to the fact that in this case the immigration is balanced by the emigration, resulting in zero net dispersal. Hence, under these assumptions spatial structure does not make any difference to the local dynamics. However, assuming random initial population sizes, we observe that the dynamics may behave differently depending on the initial values. Examples of bifurcation diagrams for sample simulations are given in fig. 3.3. In the first example (fig. 3.3(A)), initiating from a random set of initial conditions, the bifurcation diagram of the total population size, $N_1 + N_2$, is composed of period-doubling cascade from $r = 2.0$ up to $r \approx 2.85$, where the dynamics turn into two-point cycles, then stable equilibrium, and again to two-point cycles. For $r < 2.85$, the bifurcation diagram matches the bifurcation diagram of single-patch dynamics since populations in each patch are of exactly the same size at each time point, and no net dispersal occurs. For $r < 2.85$, the two patches fluctuate hand in hand while this is the case only occasionally for $r > 2.85$ (fig. 3.3(B)). The stable equilibrium for $3.1 < r < 4.3$ appears to be a combination of two two-point cycles, one in each patch, fluctuating in opposite phases (figs. 3.3(B),(C), 3.4(A). And the two-point cycles of the total population size ($r \approx 2.85$ and $r > 4.3$) turn out to be a combination of two four-point cycles (figs. 3.3)(B),(C)).

Taking another set of random initial conditions, we may get another picture of the combined dynamics. Interestingly enough, at $r > 2.5$ a unique equilibrium point replaces the period-doubling cascade, and part of the chaotic range by a single equilibrium point ($2.5 < r < 2.8$; fig. 3.3(D)). This is due to the fact that the period-doubling cascade has

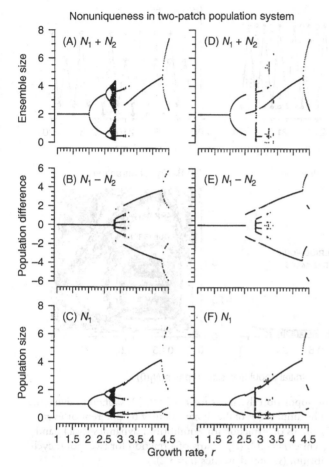

Fig. 3.3. Population dynamics in a two-patch system. (A) The bifurcation diagram obtained using random initial population values for $N_1(0)$ and $N_2(0)$, which are the same for each $r$ value. On the unstable area ($r > 2.8$), the dynamics of the total population size are two-point cycles or a stable point equilibrium. (B) For $r < 2.8$, the local population dynamics are coherent, whereas for $r > 2.8$ the populations fluctuate out of step. (C) The bifurcation diagram of the local population size differs from that of the total population size depending on whether the local populations are in step or out of step. (D) An alternative bifurcation diagram obtained using another random pair of initial population sizes, $N_1(0)$ and $N_2(0)$. A part of the period-doubling cascade is now replaced by a point equilibrium. (E) In addition to the nonuniqueness of the attractors, the coherence may also depend on the initial values of the local populations. (F) The local population dynamics may also be nonunique.

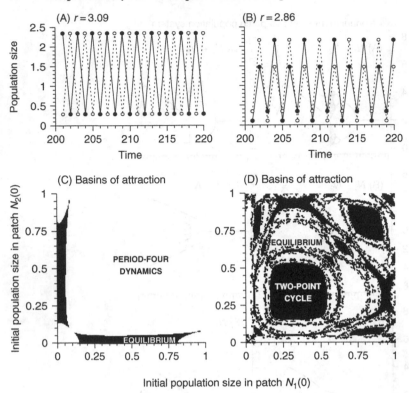

*Fig. 3.4.* (A) Incoherent population dynamics for $r = 3.09$. (B) Coherent dynamics need not be identical. The amplitudes of the dynamics may be different or even fluctuate ($r = 2.86$). (C) Basins of attraction for equilibrium dynamics (black) and period-four dynamics (white; $r = 2.6$). (D) Basins of attraction for two-point cycle (black) and stable equilibrium (white) dynamics ($r = 3.2$).

also been replaced locally by two-point cycles which are now in opposite phase, yielding a point equilibrium for the total population size. Also, exactly the same period-two fluctuation can be obtained by combining two point-four cycles (to $r \approx 4.4$, fig. 3.3(A),(B),(C)) and two point-two cycles (fig. 3.3(D),(E),(F)).

In summary, we have observed the following. The dynamics in a two-patch system may become nonunique in the sense that different initial conditions may lead to different attractors. Several different attractors may coexist, the classification of which depends on whether the population is viewed at the global or local level. For example, a stable equilibrium may be an alternative for the whole range of period-doubling cascade to chaos and even for chaotic solutions. Further, the alternative attractors seem to be connected to the coherence of the local patch dynamics. The same type of local dynamics yields different global attractors when they are coherent and when they are not. When the population dynamics become more complex, the population dynamics may be coherent, yet such that the amplitudes differ (fig. 3.4(B)). Thus, dispersal seems to have a significant role in modifying the global and local population dynamics. Generally, chaotic dynamics tend to change to periodic, indicating that dispersal may have in some cases a stabilizing effect on population dynamics. Also, there is an emergent pattern of synchrony and asynchrony of the dynamics across patches. This is an issue dealt with in more detail in the next chapter.

Extensive simulation studies in a two-patch system for logistic dynamics (Hastings 1993) and for the Ricker dynamics in a ten-patch system (Ruxton 1994) have revealed that when identical populations are coupled by redistributing individuals, two-point cyclic dynamics may continue to operate in any patch either in or out of phase, and the result will depend on initial conditions. This nonuniqueness and its ecological consequences have received rather limited attention. It is worth noting also that nonlinear interactions between age groups in age-structured populations may generate such complexities in their dynamic behavior (Caswell 2001).

Tracking the initial values leading to one or another attractor can be illustrated as basins of attractions (Hastings 1993). The prediction of the attractor based on the basin of attraction seems to work fine for, say, $r = 2.6$, where the alternative attractors are stable equilibrium and a period-four cycle (fig. 3.4(C)). The basins of attraction for both attractors are large sets that can be easily separated from each other. However, occasionally this prediction may fail completely. Hastings (1993) pointed out that the dependence of the attractor on the initial conditions may be extremely complicated. It appears that the points leading to the alternative attractors may form very complicated patterns with fractal sets and fractal boundaries (fig. 3.4(D)). Here, the basins of attraction are not well defined. In fact, they may be fractal sets without any well-defined boundaries (Grebogi et al. 1987; Hastings 1993; Kaitala and Heino 1996).

Bifurcation diagrams are convenient tools for studying the deterministic behavior of the model. One problem is, however, that the transient phase before the final attractor is reached may be extremely long (Hastings and Higgins 1994; Ruxton and Doebeli 1996; Kaitala et al. 2000). For example, for $r = 2.8$, the two alternative attractors are single-point equilibria and a two-point cycle. However, the initial, chaotic-looking transient may take thousands or tens of thousands of time steps. Thus, bifurcation diagrams, when using short time intervals, may give biased information, as is presented for $r = 3.2$ (fig. 3.4(D)), where the bifurcation diagram represents the initial transient rather than the final attractor.

Finally, we recall the close connection of the alternative attractors to the degree of coherence between the populations. Thus, we may state that the basins of attraction also represent the points from which the dynamics become synchronized or desynchronized. Nonuniqueness in two-dimensional maps has also been observed in a single-species age-structured model (Wilbur 1996; Caswell 2001) and two-species host–parasite and host–parasitoid models as examples (Kaitala and Heino 1996). Kaitala *et al.* (1999, 2000) and (Ives *et al.* 2000) give richer accounts on both host–parasitoid and predator–prey systems.

## Population dynamics in space

The simple two-patch models presented above are of course very crude representations of most biological systems. Habitat structure ranges from an entirely homogenous world (fig. 3.5) to distinct patches. Dispersal may be either local (emigrants only reaching neighboring populations), or global such that every corner of the landscape is reachable. Such different dispersal patterns may have profound effects on both local and global dynamics (e.g., Fryxell and Lundberg 1993). In the extreme case of global dispersal, distance among landscape units does not matter, which in effect means that space is only implicit. Spatially implicitly structured models have, despite the simplification, been an important first step for understanding spatial dynamics (Allen *et al.* 1993; Ruxton 1994; Bascompte and Solé 1997). A third possibility is to assume that both distance among the subpopulations and their size affect the rate and success of dispersal. That is the version to which we will primarily adhere.

### Spatially implicit models

The classical meta-population model (Levins 1969) and the extinction model by Allen *et al.* (1993) are examples of spatially implicit models. In spatially implicit models, the patches are distributed in a discrete space but their location does not affect the dispersal process. Consider $n$ local populations in an implicit space (e.g., Allen *et al.* 1993), each renewing seasonally as follows

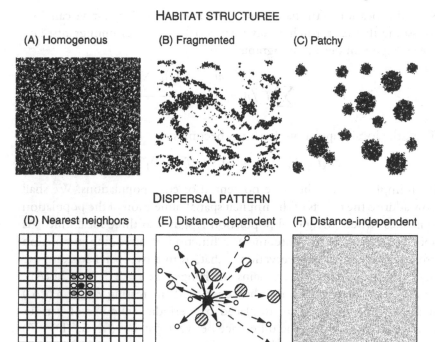

Habitat Structuree

(A) Homogenous    (B) Fragmented    (C) Patchy

Dispersal Pattern

(D) Nearest neighbors    (E) Distance-dependent    (F) Distance-independent

*Fig. 3.5.* Spatial population structures can be modeled in several alternative ways. Habitat structures may be assumed to be (A) homogenous environments, (B) heterogeneous fragmented environments and (C) patchy subpopulations. In the spatial models, dispersal may be local with only nearest neighbors linked, as in (D) the cellular automaton. In regular or irregular grids, the dispersal may be independent of the spatial setting, or the dispersal may depend on the location of the patches. When dispersal depends on the locations and inter-patch distances we refer to the models as spatially explicit population dynamics (E), otherwise spatially implicit models.

$$N_i(t+1) = N_i(t)f_i[N_i(t)] - E_i + I_i, \qquad (3.8)$$

where $i = 1, 2, \ldots, n$ and $N_i(t)$ is the population size in the $i$th subunit. One simple assumption for emigration ($E$) is that a constant fraction of the local population leaves in each time step (see, e.g., Ruxton 1996a; Palmqvist *et al.* 2000; Ylikarjula *et al.* 2000; Poethke and Hovestadt 2002 for models with density-dependent dispersal). Assuming that dispersal occurs after renewal, the number of emigrants can then be written

$$E_i = mN_i(t)f_i[N_i(t)]. \qquad (3.9)$$

Since the location of the patches does not matter, the simplest we can do is to assume that each patch receives an equal amount of immigrants, i.e., the average number of immigrants

$$\bar{I} = I_i = \sum_i E_i/n = \frac{m \sum_i N_i(t) f_i[N_i(t)]}{n}. \tag{3.10}$$

Thus, the local population dynamics becomes

$$N_i(t+1) = (1-m)N_i(t)f_i[N_i(t)] + \bar{I}. \tag{3.11}$$

In an implicit space, there are no central or edge populations. We shall now address the effects of the implicit spatial dimension of the population dynamics, where the global population is split into three local units. As above, we observe the outcome as a function of increasing population growth rate. Our first observation is that, as in a two-patch system, the point of first bifurcation remains unchanged at $r = 2.0$ (fig. 3.6(A)). Things may change radically, however, with increasing $r$ (fig. 3.6(A)) such that chaotic areas are displaced by periodic windows of various characters. Focusing on the difference between patches (fig. 3.6(D)–(F)) we see that for $r < 2.6$ all three populations obey coherent dynamics, whereas for larger values of $r$ only two out of the three populations maintain synchronous fluctuations. With different initial conditions, due to nonuniqueness, a different picture emerges (fig. 3.6(B) and 3.6(E)). Now, the major difference is that the complex part of the bifurcation diagram $(2.6 < r < 2.9)$ is replaced by period-two dynamics in the total population size.

We next reduce the value of the dispersal to $m = 0.04$. Again, the bifurcation diagrams of the total population sizes, as well as those of single populations, differ crucially from the single-patch ones (fig. 3.7(A),(C)). The local populations remain periodic until $r \approx 3$, after which the dynamics are chaotic, to return again to being periodic at $r \approx 3.3$. This appears to depend on whether the subpopulations are in synchrony or not (figs. 3.7(D)–(F)). Nonuniqueness of the attractors occurs for a wide range of parameter values. Another random set of initial conditions may yield a different bifurcation diagram (fig. 3.7(B)). For example, there may be different two-period attractors in the total population system, as is the case for $r = 2.5$ (fig. 3.7(A),(B)). In addition, chaotic dynamics may be an alternative for different periodic attractors. Again, nonuniqueness seems to be in close connection with the degree of synchronicity (Box 3.1) among the local populations.

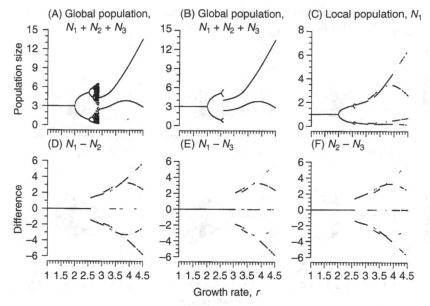

*Fig. 3.6.* Population dynamics of a three-patch system in an implicit space with the dispersal fraction $m = 0.1$. (A) Bifurcation diagram of the total population size obtained using random initial population sizes, $N_1(0)$, $N_2(0)$, and $N_3(0)$. As compared to the one- and two-patch models, the point of first bifurcation stays put, indicating that introducing one more patch to a two-patch system does not destabilize the population dynamics or the stable area of population dynamics. (B) An alternative bifurcation diagram, initiated from another set of initial population sizes, indicates nonuniqueness of the attractors. (C) A bifurcation diagram of a single patch. (D)–(F) Bifurcation diagrams of the differences between the local population dynamics. All the dynamics may be coherent or only two of them. The simulations were run for 1000 generations, and the last 50 values were used for the diagrams.

## Coupled map lattices

The "map" in *coupled map lattice* modeling is a mathematical expression to illustrate the evolution of local population size from one year to the next; "coupling" refers to the interaction of local populations via redistributing individuals, and "lattice" refers to the spatial distribution of the local populations. Note that lattice often refers to a regular grid-like structure of cells (like a square-ruled notebook page). However, for us a lattice can be either regular or irregular (population subunits are random coordinate points).

Consider now a fixed number of subpopulations distributed in an explicit space, in a regular or irregular manner. The population renewal is modeled

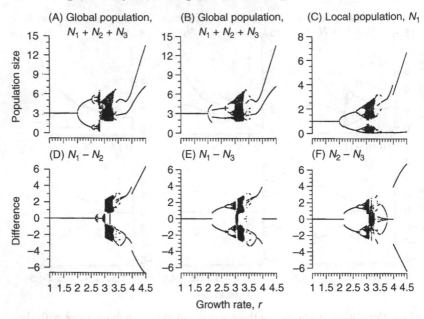

Fig. 3.7. Population dynamics of a three-patch system in an implicit space with the dispersal fraction $m = 0.04$. (A), (B) Two different bifurcation diagrams of the total population size obtained using different initial population sizes. As compared to the one-patch model, the chaotic area with high growth rates has almost completely disappeared. This indicates that a moderate dispersal may stabilize spatial population dynamics in an implicit space. (C) Bifurcation diagram of a single patch. (D)–(F) Bifurcation diagrams of the differences between the local population dynamics. All the local dynamics may be coherent, or only two of them, or none of them. The simulations were run for 1000 generations, and the last 50 values were used for the diagrams.

as in previous sections. We also assume that a constant fraction $m$ ($0 < m < 1$) of the population disperses. Then, the population dynamics are given as (Ranta *et al.* 1995a)

$$N_i(t + 1) = (1 - m)F_i[N_i(t), N_i(t - 1)] + \sum_{s, s \neq i} M_{si}(t), \qquad (3.12)$$

where $N_i(t)$ is the population size in subunit $i$ at time $t$. The last term of eq. 3.12 is the subunit-specific arrival of the redistributing individuals. Different versions of the dispersal kernel (the last term of the eq. 3.12) are given in Box 3.2.

To illustrate the effects of space on the qualitative properties of dynamics of local populations, we shall use a spatial structure with

**Box 3.1** · *Synchronicity in population dynamics*

Population dynamics are considered synchronous when population fluctuations are matching in time in two or more populations. The cross-correlation function, with various time lags, allows one to assess the degree of temporal synchrony of fluctuations in size between any two populations of matching time span. High positive values of the correlation coefficient, with time lag = 0, indicate that the pair of populations fluctuates largely in step. Values close to zero indicate no temporal match in population highs and lows, and large negative values indicate that the two populations are in the opposite phase. For more details, see Chapter 4 and Box 4.1.

---

**Box 3.2** · *Dispersal kernels*

Consider a population of a single species distributed in $n$ locations. Before dispersal, the local renewal is described as follows

$$N_i'(t) = F[N_i(t)], i = 1, \ldots, n.$$

where $N_i'(t)$ is the reproducing population size in patch $i$ at time $t$, and $N_i(t)$ is the population size before dispersal. Assume now that a fraction $m$ of individuals leaves any local patch annually. After dispersal, the local population sizes are given as follows

$$N_i(t+1) = (1 - m)N_i'(t) + M_{si}(t),$$

where $M_{si}$ is the dispersing population size arriving at patch $i$ from patch $s$.

**Spatially implicit dispersal**

In an implicit space, all the individuals are added to the joint distribution pool, and all patches receive an equal amount of dispersers

$$M(t) = \sum_i (1 - m_i)N_i'(t)/n.$$

*Kernel I*

**Spatially explicit dispersal with exponentially distributed dispersal distances**

Here we assume that the distribution of the distances determines the exchange of individuals between the patches as follows (e.g., Ranta *et al.* 1997a)

$$I(t) = \sum_{s,s \neq i} M_{si}(t),$$

where the net amount of immigrants arriving to patch $i$ from patch $s$ is given as

$$M_{si}(t) = mN_s'(t) \frac{\exp(-cd_{si})}{\sum_{j,j \neq s} \exp(-cd_{sj})},$$

where $c$ is a constant parameter, and $d_{si}$ is the distance from patch $s$ to patch $i$.

*Kernel II*

**Spatially explicit dispersal with maximum dispersal distance**

An alternative dispersal kernel for the spatially explicit kernel is as follows. Assume that the probability $S_{ij}$ that a dispersing individual survives distance $d_{ij}$ between the natal patch $i$ and the target patch $j$ is

$$S_{ij} = 1 - \frac{d_{ij}}{d_{max}}, \text{ if } d_{ij} \leq d_{max}$$

$$S_{ij} = 0 \text{ otherwise,}$$

where $d_{max}$ is the maximum feasible dispersal distance. Thus

$$M_{si}(t) = S_{si}mX_s'(t)/(n-1).$$

$n = 25$ population subunits in explicit coordinate space. Here we use dispersal kernel I in Box 3.2. We also assume that function $F$ in eq. 3.12 is the Ricker model with one added delay (Ricker 1954; Turchin 1990)

$$F[N_i(t), N_i(t-1)] = N_i(t)\exp\{r_i[1 + a_1 N_i(t) + a_2 N_i(t-1)]\}, \quad (3.13)$$

where $r_i$ is the maximum per capita rate of increase, and $a_1$, $a_2$ are parameters of the direct and delayed the density dependence, respectively. For the spatial setting, we shall use both a regular and an irregular grid.

Not much happens if the subpopulation spacing is regular (fig. 3.8 (A)–(C)). The populations tend to fluctuate cyclically as specified by the deterministic renewal process, eq. 3.13. The only visible effect of dispersal

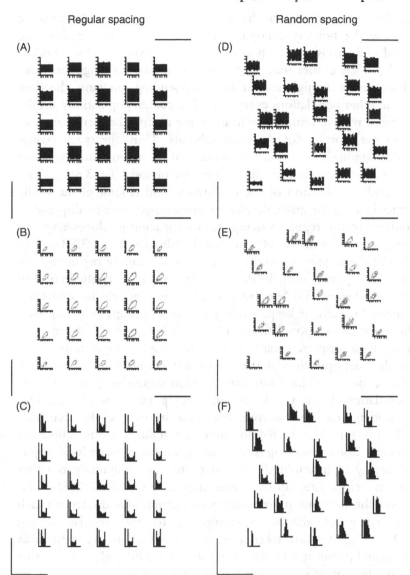

*Fig. 3.8.* Long-term dynamics of 25 local populations coupled with 10% dispersal (kernel I, p. 53). The left- and right-hand panels represent the regular and irregular spaces, respectively. (A), (D) Local dynamics (after initial transient period) for the final 1500 generations (for each panel, $y$-axis range is from 0 to 25 with 5 as subscale). (B), (E) The local population dynamics in phase space ($N_i(t+1)$ plotted against $N_i(t)$, both axes from 0 to 25 with 5 as axis division). (C), (F) The frequency distributions of population sizes for the final 1500 generations of the simulation ($x$-axis range from 0 to 25). The area of the histogram is normalized to 1 in each case.

is that the corner populations have the lowest overall population sizes; next come the bordering populations, and the highest densities are observed in the most central population (fig. 3.8(A),(B)). This is understandable, as the corner units have the least number of neighbors from which to receive immigrants, but they keep sending emigrants, the same 10% as all other populations every year. The central populations, on the other hand, receive immigrants from all populations but send fewer of them to the most remote (corner) subunits. Thus, the space creates inequality among the subunits. Common to all subpopulations is that the population size frequency distributions are bimodal (fig. 3.8(C)). This is yet another indication of the clockwork-kind ticking of the cyclic dynamics in a regular space, despite the disturbance cause by dispersal.

Contrary to the regular spacing, random spacing produces a variety of long-term dynamics of the local subunits (fig. 3.8(D)–(F)). Displacing the regular locations creates unequal inflow of immigrants into the local populations. In the example coordinate configuration in fig. 3.8, the lowest and leftmost population is relatively isolated from the others. Annually, it keeps sending 10% of its occupants as dispersers to the surrounding populations. They are, however, so distant that most of their dispersers harbor into the nearby populations. This makes the subpopulation at the lowest left corner have the smallest population density. The coordinate position dependency in the average population size is visible also in other populations (fig. 3.8(D)). This configuration also distorts the local intrinsic cyclic dynamics (fig. 3.8(D),(E)). The spatial location also influences the amplitude distribution. Sometimes population fluctuations in a given local population display high amplitude; at other times the amplitude is rather narrow, to return later on to a wide amplitude (fig. 3.8(D),(E)). This all makes the population size frequency distributions display a much greater range of shapes as compared to the regular setting (fig. 3.8(C),(F)). We have taken a closer look at the long-term dynamics of the second population in the second row (fig. 3.8(D)–(F)). Recall that the only element of stochasticity in the deterministic renewal process of the local populations is in the displacement of their coordinates. The basic pattern is that it shows the regularity of the intrinsic ten-year cyclic dynamics. Of much greater interest is that the dynamics, now and then, lose their cyclic characteristics just to reappear somewhat later on (fig. 3.9). The disappearance and return of the cycle seems to take place at rather irregular intervals. In addition, there is no obvious regularity in the changes in amplitude (fig. 3.9).

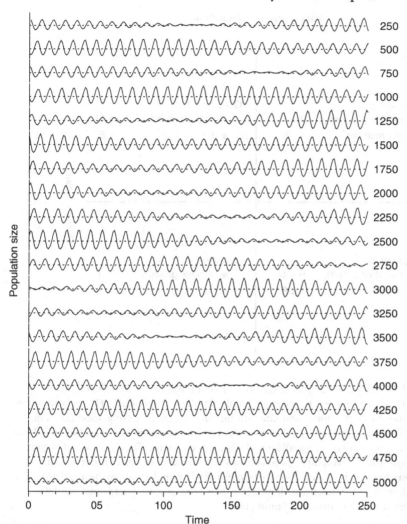

*Fig. 3.9.* Temporal fluctuations in a single population (2nd row and 2nd column in Fig. 3.8 (D)–(F) over a time span of 5000 generations (after the initial transient period). In each subpanel, the *y*-axis scale ranges from 5 to 20. The dotted line represents the long-term average population size.

## Asymmetrical dispersal

The dynamics under the above assumptions often result in asymmetries in migration rates between patches; some patches gain more individuals (on average) than they export. An interesting way to analyze the consequences

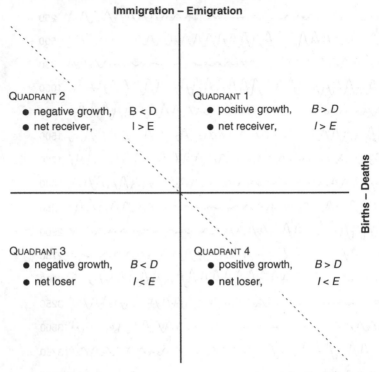

Immigration – Emigration

QUADRANT 2
- negative growth,  B < D
- net receiver,  I > E

QUADRANT 1
- positive growth,  B > D
- net receiver,  I > E

QUADRANT 3
- negative growth,  B < D
- net loser  I < E

QUADRANT 4
- positive growth,  B > D
- net loser,  I < E

Births – Deaths

*Fig. 3.10.* The demography space presented in terms of the four quadrants of $\Delta = B - D$ and $I - E$. On the right-hand side of the *y*-axis, births (*B*), exceed deaths (*D*); and above the horizontal line, immigration (*I*) exceeds emigration (*E*). The broken line across quadrants 2 and 4 indicates the combinations of *I* and *E*, and *B* and *D* giving the net rate of population change equal to zero. The dynamics of dispersal-coupled spatially structured populations make population trajectories cross the borders frequently. In the course of time, a single population may be a net producer (quadrant 1), a net loser (quadrant 3) or their combination (quadrants 2 and 4). Modified after Thomas and Kunin (1999).

of unbalanced emigration and immigration is to consider the relative importance of local (births and deaths, *B* and *D*) and regional (immigration and emigration, *I* and *E*) processes (Thomas and Kunin 1999). The approach also allows for the extremes of such asymmetries, namely source–sink dynamics (Pulliam 1988). The dynamics of the dispersal-coupled set of populations can now be understood in terms of population growth rate (births – deaths: $B - D$) and net migration (immigration – emigration: $I - E$). The balance between the two rates is described by four quadrants (fig. 3.10), defined by the inequality for the local demographic process, $B - D \neq 0$, and for the regional process, $I-E \neq 0$. The dotted line, the "compensation axis" (Thomas and Kunin 1999) in fig. 3.10, indicates the combinations of $I$, $E$, $B$, and $D$ giving the net rate of population change equal to zero. Thus, the compensation axis provides us with a way to visualize whether the population is increasing or decreasing, that is, whether $(B+I) - (D+E) \neq 0$. We will now apply the Thomas–Kunin approach to two spatial models.

**Three-patch model**

Assume three patches that are linked through dispersal. Each year, after the density-dependent renewal process, a fraction of every subpopulation leaves the natal patch to reproduce somewhere else. The entire landscape is also affected by some random external disturbing factor. We assume the patches are subject to global noise, $\mu(t)$. Each population is also separately influenced by a local disturbance $u_i(t)$. When the distance between the three patches is equal, the dispersal term in eq. 3.12 is

$$M_{si}(t) = 0.5m\overline{N}_s(t), \tag{3.14}$$

where

$$\overline{N}_s(t) = F[N_s(t), N_s(t-1), \mu(t), u_s(t)], \tag{3.15}$$

with $i = 1,2,3$. Thus, the emigrants from each patch are distributed equally between the neighboring patches. Note also that when the inter-patch distances are equal this is technically the same as implicit space. We will show here that there are often differences in the dynamics in implicit and explicit space.

We also change the spatial location of one of the patches (patch 3) keeping the others fixed. We model this by introducing a distance parameter $\chi$. When $\chi = 0.5$ all patches are equally distant from each other. When $0 < \chi \leq 0.5$, patch 3 is more distant from the two others. The smaller the value of $\chi$ the more difficult it is to reach 3 from 1 and 2 (or vice versa). We assume that the annually dispersing fraction $m$ is not affected. When the distance to patch 3 increases, fewer individuals will reach that patch. We use the following scenario, where the dispersal is defined as follows

$$M_{31}(t) = M_{32}(t) = 0.5mN_3(t)$$
$$M_{12}(t) = (1-\chi)mN_1(t); \quad M_{13}(t) = \chi mN_1(t) \tag{3.16}$$
$$M_{21}(t) = (1-\chi)mN_2(t); \quad M_{23}(t) = \chi mN_2(t).$$

However, the increasing distance has no effect on the dispersal success of individuals emigrating from patch 3.

The population size after renewal is

$$\overline{N}_i = \mu(t)\, u_i(t) N_i(t) \exp\{r[1 - N_i(t)]\}, \tag{3.17}$$

where $u_i$ is the local noise and $\mu$ is the global noise. Population-specific local noise is annually drawn from uniform random distribution (0.99, 1.01), and the global noise is drawn from a uniform random distribution on (0.95, 1.05). Emigration, $E_i(t)$, can be easily calculated from $m$ and $N_i(t)$, and from eq. 3.16 we get $I_i(t)$. Note that, on the local scale, we have not specified the values for births and deaths, $B_i(t)$ and $D_i(t)$. Instead, we calculate the difference

$$\Delta_i(t) = N_i(t+1) - N_i(t) \tag{3.18}$$

as the net change, where $N_i(t)$ is evaluated before the immigrants arrive and emigrants depart. We then calculate the net immigration rate as $I_i(t) - E_i(t)$.

When there is no difference in distance between the patches the dynamics in patches 1 and 2 are symmetrical (fig. 3.11). The variance in $\Delta$, however, is much larger than that of the net immigration, meaning that, in fact, the points are scattered around the $x$ axis. It is much more revealing, however, to look at the asymmetric case, where one of the patches is located at a distance from the others. Now, the net immigration to patches 1 and 2 is clearly positive, remaining at a nearly constant level ($I_i - E_i \approx 0.02$), whereas the net immigration to patch 3 is negative, and independent of $\Delta$ ($I_i - E_i \approx -0.04$). The compensation line crosses both sets of points, indicating that each patch acts at times as a net importer and exporter. The relative difference in the location of the net immigration level indicates, however, that the "isolated" population 3 acts more often as a net exporter than importer compared to the two others.

## A large grid

We now extend the analysis to larger landscapes, where we let $n = 25$ populations to be randomly located on a grid. Following the above recipe, global (G) noise is generated by drawing $\mu(t)$ from a uniform distribution between $1 - G$ and $1 + G$. Local (L) noise $u_i(k)$ is generated from a uniform distribution between $1 - L$ and $1 + L$. In the following, we will also use dispersal kernel I (p. 53).

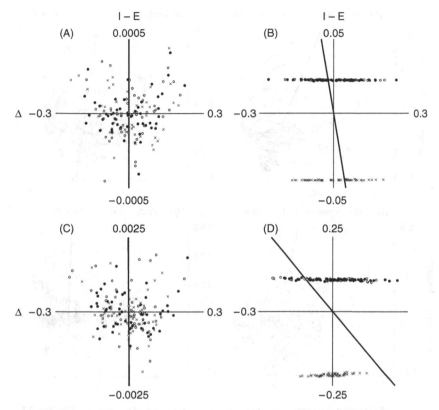

*Fig. 3.11.* The location of the three populations in the two-dimensional space of population growth rate. Population $N_1$ is indicated with ○, and $N_2$ with ●, and $N_3$ with ×. In the panels x-axis: $\Delta$; y-axis: $I - E$. The axes (thin lines) cross at the origin. Axis minima and maxima are indicated. Note that often the $(I - E)$ axis is very short compared to the $\Delta$ axis. This means that the thick compensation line indicating the combinations of $I$, $E$ and $B$ and $D$ giving the net rate of population change equal to zero often follows the $(I - E)$ axis very closely. [The compensation line can barely be distinguished from the $(I - E)$ axis in (A) and (C).] The dynamics in the local populations obey the Ricker model, with $r = 1.0$ (left-hand panels) or $r = 1.2$ (right-hand panels), both yielding stable dynamics. The populations are affected by global noise ($w = 0.05$) and by local noise (see text). (A), (B) five per cent ($m = 0.05$) of local population residents disperse from the natal patch. (C), (D) The dispersal rate is 25% ($m = 0.25$). In panels (A) and (C) the three populations are equidistant ($\chi = 0.5$), while in (B) and (D) reaching to (and reaching from) $N_3$ is much more difficult ($\chi = 0.05$; for more details see text).

Here we consider a specific example where we have selected $r = 0.53$, $a_1 = 0.05$, $a_2 = -0.1$. With these parameters, the model produces cyclic dynamics with a period of about 9–10 years (Ranta *et al.* 1997a). Although such population cycles are found in natural time series, one should be a little

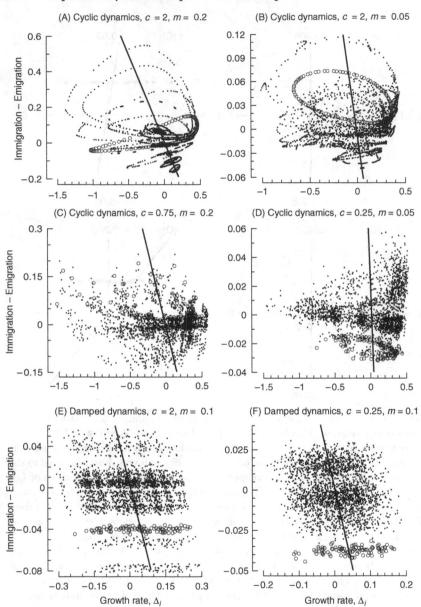

*Fig. 3.12.* Dynamics of 25 randomly located populations in a 20 × 20 co-ordinate space. The population fluctuations are presented in a space composed of growth rate, $\Delta_i$, and the difference between immigration and emigration rates, $I_i - E_i$. The dispersal parameters, $c$ and $m$, are indicated in each panel. (A), (B) Deterministic simulations. (C), (D), (E) The dynamics are disturbed with weak local ($L = 0.01$) and strong global ($G = 0.2$) noise. (F) $L = 0.05$ and $G = 0.1$. The top four panels,

wary of using cyclic dynamics to generate general insights into population dynamics. Thus, in addition to cyclic dynamics we also choose $r = 0.15$, $a_1 = -0.0035$, and $a_2 = -0.0074$, i.e., stable dynamics, for our second example (see Kaitala *et al.* 1996a for technical details). In the simulations, all populations were randomly initiated. The populations were then allowed to renew for 10 000 time steps and the last 100 steps were used as data in our analyses. Following the above analysis, the dynamics of the populations were followed in the $\Delta$ $(= B - D)$ versus $(I - E)$ plane.

The populations rarely cluster in that plane (fig. 3.12). The emerging patterns are disrupted by increasing environmental variability. With increasing dispersal rate, the paths of net immigration versus net change $(\Delta)$ in each local population become more clearly visible, although considerable overlap remains (fig. 3.12(A),(B)). This can be understood in the light of the observation that increasing dispersal makes regional (nonlocal) effects more important for the local dynamics. However, when the effect of dispersal distance is decreased (parameter $c$ of kernel I gets smaller) the outcome will become less clear (fig. 3.12(C),(D)).

The same pattern can be observed for damped dynamics (fig. 3.12(E),(F)). The population trajectories now tend to scatter in flattened horizontal clusters, however, aligned with the $\Delta$ axis. Increasing the proportion $m$ of dispersing individuals would increase the extent of the $(I - E)$ axis, as expected.

A common outcome in the two simulation experiments is that the population trajectories rarely associate along the axis indicating the combinations of $I - E$ and $\Delta$. On the compensation axis (Thomas and Kunin 1999), positive net immigration compensates for net decline in the population size due to demographic reasons, or population growth compensates for net emigration; of course this is quadrant dependent (fig. 3.10). In our simulation example, the local population trajectories hardly settle down on this axis. Rather, the dynamics are in a continuous stage of redistribution of individuals such that the sign of the net dispersal may

Caption for fig. 3.12. (*cont.*)
(A)–(D), are for cyclic dynamics with ten-year period length. The two bottom-row panes, (E) and (F), are for damped dynamics. The thick line across the panels indicates the combinations of $I - E$ and $\Delta$ giving the net rate of population change equal to zero. The line is located differently in each panel since axis scaling varies across the six panels. In each panel, a single population is marked with o to enhance inspection (note that the $x$ and $y$ co-ordinates of the 25 populations are randomly drawn in each simulation).

change in a more or less regular manner. Thus, no single population can be positioned on a temporal scale along the compensation axis (fig. 3.12). Instead, the populations tend to wander among the four quadrants defining the inequalities between births and deaths and immigration and emigration.

The dynamics of the populations in the demography space (fig. 3.12) illustrate the very nature of the dynamics of space-structured populations coupled by dispersing individuals. At times, a few populations may be located above the origin, although on average they lie below the origin. They may also be located on the left side of the $\Delta$ versus $I - E$ space, just waiting for the time to jump into the right-hand side later on. What we have stated above does not abandon the fact that there are populations that are continuously either above or below the origin (fig. 3.12). However, occasionally, a population remains permanently on the left- or right-hand side of the origin.

We may also attempt to characterize the sink–source dynamics based on average measures of the births–deaths immigration–emigration process. Concentrating on the long-term averages of the population trajectories would remove the scatter in the demographic space and would neatly align the subpopulations along the compensation axis. This result is also indicated in our simulation studies, especially for damped dynamics (fig. 3.12(E),(F)). Here, especially in figure 3.12(E), the averaging would place populations along the $\Delta$ axis. In fact, the heavy line in all the panels in fig. 3.12 goes through the population-specific long-term averages in the coordinate space.

Averaging population sizes over time certainly solves the apparent paradox of temporal imbalance in net growth and net emigration. We may question, however, the validity of averaging the population sizes. Consider, for example, the population marked with o in fig. 3.12(B). This population would have its long-term average on the compensation axis. However, during the 100-generation period it hits this axis only twice. Taking the average would be an extraordinarily good example of the fallacy of averages (Templeton and Lawlor 1981). Another example is illustrated with the similarly labeled population in fig. 3.12(A), where again the average is neatly on the compensation axis, and, in fact, precisely at the origin. This population, which is not the only one in this simulation, has an overall negative slope in the demography space. This contradicts the predictions based on the compensation axis and its suggested usage in classification of populations (Thomas and Kunin 1999).

We conclude that the usage of demography space, especially the compensation axis, in classifying spatially structured populations with dispersal linkage needs to be carefully interpreted in the light of temporal dynamics. One has to realize that populations do not stay put in the births–deaths, immigration–emigration space. The dynamics of dispersal and especially the dispersal–space interaction (Ranta *et al.* 1997a,b,c; Lundberg *et al.* 2000a) creates a complex outcome (fig. 3.12) that does not render to simple classification. Rather, the relevance of the demography space (fig. 3.12) is in rendering complex dynamics for a simple, but effective analysis.

## Summary

In this chapter we study the consequences of the population renewal processes dealt with in Chapter 2. We are especially focusing on the dynamics when the environment is spatially structured. We assume, as a first (and very useful) approximation, that the landscape a population inhabits is heterogeneous and that the "patches" the population can occupy are unambiguously delimited. We also consider various examples of landscape structures and dispersal assumption. However, the basic scenario is that, in each such patch, there is independent population growth, except for the fact that the patches are connected by dispersing individuals. The emigration rate may be density dependent or not, and the success of the dispersing individuals is dependent on the distance between patches. Models of spatial population dynamics produce a wide range of dynamic behaviors, from stable to very complex dynamics, often very dissimilar to single-patch dynamics. Also, spatial dynamics give rise to emergent properties including temporal synchrony between sub-population, and dynamics that are time variant (e.g., cycles that come and go with time). In this chapter, we also derive the different "dispersal kernels" used throughout the rest of the book.

# 4 · *Synchronicity*

Charles Elton (1924) was very well aware of the fact that many populations of a given species display large-scale temporal match in their population fluctuations. He was also among the first to propose that this, almost ubiquitous phenomenon, is due to environmental forcing, redistribution of individuals between breeding seasons, or biotic interactions of some kind. In this chapter, we shall first describe, with a set of examples familiar to us, patterns of synchronicity in various taxa. In these data, one new feature emerges that Elton did not mention: often the degree of coherent temporal population fluctuations is high among nearby populations but levels off with increasing distance. In the second part we shall address the question of how to analyze synchrony patterns. Finally we will turn to the different major explanations provided to understand large-scale synchronous fluctuations in population features of animals and plants.

Natural populations live in patchy environments. The distribution area of any given species should not be viewed as a continuous uniformly spread population, evenly painted over the landscape. Rather, the environment is composed of a network of habitable areas differing in profitability and of areas less suitable for population renewal. Even in pristine habitats, individuals are not distributed evenly all over the range. Our dogma is that natural populations are composed of local populations of varying size and quality. The independence of these units may vary: some of them can be entirely isolated while most population subunits are linked to other similar units via dispersing individuals. This view of the world suggests that neighboring individuals and individuals in proximate areas have a potential to interact more with each other than with individuals living in more distant units. It also suggests that population units nearby may share more in common than far-away units.

The independence of local populations is greatly affected by the intensity of immigration and emigration, the redistribution of individuals among the subunits in the network (Chapter 3). The two extremes are total isolation, and the mixing of individuals to such a degree that isolation plays no significance. In the two extreme situations, a single-population model is a sufficient description of the population renewal process (Chapter 2). However, most individuals live in a spatially structured world, where dispersal plays an important role. The local renewal process is, to a varying degree, due to the residents of that subunit. Thus, one has to acknowledge that dispersal linkage is significant in local dynamics. Redistribution of individuals ties local units into a regional population. If there is no redistribution, population subunits – though isolated – may share, at least to some extent, a common environment via regional climate. Here we shall focus on the temporal match in population fluctuations in systems where local populations are coupled to a network via individuals redistributing between successive population renewal occasions.

## Synchronous dynamics

Exploring temporal changes in population size in a spatially structured population network calls for population estimates replicated in space. This was first realized by Elton (1924), who brought to ecologists' attention the fact that Norwegian lemmings, Canada lynx and snowshoe hare each display synchronous population fluctuations over large geographical ranges (Lindström et al. 2001). Slightly more modern data on regional fluctuations of Finnish black grouse and mountain hare are displayed in fig. 4.1 and 4.3, respectively. The black grouse clearly had cyclic dynamics in Finland during 1964–1984 (Lindén 1989; Lindström et al. 1995), the period covered by the data (see also p. 72). Thus, it is rather easy to see that populations in different regions tend to fluctuate in step. For example in 1976–1977 there was a population low over almost all of Finland, and, likewise, most provinces had experienced a population-high phase 2–3 years before the big crash.

Visual inspection is an effective method for discovering a match of temporal phase in the dynamics of two or more populations. However, quantitative tools are also available from time series analysis techniques (Box et al. 1994; Chatfield 1999). The cross correlation function (Box 4.1) is a particularly useful method for assessing the degree of temporal coherence between any pair of population time series $X, Y$. The series is composed of $n$ pairs of values, and when the cross correlation is calculated with lag zero, $r_0$, it corresponds to the Pearson correlation, $r_{XY}$, between the two time series. Thus, large positive values of cross correlation indicate that the two populations fluctuate in synchrony. In contrast, values around zero indicate that the two time series have very little in common in their temporal fluctuations,

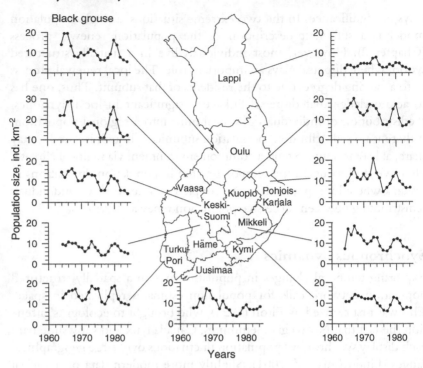

*Fig. 4.1.* Temporal (1963–1984) fluctuations in black grouse (*Tetrao tetrix*) population size in different provinces in Finland (data after Finnish Game and Fisheries Research Institute; Lindén and Rajala 1981).

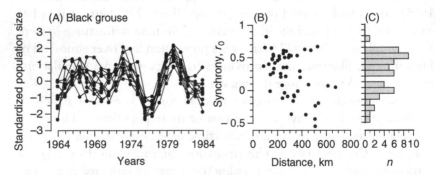

*Fig. 4.2.* (A) The 11 time series on black grouse dynamics in Finland (fig. 4.1) after linear trend is eliminated and the series are standardized into zero mean and unit variance. The graphs suggest that black grouse populations (especially from 1975 onwards) tend to fluctuate in synchrony in Finland. This is confirmed by calculating cross correlation coefficients with lag zero, $r_0$, between all provinces compared in pairs. It also appears (B) that the level of synchrony goes down with increasing distance between the provinces compared. The marginal distribution of the $r_0$ values in (B) is given in (C).

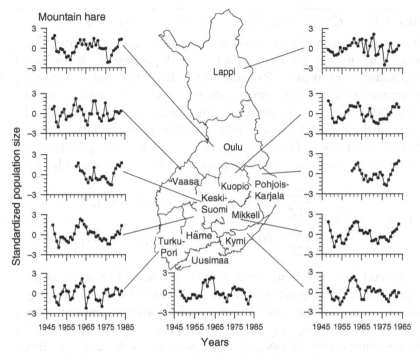

Fig. 4.3. Temporal (1948–1983) fluctuations in mountain hare (*Lepus timidus*) population size in different provinces in Finland. Linear trend is eliminated from the series and the data are standardized to zero mean and unit variance (data after FGFRI; Ranta *et al.* 1997d).

while large negative values of the correlation coefficient indicate that the populations fluctuate in synchrony, but that they are in opposite phase.

Prior to the synchrony analyses, the time series to be examined have to be de-trended (Box 4.2). Otherwise, the trend dominates the dynamics and affects the synchrony measure especially with longer lags. Prior elimination of the trend does not mean that the presence of a trend in the original data should be forgotten. De-trending is one technique in the decomposing process of the time series. Decomposing gives an understanding of the various components (trend, periodicity) of which the time series is built (Box *et al.* 1994; Chatfield 1999). Calculating cross power spectra (Box *et al.* 1994) is another tool for comparing the match in temporal structure between pairs of time series. With ecological time series, this method is seldom used, as the technique requires rather long periods of data.

The Finnish grouse data (fig. 4.1) were de-trended using linear regression against time and taking the residuals and standardizing them to zero

**Box 4.1** · *Cross correlation function: a measure of synchrony*

We assume a pair of population time series $X'$ and $Y'$ of matching length ($n = n_X = n_Y$). Often $X'$ and $Y'$ are logarithms of the original population sizes in time (Royama 1992). It is of crucial importance that prior to the synchrony analysis the series $X'$ and $Y'$ are de-trended to $X$ and $Y$. There are various time series methods of de-trending, and some of these are discussed in Box 4.2. The de-trended data have $n$ pairs of observations ($x_1$, $y_1$; $x_2$, $y_2$; ...; $x_n$, $y_n$) for both populations, for these data the estimate of the cross covariance coefficient at lag $k$ is

$$
\mathrm{COV}_{XY}(k) = 
\begin{cases}
\frac{1}{n}\sum_{t=1}^{n-k}[x(t) - \bar{x}][y(t+k) - \bar{y}]; k = 0, 1, 2, \ldots \\
\frac{1}{n}\sum_{t=1}^{n-k}[y(t) - \bar{y}][x(t+k) - \bar{x}]; k = 0, -1, -2, \ldots
\end{cases}
$$

here $\bar{x}$ and $\bar{y}$ are the sample means of the two series. The first row of the expression calculates positive lags between $X$ and $Y$, and the negative lags are calculated on the next line. The cross correlation coefficient is achieved by first calculating $s_X = \sqrt{\mathrm{COV}_{XX}(0)}$ and $s_Y = \sqrt{\mathrm{COV}_{YY}(0)}$, and then writing $r_{XY}(k) = \mathrm{COV}_{XY}(k)/s_X s_Y$, with $k = 0, \pm 1, \pm 2, \ldots$ . Box *et al.* (1994) give equations to calculate standard errors of cross correlation estimates. We shall refer to cross correlation with lag $k$ as $r_k$ and when the cross correlation is calculated over a series of lags it is referred to as CCF. Most statistics packages available today include routines to calculate CCF with corresponding 95% confidence limits (fig. B3).

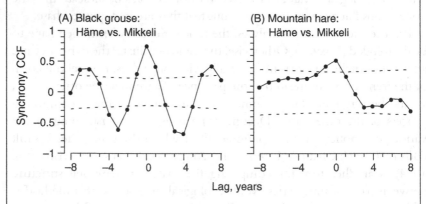

*Fig. B3.* Cross correlation functions, CCF (lags −8 to 8) for black grouse data (Fig. 4.1) and mountain hare data (Fig. 4.4) in provinces Häme and Mikkeli. The broken lines give 95% confidence limits.

---

**Box 4.2** · *De-trending of time series*

The two most often used de-trending methods with ecological data are elimination of linear trend and differentiation (for other methods, see, e.g., Chatfield 1999). Linear trend can be removed simply by fitting a first-order polynomial (linear regression) between time and the time series values and by taking residuals as the new values for the time series: $X = X' - (a_X + b_X X')$. Here $a$ is a constant and $b$ is the slope of the linear equation capturing the linear trend (if $b \approx 0$, $a$ is the long-term average of the population size). For curved trends second-order or higher-order polynomials can be used but one has to use caution here so as not to introduce arbitrary dynamics into the residuals by fitting too complex a polynomial model into the original data.

Differencing, often done for $\log(X')$ transformed time series, is, as the name implies, the difference between two subsequent observations. In ecological literature, log-transformed differentiation is referred to as $R(t)$, population growth rate. Differentiation is a simple and powerful tool to get rid of any trend. However, one has to keep in mind that in population data differentiation also changes the topic of interest from population size fluctuations into fluctuations of population growth rate.

De-trending, principally a straightforward process, can become a tricky business. A few examples are shown in fig. B4. The trend elimination for the black grouse in the province Keski-Suomi (fig. 4.1) is done both by taking residuals of a linear regression of population size against time (fig. B4(A)), and by differentiating (fig. B4(B)). As can be seen, the cyclic character of the grouse dynamics is retained in both techniques. We emphasize that fig. B4(A) displays periodic population fluctuations while fig. B4(B) displays cyclic fluctuations in $R(t)$.

The data in fig. B4(C) are generated using an AR(1) process with $\phi_1 = 0.5$ and $\delta = 12$. Both time series are superimposed with a trend, the slope being either increasing, $X'_P$, (slope $b_P = 0.2$ units per time step) or decreasing, $X'_N$, ($b_N = -0.2$). Both series are subjected to local noise ($b$ being multiplied by random numbers between 0.5 and 1.5 having an expectation of 1). Calculating CCF for the nontreated time series gives negative $r_k$ values with all lags, while the CCFs of the series where linear trend is removed and residuals are standardized to zero mean and unit variance, and where the de-trending technique is differentiation of the log-transformed original series give the highest synchrony measures at lag zero, as expected (fig. B4(F)).

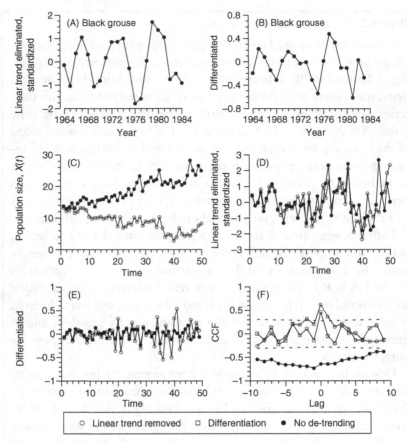

*Fig. B4.* (A) Black grouse data from the province Keski-Suomi (Fig. 4.1) after linear trend is eliminated and the series are standardized into zero mean and unit variance. (B) The same data after using differentiation (of the log-transformed data). (C) Two time series, one with increasing trend, $X'_P$, and the other one with decreasing trend, $X'_N$. The time series $X'_P$ and $X'_N$ after the trend is eliminated with a linear regression technique (D) and after using differentiation of log-transformed data (E). The cross correlation functions $CCF$ (lags $-9$ to 9) for the original $X'_P$ and $X'_N$ data, and after the de-trending with the two methods (F). The broken lines give 95% confidence limits. The legend below panels (E) and (F) is for (F).

The nontreated ($r_0 = -0.63$) and de-trended (linear trend removed $r_0 = 0.62$, differentiated log-transformed series $r_0 = 0.47$) time series differ greatly in terms of CFF and synchrony at lag 0. One may suspect that $X'_P$ increases, e.g., because of increasing carrying capacity, while $X'_N$ decreases due to deterioration. However, the long-term trend is not the focus of the study, rather, the interest is in the residuals fluctuations, which may be caused by factors such as climate fluctuation. Thus de-trending is the transformation called for, and we have learned that over a longtime $X'_P$ increases and $X'_N$ decreases, while $X_P$ and $X_N$ fluctuate in rather high synchrony.

The first step in time series analysis is always to graph the data (Chatfield 1999). Visualization effectively reveals characters of the time series (trend, possible periodicity, discontinuities, etc.). For example, black grouse populations display a clear decreasing trend in 1964–1984 in most provinces in Finland. Of course, time series are children of their time: they are just samples in time of the temporal behavior of the target population. For example, the two series with opposing trend in fig. B4(C), may both possess slow sine-wave-like dynamics in opposite phase, but the 50 year sample does not reveal it. Thus, the time series data are to be taken as they are. One has to bear in mind that de-trending, easy as it is to do, may be a cause of introducing artifacts into the data.

mean and unit variance (fig. 4.2(A)). With black grouse, displaying cyclic dynamics with a period of 6–7 years (Lindén 1989; Lindström et al. 1995), a casual look at fig. 4.1 reveals that populations of this species tend to fluctuate in temporal match over most of Finland. De-trending and standardization enhance this impression (fig. 4.2(A)), and when the level of synchrony is assessed by calculating the cross correlation coefficient with lag zero, $r_0$, the impression becomes quantified (fig. 4.2(B),(C)). There are data for 11 provinces; therefore, one can calculate 55 cross correlation coefficients in pairs, i.e., $[(n-1)n]/2$.

A few comments are worth noting here with respect to the black grouse data. First, the marginal distribution of the $r_0$ has a weight on positive values ($r_0 > 0$ in 73% of all cases). Second, when the synchrony values are graphed against the distances among the geographical mid-points of the Finnish provinces, one finds the relationship to be negative, the correlation being $r_D = -0.46$. Third, assessing statistical significance for the overall synchrony level in black grouse population fluctuations in Finland, or how the degree of synchrony levels off against distance (fig. 4.2) is a tricky task. This is because in all possible comparisons in pairs the resulting synchrony measures are not entirely independent of each other. Re-sampling techniques in statistical analysis (Efron and Tbshirani 1983) come to the rescue (Ranta et al. 1995b; Koenig and Knops 1998; Buonaccorsi et al. 2001). For example, the re-sampled average synchrony (with 95% confidence limits) was 0.27 (0.18–0.35) for the 1964–1984 black grouse, and the re-sampled value for the $r_D = -0.40$ (Lindström et al. 1996).

When correlations are calculated, the question is always raised about the statistical significance of the estimate. Such is also the case with the $r_k$. Calculating the standard error for the cross correlation estimate with lag $k$ is a straightforward process (e.g., Box et al. 1994). However, when more than one synchrony measure is calculated between a number of time series in pairs, the number of resulting correlation coefficients increases rapidly, hence test statistics based on the synchrony estimates and their error terms will be devalued by the multiple-testing effect. Specially tailored statistical tests are called for when one needs to assess the statistical significance of the relationship between synchrony and it leveling off with distance. For this purpose, Koenig and Knops (1998) provide a novel spatial autocorrelation method. This method, and the others used in assessing statistical significance of the synchrony measures, are reviewed by Buonaccorsi et al. (2001). In the pattern seeking, it often suffices to score on which side of zero the weight of the synchrony measures lies, and the sign of the $r_D$. However, when specific hypotheses about the causes and consequences of population synchrony are to be tested, proper statistical tools are of great value.

There is also a caveat in calculating average synchrony over a number of population time series. This can be best illustrated by an example. Assume two sets of population time series $A$ and $B$, both consisting of long-term observations of a number of local populations. All populations in set $A$ are in high synchrony with each other, also in $B$ all populations fluctuate more or less in step. However, the sets $A$ and $B$ are fluctuating in opposite phase. Now, when the synchrony measures are calculated in pairs between all populations, the emerging frequency distribution of the $r_0$ values is bimodal. One mode peaks with high positive $r_0$ values (populations $A$ against each other and populations $B$ against each other), the other one with high negative synchrony values (populations in $A$ against populations in $B$). Averaging over a bimodal frequency distribution like this would yield an overall synchrony measure close to zero (depending on the numbers of populations in the groups $A$ and $B$). Thus, it is advisable to have a look at the marginal distribution of the $r_0$ values (as displayed in figs. 4.2 and 4.4).

The 1964–1984 black grouse population fluctuations in Finland are cyclic without doubt (Lindström et al. 1995, 1999). Elton's (1924) examples were also on species with pronounced periodic fluctuations. However, cyclic dynamics are by no means a necessity for synchrony to

FINNISH DATA (A–F)

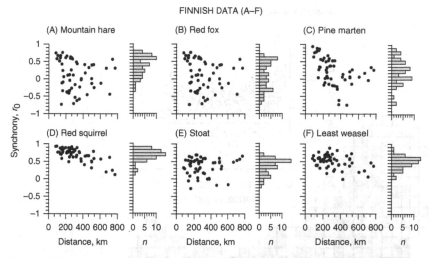

*Fig. 4.4.* Data on synchronous fluctuations in (A) mountain hare (*Lepus timidus*), (B) red fox (*Vulpes vulpes*), (C) pine marten (*Martes martes*), (D) red squirrel (*Sciurus vulgaris*), (E) stoat (*Mustella erminea*) and (F) least weasel (*Mustella nivalis*) populations in 11 provinces in Finland. For each species the scatter plot diagram gives the level of synchrony, $r_0$, against the distance between the pair of provinces compared, and the histogram gives the marginal distribution of the $r_0$ values. Linear trend is eliminated from the series and the data are standardized to zero mean and unit variance ((A)–(F): data after FGFRI; Lindén 1988).

emerge in regional population dynamics (Ranta *et al.* 1995a, 1998, 1999a). Our second example of regional data, also courtesy of Finnish Game and Fisheries Research Institute (FGFRI), is the 1948–1983 fluctuations of Finnish mountain hare (fig. 4.3). There is no good evidence for cyclic fluctuations in mountain hare (Ranta *et al.* 1997d). This is well exemplified by calculating the cross correlation function with lags from −8 to 8 for both black grouse and mountain hare data in provinces Häme versus Mikkeli (fig. B3 in Box 4.1). The sine-function-like cross correlation function clearly reveals the periodicity of the black grouse fluctuations, while no such pattern emerges for the mountain hare populations in these neighboring provinces (for better means to find periodicity in time series see p. 32 and Lindström *et al.* 1997a). Yet, for both species the synchrony measure is at its highest with zero lag. When the mountain hare data are analyzed in the same way as the black grouse data, similar features in the synchronicity pattern emerge: at the national level there are evidently synchronous dynamics in hare population fluctuations, and the level of synchrony decreases with increasing distance between the

SNOWSHOE HARE IN CANADA

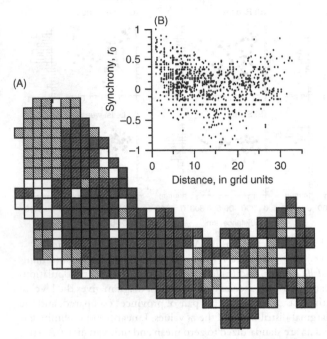

*Fig. 4.5.* Spatial synchrony for snowshoe hare in Canada (1931–1948; Smith 1983). In the map (A), four spatial clusters with matching dynamics averaged over the study period are indicated with matching shading of the grid units (each covering about 900 square miles). (B) Synchrony against distance in grid units of the map (redrawn from Ranta *et al.* 1997d).

provinces (fig. 4.4). In fact, the FGFRI data on other small game animals also display a matching pattern (fig. 4.4). However, when comparing large enough geographical areas, like that of the population fluctuations of snowshoe hare in Canada (fig. 4.5; Ranta *et al.* 1997d), one might also find that the degree of synchronous fluctuation first levels off with distance, to increase again when distance, among the areas compared becomes long enough (fig. 4.5). Similar observations are also found for the Canada lynx, the major predator of the snowshoe hare (Ranta *et al.* 1997a,b; Krebs *et al.* 2001).

Periodic masting, or mass fruiting, by temperate trees (Kelly 1994; Koenig and Knops 1998, 2000a,b) is an example of almost continent-wide synchrony of sessile organisms. A matching example is synchronous tree-ring growth in various tree genera in North America and Europe (Fritts 1976; Koenig and Knops 2000a). Also here the level of temporal synchrony fades away with increasing distance. However, synchrony levels tend to be rather high, up to ranges of several hundred kilometers (Koenig and Knops 2000a).

After this short exploration of synchrony in population fluctuations, questions arise about the within-species synchrony. First, how common is the finding that populations of various species fluctuate in step over large geographical ranges? Second, if one finds large-scale synchronicity, does the degree of synchrony level off with increasing distance as the data in fig. 4.4 are suggesting?

## Explanations of synchrony

### Moran's theorem

The question of why two or more populations of a given species fluctuate in temporal match was first addressed in quantitative terms by an Australian statistician, P. A. P. Moran (1953a,b). He proposed that if two populations $X$ and $Y$, sharing a common structure in the density-dependent feedback of the renewal process, are disturbed by external forces $\varepsilon$ and $\eta$ that are correlated $\rho(\varepsilon, \eta)$, populations $X$ and $Y$ will also start to fluctuate in synchrony, $r_0(X, Y)$. For his argument, Moran used the second-order linear autoregressive model, AR(2), where the common structure is via matching coefficients $a_1$ and $a_2$

$$X(t + 1) = a_1 X(t) + a_2 X(t - 1) + \varepsilon(t)$$
$$Y(t + 1) = a_1 Y(t) + a_2 Y(t - 1) + \eta(t). \tag{4.1}$$

In fact, Moran (1953b) said that with eq. 4.1 the correlation should be $r_0(X,Y) = \rho(\varepsilon,\eta)$. The explanation was later coined with the name "Moran's theorem" (Royama 1992). It is a widely spread phenomenon that populations of a given species tend to fluctuate in step over large geographical ranges (Elton 1924; Elton and Nicholson 1942; Butler 1953; Smith 1983; Marcström et al. 1990; Steen et al. 1990; Pollard 1991; Thomas 1991; Hanski and Woiwod 1993; Ranta et al. 1995a,b; Sutcliffe et al. 1996; Myers et al. 1997a; Ranta et al. 1997a,b,d,e; Myers 1998; Bjørnstad et al. 1999a,b; Ranta et al. 1999a; Rusak et al. 1999; Paradis et al. 2000; Williams and Liebhold 2000; to list just a few), so it is surprising that Moran's theorem was left almost unnoticed until relatively recently (Royama 1992). In this respect, Leslie (1959) is a clear outlier of his time as he soon took further Moran's idea, and proposed a matrix-modeling approach to the Moran effect (Leslie 1959). Leslie assumed two unconnected populations (both with four age groups) living in limited environments. These populations were expected to be proximate enough to share the effect of some external random and density-independent factor. Under these conditions, Leslie wrote the population renewal process to be

$$\mathbf{N}_i(t + 1) = \mathbf{L}\mathbf{R}^{-1}(t)\mathbf{N}_i(t),$$ (4.2)

where $\mathbf{N}_i(t)$ is the column vector indicating the number of individuals in each age group of the $i$th population, $\mathbf{L}$ is the Leslie matrix (p. 12), while $\mathbf{R}$ is a diagonal matrix containing the density-dependent components influenced by the Moran noise (Leslie 1959). Leslie showed that, in this system, populations initially fluctuating out of phase would soon become synchronized in their oscillations (Lindström et al. 2001).

### The Moran effect

Moran's theorem is now perhaps better known by more liberal names, such the "Moran effect," "Moran disturbance," and "Moran noise." Other synonyms are "external disturbance," "environmental noise," or just "disturbance" and "noise," or even "stochasticity." The most notable difference between the liberal versions and the original theorem is that the modern varieties do not assume a strict linear (or linearizable) structure in the population dynamics, or in the way the external disturbance is implemented to influence the renewal process. To make this distinction clear, and to avoid further confusion, we shall suggest the name Moran's theorem to be used only for systems strictly confirming to the original description given by Moran (1953b), otherwise is better to use more liberal epithets for the noise that modulates the dynamics of populations. Moreover, departing from the structure of the original model is likely to lead to situations where the Moran theorem, i.e., $r_0(X,Y) = \rho(\varepsilon,\eta)$, no longer holds.

To study in more detail the synchronizing potential of the external modulator we shall now rephrase eq. 4.1 in a general form

$$X(t + 1) = F[X(t), X(t - 1), \ldots]\mu_X(t)$$
$$Y(t + 1) = F[Y(t), Y(t - 1), \ldots]\mu_Y(t)$$ (4.3)

Here $F$ is a general population renewal process as a function of past population sizes with various lag terms. Note that − in contrast to Moran (1953b) − the function $F$ is not necessarily assumed to be a linear one, or a function that can be linearized. The disturbance, the Moran effect $\mu$,

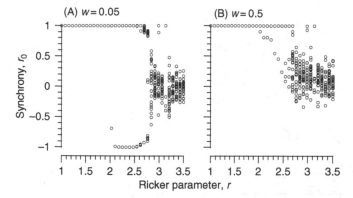

Fig. 4.6. Synchrony levels between two populations both obeying Ricker dynamics (with $K = 1$). The color of the external disturbance is white, but in (A) ($w = 0.05$) it is much weaker than in (B) ($w = 0.5$). In the region of stable dynamics ($r < 2$), the two populations will become synchronized, but not always in the region of periodic dynamics ($2 < r < 2.76$) and very seldom in the region of chaotic dynamics ($r > 2.76$). For a bifurcation diagram of the Ricker dynamics see, fig. B1 in Box 2.2 (p. 19).

is taken here as a multiplicative effect from a uniform distribution of random numbers between $1 - w$ and $1 = w$, where $0 < w < 1$ (Ranta *et al.* 1995a). The term $w$ is the strength of the external impact.

Consider now a two-population system where the renewal function $f$ obeys the Ricker dynamics with $1 \leq r \leq 3.5$. To generate the Moran effect we shall let the impact strength have two differing values, $w = 0.05$ and $w = 0.5$. The population renewal process (initiated with random numbers between 0 and 1) was left running for 1100 generations. The synchrony between $X$ and $Y$ was calculated from the final 100 time steps. Unlike the AR process in eq. 4.1, the Ricker dynamics will not become synchronized very well in the periodic and chaotic region ($r > 2$), and if the strength of the impact $w$ is weak for $2 < r < 2.76$ the two populations are either in perfect synchrony or entirely out of phase in 50% of the trials (fig. 4.6). It appears that the Moran effect is not completely capable of synchronizing complex nonlinear dynamics (Ranta *et al.* 1997e,f).

In eqs. 4.1–4.3 the Moran effect forcing the two populations is global, i.e., the noise influencing $X$ and $Y$ is perfectly correlated. This is not always the case. The different populations may share a common environment, i.e., they are all disturbed by a global noise, but they may be differently influenced by local noise. To explore this we selected a system of $n = 10$ unconnected populations variously affected by local and global disturbances ($w_{LOCAL}$ and $w_{GLOBAL}$ ranges being from 0.01 to 0.25). In the

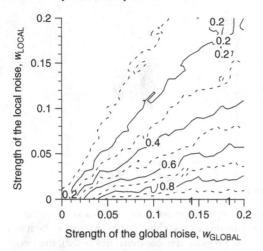

*Fig. 4.7.* Average synchrony isoclines ($r_0$) between ten nonlinked populations obeying delayed Ricker dynamics with parameters selected ($r = 0.53$, $a_1 = 0.05$, $a_2 = -0.1$) so that the emerging dynamics display cyclic dynamics with approximately ten-year periodicity. The populations are disturbed with global and local perturbations (equal range in both). The results are averaged over 50 replications of all parameter combinations.

exercise, local disturbance is by definition population specific, while the global noise affects all $n$ populations likewise. The populations were set to obey the delayed Ricker dynamics with a ten-year cycle (p. 23; the simulations were left running for 1100 generations of which the final 100 were used to score the average level of synchrony $\bar{r}_0$ among the populations). Not unexpectedly (fig. 4.7), the level of synchrony achieved under such a situation depends on the relative strength of the global and local effects (Ranta *et al.* 1997f). The global perturbations have to dominate the local ones in order to achieve reasonable levels of synchrony without any other effects called into action. This observation underlines the relative impact of local and global processes in affecting the dynamics of local population subunits. The delayed Ricker nonlinear second-order dynamics do not render so easily to perfect synchrony (fig. 4.7), except when exposed to relatively high global disturbing forces. This is in contrast to the linear AR models of the Moran theorem that will be synchronized once the global effect takes over the local one.

A recent debate (Blasius and Stone 2000; Grenfell *et al.* 2000) concerns synchronicity in the fluctuations of numbers of two feral sheep populations (Grenfell *et al.* 1998). The populations on two islands in the

St. Kilda archipelago are surrounded by rough sea, effectively cutting off dispersal; thus, making up a Moran's dream system. The discussion concerns the fact that the environments shared between the two islands correlate more strongly than the dynamics of the two sheep populations, $r_0(X, Z) < \rho(\varepsilon, \eta)$. The conclusion (Blasius and Stone 2000) is that the Moran theorem seems not to hold. The argument goes that the nonlinearity observed in the sheep dynamics (Grenfell *et al.* 1998, 2000) is responsible for this mismatch (but see Ripa 2000). Overall, observing such apparent inconsistency between data and theory does not nullify Moran's achievement. What is badly needed is a general picture of the population synchronies when the circumstances do not match exactly the assumptions in Moran's theorem. Possible deviations from Moran's assumptions include nonlinearities (Grenfell *et al.* 1998), temporally autocorrelated noise, and different local endogenous dynamics. Thus, despite efforts over the past 50 years, the population level consequences of the Moran effect are far from being completely understood.

Seed masting and tree-ring growth both show large-scale synchrony; in addition they demonstrate that synchrony levels off with increasing distance (Koenig and Knops 1998, 2000a). Here the synchrony scale is several hundreds of kilometers. Climate events (rainfall, temperature) are the most likely candidates for the external agent synchronizing masting and tree-ring growth (Koenig and Knops 2000a). With trees, dispersal is thought to be of no relevance (but for pollen flow, dispersal is an agent of synchrony, see Satake and Iwasa 2000) in synchronizing the life history events. Surprisingly enough, the spatial pattern in masting or growth and in various climate indices is different (Koenig and Knops 2000b). Hence, in this case climatic events cannot be unambiguously advocated as the Moran effect making the life history events take place in synchrony. However, elsewhere (Post and Stenseth 1999; Post *et al.* 2001) it has been shown that climate effects influence plant life histories over large geographical ranges.

We shall close this section by pointing out that when the external noise modulates the dynamics of the focal populations, it might turn out that the underlying dynamics become unrecognizable. This has been shown to be the case with both Ricker and delayed Ricker dynamics when the disturbance is frequent and strong enough (Ranta *et al.* 1998). Thus, the Moran effect has a dual face: it may generate synchronicity in population fluctuations over large ranges, but it may also mutilate the underlying skeleton of the dynamics to the extent that it becomes unidentifiable. This is the "visibility" problem (Ranta *et al.* 1997f, 2000a; Kaitala and Ranta

2001; Laakso *et al.* 2001; Lundberg *et al.* 2001; Jonzén *et al.* 2002a) discussed in Chapter 2.

## Dispersal

Elton (1924) first proposed that redistribution of individuals between breeding seasons might synchronize dynamics of populations. However, to substantiate this suggestion took half a century. Maynard Smith (1974) demonstrated – using theoretical machinery – that dispersal, indeed, synchronizes fluctuations of populations. Since then, many authors have suggested dispersal as one of the key agents responsible for synchronicity (e.g., Elton 1924; Butler 1953; Smith 1983; Steen *et al.* 1990; Hanski and Woiwod 1993; Ranta *et al.* 1995a; Bjørnstad *et al.* 1999a; Cattadori *et al.* 2000). To illustrate the synchronizing power of redistributing individuals let us first return to the two-population Ricker dynamics (p. 44, fig. 4.6), which the Moran effect could not synchronize. We shall now remove the external perturbation but will let 10% of the residents in $X$ take leave to $Y$ and the other way round. It turns out that with periodic Ricker dynamics dispersal is capable of synchronizing fluctuations in the two populations (fig. 4.8) but not when the dynamics are complex (Ranta and Kaitala 2000). It is worth noting, however, that dispersal modulates the originally complex dynamics into a two-point periodicity.

Ranta *et al.* (1995a, 1997c) did a detailed analysis of the significance of the Moran effect and dispersal on synchrony level among local populations. Here we shall revisit part of the analysis (Box. 4.3) with a $2^2$-factorial treatment: Moran effect (yes, no) and dispersal (yes, no) as factors. The four differing combinations also give differing results as to the degree of temporal match in coherence of fluctuations among the $n$ populations. Obviously, if there is no external disturbance and no dispersal, the populations, if initiated out of phase, will also keep on fluctuating out of phase (fig. 4.9(A)). The Moran effect alone is capable of raising the level of synchrony $r_0 \approx 0.5$ (fig. 4.9(B)). The distance-dependent dispersal enhances the level of synchrony between closely located populations but less so for more distantly related populations (fig. 4.9(C)). It is worth noting that negatively distance-dependent dispersal can alone yield synchrony patterns matching those observed with real population systems (figs. 4.2, 4.4). When both the Moran effect and dispersal are in action one gets enhanced synchrony values that also level off with distance (fig. 4.9(D)). In fact, as already pointed out by Ranta *et al.* (1995a), with real data the two cases, distance-dependent

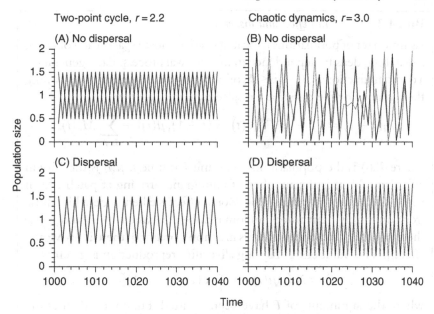

Fig. 4.8. Population trajectories of a two-population system (thick and thin line) obeying periodic ($r = 2.2$) and chaotic ($r = 3.0$) Ricker dynamics ($K = 1$). Panels (A) and (B) indicate the initial dynamics, while the corresponding dispersal-modulated dynamics are given in (C) and (D). Dispersal (10% of individuals moving between the two populations) is capable of synchronizing Ricker dynamics in the periodic region but not in the chaotic region of $r$. In the chaotic region with dispersal, the fluctuations in population size will be modulated into two-point cycle. Modified after Ranta and Kaitala (2000).

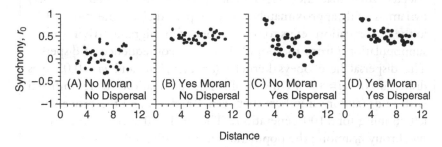

Fig. 4.9. Synchrony levels against distance among nine populations obeying delayed Ricker dynamics with parameters selected so that the emerging cyclic dynamics have a ten-year periodicity. The four panels indicate results of a system where the Moran effects (no, yes) and negatively distance-dependent dispersal (no, yes) were superimposed on the population renewal process in various combinations.

**Box 4.3** · *Moran effect and dispersal*

Assume a set of populations $n$ located in co-ordinate space, each obeying a matching density dependence in the renewal process. Each generation a constant fraction $m$, $0 \leq m \leq 1$, of each population redistributes among the $n$ units. For the dispersal-coupled dynamics we have

$$X_i(t+1) = (1-m)F[X_i(t), X_i(t-1), \mu(t)] + \sum_{s,s\neq i} M_{si}(t),$$

where $X_i(t)$ is the population size in unit $i$ at time $t$, $\mu(t)$ is the Moran effect and $M_{si}(t)$ is the number of immigrants arriving at patch $i$ from patch $s$ because of dispersal. The Moran effect is taken to be white noise from uniformly distributed random numbers between $1-w$ and $1+w$ [here $w = 0.1$ (the Moran effect is in action), or $w = 0$ (there is no Moran effect)]. The number of offspring alive after reproduction is given as

$$F = X_i(t)\mu(t)u_i(t)f[X_i(t), X_i(t-1)],$$

where the arguments of $F$ have been omitted. Function $f$ defines the delayed density-dependent per capita reproductive rate. The term $u_i(t)$ is local noise (drawn from uniform random numbers between 0.95 and 1.05). For the Ricker dynamics with delayed density dependence we write

$$f[X_i(t), X_i(t-1)] = \exp\{r[1 + a_1 X_i(t) + a_2 X_i(t-1)]\},$$

where $r$ is maximum per capita rate of increase, and $a_1$ and $a_2$ are parameters determining density dependence. The parameters were selected such that they yield cyclic ($r = 0.53$, $a_1 = -0.05$, $a_2 = 0.1$) dynamics with approximately ten-year periodicity. The term $M_{si}(t)$ in the first equation refers to the number of immigrants arriving at the subpopulation $i$ from the population $s$ as a consequence of dispersal. The dispersal here obeys kernel I ($c = 2$; p. 53), and $m = 0.05$ (yes dispersal) or $m = 0$ (no dispersal). The populations were initiated with i.i.d. random numbers (between 1 and 10) and the process was left running for 1100 generations. The final 100 were used to score synchrony $r_0$ among the populations in pairs (fig. 4.9).

dispersal alone or distance-dependent dispersal acting in concert with the Moran effect, are hard to tell apart.

We shall now briefly return to the system of $n$ unconnected populations that are variously affected by local and global disturbances ($w_{\text{LOCAL}}$ and

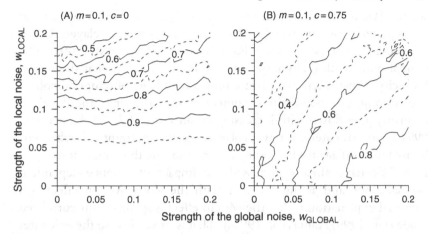

*Fig. 4.10.* Average synchrony isoclines ($r_0$) between ten dispersal-linked populations (spatially implicit structure) obeying delayed Ricker dynamics with parameters selected ($r = 0.53$, $a_1 = 0.05$, $a_2 = -0.1$) so that the emerging dynamics display cyclic dynamics approximately with ten-year periodicity. The populations are disturbed by global and local perturbations (equal range in both). The results are averaged over 50 replications of all parameter combinations (see also fig. 4.7).

$w_{GLOBAL}$). In this example the populations follow the delayed Ricker dynamics, (Box 2.5), with ten-year cycle period length, and, precisely as previously, the simulations were left running for 1100 generations of which the final 100 were used to calculate in a pairwise manner the average level of synchrony in $n$ populations. The only deviation to the analysis reported in fig. 4.7 is that we allowed 10% ($m = 0.1$) of the resident individuals to disperse among the population subunits. If dispersal distance plays no role ($c = 0$ in the kernel I (p. 53), dispersal becomes independent of distance among the subunits (i.e., one has spatially implicit population structure)) in the redistribution of individuals, the synchrony levels achieved become dependent only on the intensity of the local disturbance. With more localized dispersal ($c = 0.75$) the synchrony level isoclines bend from horizontal ones (fig. 4.10(A)) towards 45° (fig. 4.10(B)). Comparison with fig. 4.7, however, shows that even a modest dispersal over short distances enhances the overall level of synchrony in population fluctuations.

## Spatially autocorrelated Moran effect

The outcome of the experimentation with the Moran effect and dispersal as synchronizing agents (fig. 4.9) is not too sensitive to the underlying population renewal process. In their original analysis Ranta *et al.* (1995a) used also Ricker dynamics and age-structured population dynamics. Later on this pattern was confirmed with damped and complex delayed Ricker dynamics and with AR(1) and AR(2) dynamics (Ranta et al. 1999a). The essential point with many population series from nature that show synchronicity is that often there is not only a high level of temporal coherence among the populations compared but also that the level of synchrony goes down with increasing distance (fig. 4.2, 4.4; Ranta et al. 1999a). Unfortunately, a few explanations may account for why synchrony levels off against increasing distance among the populations compared. We have already discussed the impact of distance-dependent dispersal alone and acting together with the global Moran effect. An additional explanation is that the Moran effect is spatially autocorrelated (Lande et al. 1999; Ranta et al. 1999a). That is, areas close to the epicenter of the disturbance receive a higher level of perturbation than areas far away from the center point of the disturbance.

One way of implementing spatial autocorrelation in the Moran effect is (Ranta et al. 1999a)

$$X_i(t + 1) = (1 - m)F[X_i(t), X_i(t - 1), \mu(d_i, t)] + \sum_{s, s \neq i} M_{si}(t), \quad (4.4)$$

with a slight modification of the first equation in Box 4.3. The last term of eq. 4.4 is for the dispersal kernel I. Here the $\mu(d_i)$ is the Moran effect, now characterized by its intensity, $\mu_i$, for each subpopulation as a function of the distance $d_i$ of each subpopulation $i$ from the place of the hit, e.g.,

$$\mu_i(d_i, t) = 1 - \exp(-c_{MORAN}d_i). \quad (4.5)$$

If $\mu_i(d_i) < 0.2$ we set $\mu_i(d_i) = 0.2$ (that is, a maximum of 80% of a subpopulation located precisely at the place of the effect's hit is wiped off, or if the Moran effect is global, 80% of individuals in the various populations will be eliminated). The parameter $c_{MORAN}$ takes care of the spatial extent of the Moran effect: $c_{MORAN} = 0$ equals global disturbance, while the spatial coverage of the Moran effect goes down with increasing value of $c_{MORAN}$. This way of implementing the spatially autocorrelated Moran effect implies that the location of the strongest effect may vary from year to year (Ranta et al. 1999a).

To demonstrate the impact of spatially autocorrelated Moran effect on synchrony we shall take a reduced subset of the analysis done by Ranta et al. (1999a). Following them, we shall assume delayed Ricker dynamics with ten-year cycles. A set of $n = 25$ population subunits is

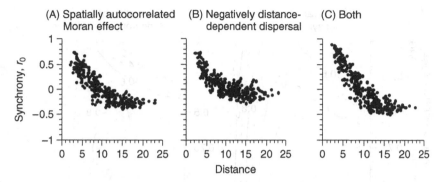

Fig. 4.11. Synchrony levels against distance among pairs of populations ($n = 10$) with (A) spatially autocorrelated Moran in action ($c_{MORAN} = 1.0$; kernel I, p. 53), with (B) distance-dependent dispersal in action ($m = 0.1$, $c_{DISP} = 0.75$), and when (C) both are operating in concert. The dynamics are ten-year cycle length after the delayed Ricker equation. The synchrony levels, $r_0$, are assessed for 100 time units after elapse of an initial phase of 1000 generations.

randomly placed into a $20 \times 20$ co-ordinate space, and each population initiated in random phase with random numbers from uniform distribution between 1 and 10. Here, any dispersal follows kernel I (p. 53), while the spatially autocorrelated Moran effect, when in action, obeys eqs. 4.4 and 4.5.

As anticipated, the relationship between synchrony and distance is much the same regardless of whether we have only the spatially auto-correlated Moran effect in action (fig. 4.11(A)), or dispersal alone that is negatively distance-dependent (fig. 4.11(B)), or both of them operating simultaneously (fig. 4.11(C)). As the landscape in our system is arbitrary, one should perhaps not pay too much diagnostic attention to the subtle differences in the shape in which the data points are patterned in the synchrony versus distance space in fig. 4.11. However, what is relevant here is that, even with the ten-year cyclic dynamics, there is a substantial parameter space ($c_{DISPERSAL}$, $c_{MORAN}$) that will produce matching patterns in overall synchrony and how the synchrony relates to distance among the populations compared (fig. 4.12). Localized disturbance and dispersal (i.e., large values of $c_{MORAN}$ and $c_{DISPERSAL}$; because they are coefficients of a negative exponential function) tend not to synchronize the dynamics of populations well. The outcome is in details that are dependent on the kind of underlying dynamics, as shown by Ranta *et al.* (1999a) in their more extensive analysis. This warns against making conclusions too hastily based on a simulation of a single kind of population renewal process alone. However, the issue is far from being completely understood, partly because

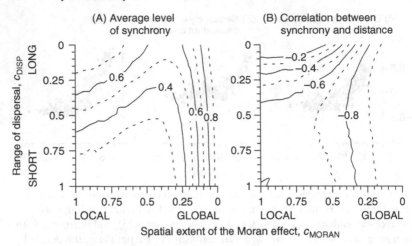

*Fig. 4.12.* Isoclines of (A) average level of synchrony and (B) correlation between synchrony level and distance among the populations compared. In these explorations the spatial extent of the Moran effect and dispersal are varied. The dynamics are ten-year cycle length after the delayed Ricker equation.

of a lack of proper data on the spatial extent of the disturbance effect, but also due to a lack of theoretical research.

## Predation

External disturbance and dispersal are not the sole agents proposed to be responsible for the observed large-scale synchronicity in population fluctuations. Trophic interactions, most notably predator–prey interactions, have frequently been raised as a candidate cause of synchrony (Elton 1924; Butler 1953; Watt 1968; Maynard Smith 1974). In more recent years the significance of predation as a synchronizing agent for their prey species has been raised by Ydenberg (1987), Ims and Steen (1990), Korpimäki and Norrdahl (1991), Small *et al.* (1993). The hypothesis calls for specialist predators that can cover large regions in a short time. Recently Ims and Andreassen (2000) collected the first experimental field evidence to prove that predators indeed synchronize the dynamics of several vole species. Using predation-excluded cages the authors were able to demonstrate that migrating owls (*Asio otus, A. flammeus*) and raptors (*Falco tinnunculus, Buteo lagopus*) were capable of reducing experimental vole populations down to the level observed simultaneously in nature. That predators have a synchronizing impact indicates a tight enough predator–prey interaction, to the extent that predators are the ones that control prey numbers.

## Synchrony across species

To our knowledge there are a few data sets displaying a relatively high degree of synchronicity in population fluctuations across species. The oldest one comes from Butler (1953) who showed, with the Hudson Bay Company's fur records, that Canadian mammal species such as Canada lynx, red fox and fisher, or mink and muskrat as an other set of species (fig. 4.13) tend to have their population peaks across Canada in closely matching years. It is interesting to note that different grouse species tend to fluctuate in step in Finland (Lindén 1988; Ranta et al. 1995b), Scotland (Mackenzie 1952; Hudson 1992) and the Italian Alps (Cattadori and Hudson 1999; Cattadori et al. 2000). The same goes for Scandinavian vole species (Henttonen 1985; Korpimäki and Norrdahl 1998; Stenseth 1999), and to some extent also for British aphids and moths (Hanski and Woiwod 1993).

To exemplify synchrony across species we shall use the 1967–1983 records of the dynamics of voles and six small game species and four grouse species in 11 provinces in Finland (Lindén 1988). To assess the overall level of synchronicity in dynamics of the 11 taxa in Finland, we made a principal component analysis with the species as variables and province-specific de-trended data as observations. Their results suggest that the tightest coupling seems to be among the grouse species, capercaillie, black grouse and hazel grouse (fig. 4.14), which (using different census data; p. 68) are known to fluctuate in rather tight temporal match (Lindén 1988; Ranta et al. 1995b). The two other groups of species (stoat versus least weasel, mountain hare versus red squirrel) are more heterogeneous in terms of population renewal, yet they display a relatively high level of synchrony in large areas in Finland.

It is not entirely clear to us why some of the taiga forest fauna in Canada and in Finland, or the grouse in Scotland and Italian Alps, or moths and aphids in England display synchronous dynamics across species. Thus, we shall begin with a parsimonious explanation: synchrony across species is due to the Moran effect. At this stage the only factor that is assumed is that the focal pair of species $X$ and $Y$ share a common environment. For the population renewal functions, we have selected here the Ricker dynamics and Maynard Smith–Slatkin dynamics in pure and mixed combinations. The external disturbance (white noise; Box 4.3) is assumed to affect simultaneously populations of the two species, say $X$ and $Y$, with a similar force $\mu(t)$.

With both the Ricker and Maynard Smith–Slatkin dynamics (Box 2.1), we found that the Moran effect is capable of synchronizing the dynamics

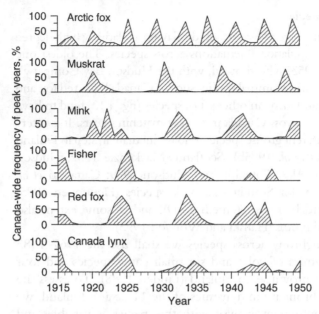

*Fig. 4.13.* Nationwide frequency of peak years in population fluctuations of Canadian mammals (drawn from data in Butler 1953). Butler divided Canada into 63 zones and scored from the Hudson Bay Company's records in how many of the zones the different species had their population high year during the winters 1915–1916 to 1950–1951. For example, with perfect synchrony the Canada lynx peak years should all have coincided at perfectly matching ten-year intervals. That there is some spread (e.g., from 1921 to 1926 with the high in 1924) indicates that the synchrony is not perfect and that it may level off with increasing distance. Note also the pronounced match across species. For Butler (1953) these data show three clusters of species: (i) Canada lynx, red fox and fisher, (ii) mink and muskrat, and (iii) arctic fox.

across species in the area of stable dynamics (4.15(A),(B)), and also in scattered areas with periodic dynamics, but not when the dynamics are complex (except when $X$ and $Y$ have matching growth rates). To some extent, the hybrid renewal process pairing (Ricker versus Maynard Smith–Slatkin) echoes (fig. 4.15(C)) what was found when the populations were renewed according to identical models.

Our findings show that the Moran effect is a possible synchronizing agent also for across-species dynamics despite their different density-dependence structures. Note that we did not assume any interaction between the two species, just that they shared a common environment with its external noise. This finding is intriguing, because the simulations also show that growth rates or autocorrelation structures must not differ too much in order to produce synchronous dynamics across differing population renewal

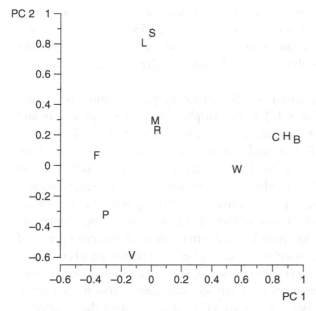

*Fig. 4.14.* Ordination of 11 bird and mammal species (C = capercaillie, B = black grouse, H = hazel grouse, W = willow grouse, S = stoat, L = least weasel, M = mountain hare, F = red fox, P = pine marten, R = red squirrel, V = voles) in a principal component space with two components (PC 1 and PC 2; these components, eigenvalues 3.21 and 1.78, scored 74% of the total variation in the data). The ordination is based on 1967–1983 fluctuations of the 11 species in 11 provinces in Finland. Species located closely in the ordination space (capercaillie, black grouse, hazel grouse; least weasel, stout; mountain hare, red squirrel) fluctuate more in synchrony than distantly spaced species.

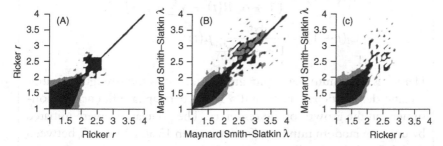

*Fig. 4.15.* Synchrony profiles for two populations $X$ and $Y$ obeying (A) the Ricker dynamics, (B) Maynard Smith–Slatkin dynamics and (C) Ricker dynamics against Maynard Smith–Slatkin dynamics. The relevant parameter values are indicated on the $x$ and $y$ axes. The darker the shading the higher the level of synchronous dynamics (white: $r_0 < 0.33$, gray: $0.34 < r_0 < 0.66$, black: $0.67 < r_0 < 1$). Note that in (A) and (B) the positive diagonal is for two populations with matching parameter values.

processes. When the types of density dependence are mixed, the synchronizing capacity of the Moran effect becomes less obvious. However, as long as the deterministic dynamics are stable, synchronous dynamics via the Moran effect are possible even with different demography of the two species.

An alternative explanation for the synchrony pattern found in the two data sets (fig. 4.13 and 4.14) is that trophic interactions (predators and disease) are important (Ydenberg 1987; Ims and Steen 1990; Korpimäki and Norrdahl 1991). For example, many members of the prey community in Finnish forests are often positively correlated among themselves, as are many of the dominating predators. Among predators and prey, however, there are rather strong negative correlations with lag zero (Lindén 1988). To address the extent to which trophic interactions among species are capable of synchronizing population fluctuations of an interacting pair of species, we shall use a model on dynamics between resource and consumer species (Leslie and Gower 1960; Box 4.4). First, in a one-resource–one-consumer species system we shall explore synchrony across species when the dynamics of the two species are affected by external disturbance in varying degrees. The Moran effect is superimposed, in turn, on only the resource population, the consumer population, or on both of them (fully

---

**Box 4.4** · *Discrete-time resource–consumer dynamics*

The Leslie and Gower (1960) model of resource ($R$) and consumer ($C$) interactions is

$$R(t+1) = \left\{ \frac{\lambda_R R(t)}{1 + \alpha_R R(t) + \gamma \sum C_i(t)} \right\} \mu(t),$$

$$C_i(t+1) = \frac{\lambda_{C,i} C_i(t)}{1 + \alpha_{C,i} \frac{C_i(t)}{R(t)}} \mu(t).$$

Here $\lambda$ is maximum per capita growth rate, $\alpha$ is the strength of intraspecific density dependence and $\gamma$ is the per capita influence of consumers on the growth of the resources. The populations were initiated by uniform random numbers (for $R$ between 15 and 20, for $C_i$ between 2 and 7), and were let to renew after the equation above for 600 generations; the final 100 were used to calculate the level of synchrony in dynamics between $C_1$ and $C_2$ and between $R$ and $C_i$. The process was repeated for 1000 times and the averages are reported in fig. 4.16.

## LESLIE–GOWER RESOURCE–CONSUMER DYNAMICS

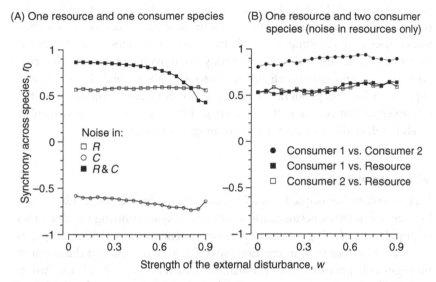

(A) One resource and one consumer species

(B) One resource and two consumer species (noise in resources only)

*Fig. 4.16.* Synchrony across species in Leslie–Gower resource–consumer dynamics. (A) Synchrony across trophic levels with one resource $R$ and one consumer $C$ species with the Moran disturbance variably affecting $R$, $C$ or both. (B) Synchrony across species with one resource (influenced by external noise) and two consumer species. We used the original parameter values by Leslie and Gower (1960): $\lambda_R = 1.25574$, $\lambda_C = 1.1892$, $\alpha_R = 0.0005148$, $\alpha_C = 0.2838$, $\gamma = 0.0018018$. In the system with two consumer species $C_1$ and $C_2$, the Moran effect influenced only the resource dynamics (the parameter values for resource species were as above, while the values for $\lambda_{C1}$ and $\lambda_{C2}$ were drawn from uniform random distribution between 1 and 3, and $\alpha_{C,1}$ and $\alpha_{C,2}$ from uniform random distribution between 0.1 and 0.3).

correlated noise). Second, we extend our system to include two consumer species that interact only via their common resource. In this system, the noise influences only the dynamics of the resource.

The Leslie–Gower dynamics, a ratio-dependent predator–prey interaction, yield stable dynamics if the system is left running without any stochastic influence (Leslie and Gower 1960). When the one-resource–one-consumer Leslie–Gower system is disturbed with external noise, the achieved synchrony level across trophic levels now depends on which component of the interacting species is influenced by the external noise (fig. 4.16(A)). The synchrony between the resource and consumer species is rather high ($r_0 > 0.5$) when the noise is only affecting the resource species, or when the resource and consumer species both receive perfectly correlated noise. However, when the modulating noise acts only upon the dynamics of the consumer the two populations start to fluctuate out of phase ($r_0 < -0.5$; fig. 4.16(A)). The next question is the degree to which the resource–consumer interaction can synchronize the dynamics of two consumer species interacting via a shared resource species. The answer is straightforward: the level of synchrony in dynamics between the two consumer species depending on one resource is very high ($r_0 > 0.7$) and appears to be independent of the level of disturbance modulating the dynamics of the resource (fig. 4.16(B)). Thus, synchrony across trophic levels is achievable in a resource–consumer interaction.

## Where are we?

Half a century has elapsed since Moran outlined his ingenious idea about how environmental perturbation is capable of synchronizing the dynamics of populations sharing a common environment (Moran 1953b). His idea was a direct response to interactions with Elton, who initiated the research on large-scale phenomena in population dynamics (Elton 1924; Lindström et al. 2001). Now, we can attempt to answer the following question: what have we learned about population synchronicity during the past five decades (Moran 1953b), or rather, during the last three-quarters of a century (Elton 1924)? This question has been the subject of a few reviews (Ranta et al. 1997c; Hudson and Cattadori 1999; Bjørnstad et al. 1999a, Liebholt et al. 2004), and one can conclude that by now we have much more data on the taxonomic and geographic extent of synchronicity both within and across species (references in the previous sections of this chapter). We have an enriched palette with a number of agents that have the potential to force populations, regulated by density-dependent feedback, to fluctuate in step: the Moran effect, redistribution of individuals between breeding seasons and predation/disease. These may act in solo, but more likely in concert.

Accepting the Moran effect and dispersal in action, Kendall et al. (2000) and Ripa (2000) both used linear models for the population renewal process (and worked independently of each other). Kendall et al. (2000) addressed the joint effect of the spatial correlation in the environment and dispersal. They come to the following conclusion: there is always interaction between the two. The interaction is naturally small when one or both effects are small. Most interestingly, their analysis indicates that the interaction between the external disturbance and dispersal is opposite in sign to the environmental correlation. It is generally assumed that $\rho\,(\varepsilon, \eta) > 0$ (the populations share a common environment). Kendall et al. (2000) concluded that population synchrony will be lower than predicted by simply adding the effects of dispersal and environmental correlation. Hence, the effects of dispersal and environmental correlation are subadditive. Ripa, also using a theoretical approach with linear models, argued that external disturbance (after Moran 1953b) sets a baseline for synchrony level among populations, and that dispersal can only enhance the level of synchrony. For Ripa, dispersal is only an effective synchronizing mechanism when the population renewal process is close to unstable. Both Kendall *et al.* (2000) and Ripa (2000) agree that our interpretation of population synchrony can vary depending on the timing of the population census. Simultaneously with the research described, a third team, Lande *et al.* (1999), was working on the same issue. They used stochastic nonlinear modeling to address synchrony with the Moran effect and dispersal. The major conclusion is simply that, relative to $\rho(\varepsilon,\eta)$, the contribution of dispersal to the spatial scale of synchrony is ". . . magnified by the ratio of dispersal rate to the strength of density regulation. Thus even if the scale of dispersal is smaller than that of environmental correlation, dispersal can substantially increase the scale of population synchrony for weakly regulated populations" (Lande *et al.* 1999, p. 271). Thus, Lande and his associates appear to agree with Ripa. This is also in harmony with the simulations shown on the preceding pages. Nonetheless, the three teams agree, as do our simulation results in this chapter, that the strength of density–dependent feedback plays a significant role in the synchronizing process.

Greenman and Benton (2001) used unstructured nonlinear renewal processes. They return to the roots by being interested in how two independent populations, influenced by the Moran effect, will become synchronized. This matches the feral sheep setting (Grenfell *et al.* 1998) on two oceanic islands with no dispersal whatsoever. In this case, the environments shared between the two islands correlate more strongly than the dynamics of the two sheep populations. Greenman and Benton (2001) showed that for a great variety of nonlinear models there is almost always $r_0(X,Z) < \rho(\varepsilon,\eta)$. Thus, their results agree with Grenfell *et al.* (1998, 2000). Greenman and Benton (2001) also emphasized a general point often forgotten in synchrony research. Even though a rich variety of environmental noise structures can generate synchrony, it is hard – if not impossible – to identify correctly the noise that synchronized given population data. We cannot agree more (Ranta *et al.* 1995a, 1998, 1999a). Any set of populations obeying a given renewal process can be synchronized in a rich variety of ways. Thus, finding out who is the Moran in the Moran effect becomes a futile task. This is the reverse of the "visibility" problem (p. 34, Ranta *et al.* 2000a; Laakso *et al.* 2001).

Spatial scales for investigating synchrony range from meters to kilometers (Thomas 1991; Sutcliffe *et al.* 1996) to hundreds of kilometers (Hanski and Woiwod 1993; Ranta *et al.* 1995b; Koenig and Knops 2000a), and even up to thousands of kilometers (Ranta *et al.* 1997a,b,d; Hawkins and Holyoak 1998; Paradis *et al.* 2000). The relative importance of dispersal and the external perturbation as synchronizing agents is often argued to depend on the scale: dispersal at a local scale and the Moran effect at larger ranges. This is not so. It is a relatively easy task to show that local dispersal yields global synchrony given a long enough time. For example, we took a set of 10 000 populations, arranged in a string, and let 10% of the annual population size disperse to the neighboring cell on the right and left (at the ends only to the right, or to the left). The populations obeyed delayed Ricker dynamics with ten-year period length, and they were initiated in random phase. Checking the match of population fluctuations, after 5000 generations elapsed, showed that the system was in perfect synchrony all over. Experimentation with various other renewal processes gives matching results.

With synchronicity, we often find that the level of synchronous fluctuations goes down as the distance increases between the areas from where the data are derived. This pattern can be generated in different ways: dispersal that is negatively distance-dependent, dispersal and global Moran effect acting together, and with a spatially autocorrelated Moran effect (Ranta *et al.* 1999a; Lande *et al.* 1999). Many of the climate parameters (rainfall, temperature) are spatially autocorrelated (Burroughs 1992), e.g., in Finland the radius of reasonably strong spatial autocorrelation for temperature is 200 km (Heino 1994). Finding spatial autocorrelation in population time series tempts one to start looking for the match between climate variables and population data. However, matching spatial autocorrelation profiles might not necessarily reveal that one of them (population fluctuations) is due to the other one (spatial fluctuations in precipitation or temperature). Our argument is that the identity of the Moran effect might be hard to find out. It also might be that the external disturbance comes from different sources in different years (Ranta *et al.* 1995a; Lindström *et al.* 2001). One year it may be a cold spell in a critical phase of life history of the focal species; another time it might be exceptional rainfall or a dry spell. The issue is that natural populations live with redistribution of individuals and are impacted by a number of various stochastic processes. To pinpoint one cause of synchronicity might be too far off the mark. However, one has to agree with Paradis *et al.* (2000, p. 2123), who say that "synchrony in natural populations seems to be determined by complex interactions between abundance, population variability, species characteristics, and demographic mechanisms." Despite the progress made since Moran (1953b), there still seems to be a great deal of research to do in order to fully understand the causes and consequences of synchronous population fluctuations. There are reasons to be worried: species with synchronous populations are likely to face greater risks of extinction because density crashes can occur simultaneously in all populations (Heino *et al.* 1997a). A covering review of various topics in population synchronicity is provided by Liebhold *et al.* (2004).

## Summary

In this chapter, we showed that, in a great many species, populations tend to fluctuate in a rather good temporal match across large geographical ranges. Large-scale synchrony is also observed in plant life history events (seed set, growth). External disturbance, the Moran effect, is the first mechanism raised to explain this pattern. In many cases, we also see that the level of synchronicity goes down when the distance between the populations compared is increased ($r_D < 0$). The second explanation for synchronous fluctuations, dispersal of individuals, can account for this pattern. Predation/diseases are the third mechanism suggested to account for synchrony. That $r_D < 0$ can also be achieved by the Moran effect being spatially autocorrelated. It is most likely, however, that the various synchronizing agents work in concert. Many different noise signals can synchronize the dynamics of populations obeying common density-dependent feedback. As to the external disturbing agent, it may very well be that, in different years or at different places, it is a different cause that generates population synchrony. It may also be that in structured populations different life stages may be differently vulnerable to the external modulator(s). This makes it difficult to identify the exact nature of the external disturbance. Much more experimental and theoretical research is needed before the stochastic processes influencing the dynamics of populations are properly understood. This is especially true for nonlinear population renewal processes. Furthermore, there is substantial evidence of synchronous population fluctuations across species. Yet, a theory of synchronicity across species with various interactions is almost nonexistent. Despite a substantial amount of extant data, we still lack information from which various aspects of geographical scale and synchronicity could be properly assessed.

# 5 · *Order–disorder in space and time*

In this chapter, we shall continue exploring how large-scale ecological processes may be fundamentally important for our understanding of emergent phenomena in various population systems. The modern ecological literature on spatial population dynamics has drawn our attention to various captivating configurations, such as traveling waves, suggesting that spatially structured populations may become self-organized. Spatial interactions in population dynamics can create a variety of spatial-temporal patterns. Spatial self-organization was first demonstrated in dispersal-coupled predator–prey and host–parasitoid models and we will discuss them first. We then proceed to expand the self-organization on a large scale of redistribution-coupled population processes. We shall finish by discussing data that support some of the theoretical findings.

Spatial interactions in population dynamics can generate a rich ensemble of spatial patterns. One example is traveling waves (Shigesada *et al.* 1986; Kot 1992; Ranta and Kaitala 1997; Shikesada and Kawasaki 1997; Kaitala and Ranta 1998) and they may appear in the form of wave fronts, periodic waves, spirals, and rings. Other eye-catching patterns are represented by crystal lattices, patches, and spatial chaos (Hassell *et al.* 1991, 1994; Solé and Valls 1991; Comins *et al.* 1992; Solé *et al.* 1992a,b; Solé and Bascompte 1993; Hassell 2000; Bjørnstad and Bascompte 2001). There are no strict definitions for the different spatial configurations emerging due to spatially coupled population renewal, although some attempts to derive more formal approaches to identify the patterns have been presented (Bjørnstad and Bascompte 2001; Bjørnstad *et al* 2002a; Kaitala 2002). These classifications refer to aggregated population highs and lows, which may take the form of a wave front and which may or may not move to a certain direction in space (Bjørnstad *et al.* 2002b).

Invasion of a species to new distribution ranges may cause a moving wave front (Shikesada and Kawasaki 1997; Sherratt 2001). Another form of a traveling wave is repeated traveling waves that, according to definition, will reoccur at more or less regular intervals. This phenomenon is comparable to the Mexican wave, often observed in football arenas, where hand raising by people in the audience generates waves circling around the stadium (Sherratt 2001). Traveling waves due to ecological interactions are the particular feature that we shall address with the few extant examples. In crystal lattices, yet another spatial pattern, the aggregations of local population sizes stay put in space, but locally population abundances may differ. In spatial chaos, local populations show chaotic dynamics and the population highs may, but need not, be aggregated (locally in synchrony).

In recent years, various spatial patterns have also been reported from natural populations, especially regarding traveling waves (Bulmer 1974; Ranta and Kaitala 1997; Kaitala and Ranta 1998; Lambin *et al.* 1998; Savill and Hogeweg 1999; Moss *et al.* 2000; Mackinnon *et al.* 2001; Sherratt 2001; Bjørnstad *et al.* 2002b; Sherratt *et al.* 2002), and from epidemics (Grenfell *et al.* 2001, 2002; Bjørnstad *et al.* 2002a). We are thus right at the beginning of a stronger matching between empirical and experimental data, and abstract theory. However, the theory and data have to be further developed before the role of self-organization in natural systems can be better established. Spatial self-structuring is not only the fancy of theoreticians; it has potentially important bearings on population management, conservation biology as well as on improving the understanding of evolutionary processes.

Self-organized structures have often only been assessed by visual inspection of the snapshot patterns emerging in simulations. This is a sign of how young the research on spatial self-structuring is: if an interaction generates, e.g., symmetrical spatial properties or wave patterns in a lattice, identifiable to human eye, these patterns are likely to be a product of self-structuring processes occurring in the population. Obviously, visual inspection is not enough. Further insight is needed, preferably in terms of statistical indices that can be objectively assessed. Another problem is that data are often not collected for the purpose of testing this theoretical work.

Since the spatial and temporal dynamics are two sides of the same coin, spatial self-organization should also be reflected as temporal self-organization; for example, in the frequency structure of population dynamics (p. 31). It becomes evident that spatial order should be treated together with temporal order. This is what we are after.

## Spatial host–parasitoid dynamics

### The template

Self-organization of population dynamics in space can be easily illustrated in a system where $n$ dispersal-coupled population subunits are arranged into a regular grid of cells (lattice). Pattern formation on a lattice was first studied by Solé and Valls (1991) for predator–prey models, and by Hassell et al. (1991, 1994) for host–parasitoid models. Solé and Valls (1991) showed that order may arise in a configuration starting from random initial population sizes. They were able to show that various spatial patterns, such as spiral-like waves, patches, and rings, will emerge when local population subunits are coupled by dispersing individuals. Moreover, they observed that such patterns might exist under complex and even under chaotic population dynamics.

To exemplify the self-emergence of spatially organized patterns we shall re-work in some detail the host–parasitoid model used by Hassell and others (Hassell et al. 1991, 1994; Hassell 2000). For these purposes, consider first a single-population (or isolated-population) version of the model, where the dynamics are given as

$$
\begin{aligned}
N(t+1) &= \lambda N(t) f[N(t), P(t)] \\
P(t+1) &= q N(t) \{1 - f[N(t), P(t)]\}.
\end{aligned}
\tag{5.1}
$$

The host population size at time $t$ is $N(t)$ while $P(t)$ is the corresponding size for the parasite population, $\lambda$ and $q$ are parameters, and $f$ represents the functional response of parasitoids against the number of its host. Following Hassell et al. (1991), we assume that the functional response is given as $f(N,P) = \exp(-aP)$, where $a$ is a parameter. This is the classical Nicholson–Bailey host–parasitoid model (Nicholson and Bailey 1935), where the underlying assumption is that parasitoids search and attack their hosts randomly. For the single-patch model, the dynamics become

$$
\begin{aligned}
N(t+1) &= N(t)\exp[r - aP(t)] \\
P(t+1) &= q N(t) \{1 - \exp[-aP(t)]\},
\end{aligned}
\tag{5.2}
$$

where $r = \ln(\lambda)$, i.e., the population growth rate of the host, and $a$ is a constant. The equilibrium population sizes for the host and parasitoid are now given as

$$N^* = \frac{r}{aq[1 - \exp(-r)]}$$
$$P^* = \frac{r}{a},$$

(5.3)

respectively. The equilibrium of the Nicholson–Bailey model is unstable for all parameter values (Hassell and May 1973), hence persistent populations cannot be generated with eq. 5.2.

Assuming nonglobal dispersal both in the host and in the parasitoid, we can extend our explorations of the host–parasitoid dynamics in a lattice. The population dynamics proceed in two steps as follows. First, both species reproduce at the same time. This is described by the local two-species interaction on cell $i$

$$N_i'(t) = \lambda N_i(t)\exp[1 - aP_i(t)]$$
$$P_i'(t) = qN_i(t)\{1 - \exp[-aP_i(t)]\}.$$

(5.4)

A fraction $m_N$ and $m_P$ of the renewed host $N'$ and parasitoid $P'$ population disperses resulting in the new population sizes in each cell in the next time step as follows

$$N_i(t+1) = (1 - m_N)N_i'(t) + \sum_{j \in N_i} m_N N_j'(t)$$
$$P_i(t+1) = (1 - m_P)P_i'(t) + \sum_{j \in \Pi_i} m_P P_j'(t),$$

(5.5)

where the latter term in each equation gives the immigration of both species to grid $i$ from its neighborhoods $N_i$ and $\Pi_i$. Here we assume that for both species dispersal is local and that the neighborhood is defined as the eight neighboring cells. Note that dispersal may be species-specific and it can be density-dependent, such that the predator or parasitoid dispersal is determined by the host density (Rohani and Miramontes 1995b).

Hassell et al. (1991) showed that the spatial patterns tend to form different configurations depending on the parameter values, especially on the dispersal parameters $m_N$ and $m_P$ (see also Wood and Thomas 1996; Bolker and Pacala 1999; Bjørnstad and Bascompte 2001). Roughly, three different classes of spatial patterns are identifiable in the simulations: crystal lattices, spiral waves, and spatial chaos. Crystal lattices are spatially standstill patterns where the population sizes may still go up and down. In the two other patterns, spiral waves and spatial chaos, the patch-like population aggregations move around the space and they may be formed

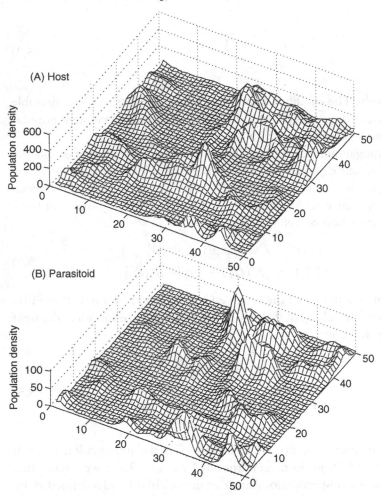

*Fig. 5.1.* An example of the self-organization behavior of the dispersal-coupled Nicholson–Bailey host–parasitoid model (a 50 × 50 lattice, $\lambda = 2$, $q = 0.5$, $m_N = 0.2$, $m_P = 0.8$, $a = 0.09$). The system was initiated with uniform random numbers (between 1 and 40) for both the host and the parasitoid subpopulations. The snapshot situation is given at $t = 500$. Note that both populations are spatially aggregated, and, to some extent, the spatial configuration of the parasitoid population follows that of the host (spatial correlation between host and parasitoid numbers $r = 0.425$).

and lost repeatedly. Such a spatial pattern formation can be easily recognized by eye (fig. 5.1). A major challenge for population ecology is, however, to proceed further with describing the population processes by more objective tools than aesthetic evaluation of what we see.

(A) Host

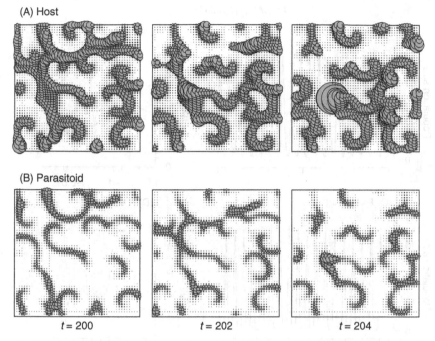

(B) Parasitoid

$t = 200$ $t = 202$ $t = 204$

*Fig. 5.2.* Examples of spiral waves in host (A) and parasitoid (B) dynamics in a $50 \times 50$ lattice ($\lambda = 2$, $q = 0.5$, $m_N = 0.4$, $m_P = 0.4$, $a = 0.09$). The dot size indicates local population size; both (A) and (B) are drawn to the same scale. The three snapshots are at $t = 200$, $t = 202$ and $t = 204$.

## The play

We shall now deviate from the approach presented by Hassell *et al.* (1991). In particular, we pay attention to the temporal patterns in the connection of spatial pattern formation. Hassell *et al.* (1991) showed that crystal lattices will be obtained for relative small areas of the dispersal parameters $m_N$ and $m_P$. For example, when $\lambda = 2$, this will occur when $m_N$ is small and $m_P$ is close to 1. On the other hand, it is much easier to find parameter areas producing the two other patterns, spiral waves and spatial chaos. Let us concentrate on an example of spiral waves (fig. 5.2). Spiral waves should be understood as a general expression for different wave patterns arising and proceeding in varying directions. On a large lattice, many waves may be seen at a time. A single wave can occasionally be split into two wave fronts, each moving in a different direction. As Solé and Valls (1991) comment, spiral waves on a lattice, produced by

LOCAL DYNAMICS

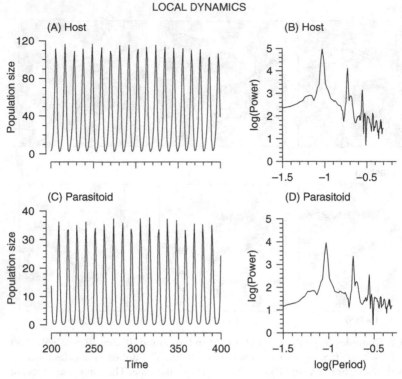

*Fig. 5.3.* Dynamics of the host (A) and the parasitoid (C) in one randomly selected cell in a 50 × 50 lattice (fig. 5.2). The panels (B) and (D) give the corresponding power spectra. Note that approximately 10-year cycles heavily dominate population fluctuations in both the host and parasitoid.

ecological models, may not be spirals in the tight sense. Rather, they are more or less spiral-like, moving population density formations.

Spiral waves will be observed, e.g., for $\lambda = 2$, $m_N = m_P = 0.4$ (on a 50 × 50 lattice, dynamics run for 1000 time steps). The aggregated population size over all cells in the lattice as well as local population size of a randomly chosen cell fluctuate in a regular manner, as expected from the Nicholson–Bailey dynamics (figs. 5.3, 5.4). The power spectrum indicates that the local host and parasitoid dynamics are periodical with the period length being about ten time steps. However, this is not the case with the global (aggregated over all the lattice cells) dynamics: global temporal fluctuations in the host and in the parasitoid fit better to a $1/f$ power law (Halley 1995). We also observe that, in the spatial scale, both the host and parasitoid populations exhibit spiral-looking fronts (fig. 5.2).

GLOBAL DYNAMICS

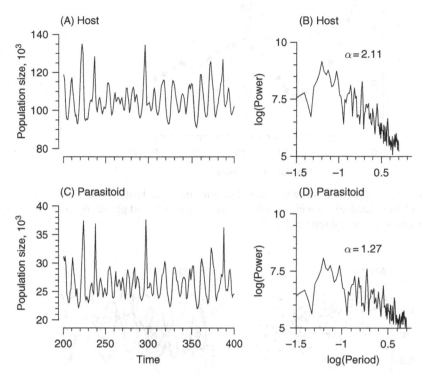

*Fig. 5.4.* Global dynamics of the host (A) and the parasitoid (C) in a 50 × 50 lattice (fig. 5.2). The panels (B) and (D) give the corresponding power spectra. Note that the locally periodic (approximately 10-year cycle length) has been forced, due to spatial coupling via redistributing individuals, into red power law dynamics (slopes inserted).

A further inspection shows that the fronts of the parasitoid populations follow behind the fronts of the host. This observation becomes understandable by noting that the local populations of parasitoid lag behind the local populations of the host in a very clear-cut way (fig. 5.5), the lag being two to three generations, while at the global scale the lag is shorter, and soon fades away with increasing time. Looking at the spatial patterns (figs. 5.1, 5.2) indicates that the synchrony levels both of host and parasitoid populations go down with increasing distance between cells. This was verified by randomly selecting 20 cells and sampling host and parasitoid dynamics for the final 100 time steps (fig. 5.6). For the same cells, we also calculated synchrony in pairs by using a sliding time-window technique (Ranta *et al.* 1997a, 1999b; Kaitala *et al.* 2001a). A window of 20 years was passed though the final 700 time units of the data with

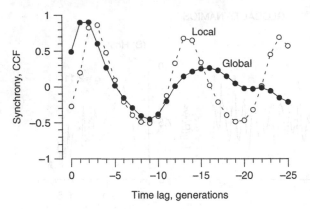

Fig. 5.5. Synchrony (cross correlation function) between host and the parasitoid population fluctuations with various lags. Local (fig. 5.3) and global (fig. 5.4) dynamics are treated separately.

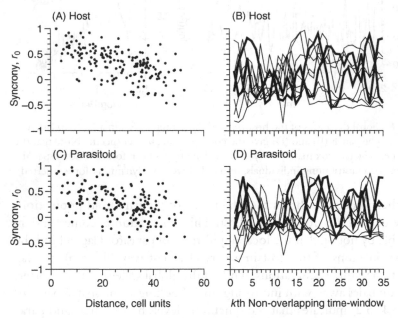

Fig. 5.6. (A), (C) The synchrony level in the dynamics of the host and the parasitoid populations (fig. 5.1, 5.2) goes down with increasing distance, as indicated by randomly selecting 20 cells and sampling the synchrony of their dynamics for the final 100 time steps. (B, D) The synchrony in pairs was calculated for the same cells by using a sliding time-window technique. There are pairs of local host and parasitoid populations, respectively, fluctuating in synchrony at times, drifting out of synchrony to fluctuate in opposite phase, and then returning back to fluctuate in step.

(A) Host

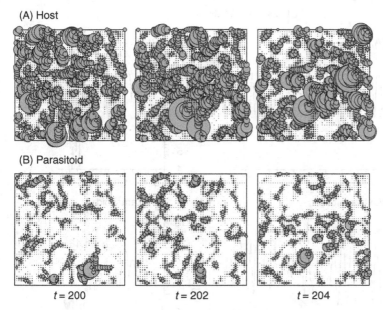

(B) Parasitoid

$t = 200$          $t = 202$          $t = 204$

*Fig. 5.7.* Examples of spatial chaos in host (A) and parasitoid (B) dynamics in a
$50 \times 50$ lattice ($m_N = m_P = 0.1$). The dot size indicates local population size; both (A)
and (B) are drawn to the same scale. The three snapshots are at $t = 200$, $t = 202$ and
$t = 204$. The local dynamics of the host and the parasitoid are cyclic with
approximately 10-year period length.

nonoverlapping sections. The results indicate that there are pairs of
populations, both in the host and in the parasitoid, that fluctuate in
synchrony for a while, drift out of synchrony to fluctuate in opposite
phase, and then return back to fluctuate in step (fig. 5.6). A similar finding
of time-variant synchrony is documented for the Canada lynx (p. 128;
Ranta *et al.* 1997a; Kaitala *et al.* 2001a).

We next turn to explore spatial chaos in the Nicholson–Bailey host–
parasitoid dynamics. Spatial chaos can be generated by using eqs. 5.4 and
5.5 with, e.g., $\lambda = 2$, $m_N = m_P = 0.1$ (fig. 5.7). Again, the local dynamics of
the host and the parasitoid are cyclic with approximately 10-year period
length. However, at the global level spatial coupling breaks down the
regular cycle. The power spectra show that long-term dynamics of the
host and the parasitoid appear to be a $1/f$ power law process (p. 104) (fig.
5.8). In addition, in this case we find a rather high level of synchrony among
the host and among the parasitoid populations. Again, the level of synchrony goes down against distance, and the synchrony is time-variant.
However, these results are not displayed, as they would merely repeat

GLOBAL DYNAMICS

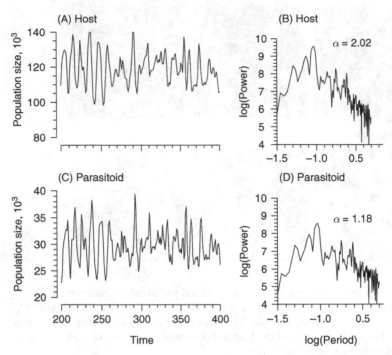

*Fig. 5.8.* Global dynamics of the host (A) and the parasitoid (C) and the corresponding power spectra (B, D) in a 50 × 50 lattice (fig. 5.7). The long-term dynamics of the host and the parasitoid appear to obey the $1/f$ power law (slopes inserted).

those in fig. 5.6. We encourage research along these lines using different models on interspecific interactions.

## Traveling waves

We shall now move on to show that self-organization may also be observed in single-species population dynamics in a spatial setting. It will be shown that semi-independent population subunits, coupled via redistributing individuals, are capable of supporting the evolution of order in their dynamics (Kaitala and Ranta 1998). In this system, all the patches are of matching quality. They differ only in their spatial co-ordinates. Here we shall maintain the regularity of the spatial grid. In addition to characterizing self-organization visually, we also analyze the emerging patterns quantitatively through the synchronicity of local dynamics. We shall concentrate on the temporal and spatial patterns of synchrony.

Consider a set of populations using the Ricker model with delayed density dependence (Box 2.5). Each time step a constant fraction $m$ ($0 < m < 1$) of each subpopulation will emigrate from the natal patch. All the dispersing individuals are assumed to survive [we are using dispersal kernel I (p. 53)]. In a single-population system, or in the absence of dispersal ($m = 0$), the population dynamics are stable for $r < 2$ when $a_1 = a_2$. At $r = 2.0$, a bifurcation occurs such that for $r > 2$ the dynamics will be a 4-year cycle. Simulations, power spectral analyses, and dominant Lyapunov exponents (for their calculation, see von Bremen *et al.* 1997) for $a_1 = a_2 = 0.05$ indicate that a further increase in the growth rate $r$ ultimately leads, through a period-doubling range, to chaotic population dynamics with periodic windows ($r > 2.7$).

The coupled map lattice structure will be constructed by locating 25 populations regularly on a $20 \times 20$ grid such that the neighboring populations are four grid units apart. As in Kaitala and Ranta (1998), we used the parameter values for the delayed Ricker dynamics that will yield 4-year cycles [as in Scandinavian vole dynamics (Hansson and Henttonen 1985)]. We used $r = 2.5$ and $r = 3.5$ and the simulations were initiated with random population densities for the first 2 years (i.i.d. random numbers from uniform distribution between 0 and 50). To remove the effects of the initial transients the simulations were run 1500 generations before analyses.

Self-organization in this system (fig. 5.9) is observed in the repeated pattern that arises from random initial conditions. In the first sampling year (fig. 5.9(A)), the population levels are low in all 25 patches, while in the next time step (fig. 5.9(B)), six adjacent patches reach peak densities in the north-west corner of the space. The following year, the peak densities have moved south. Now the populations show a band of peak densities spanning from southwest to northeast (fig. 5.9(C)). Finally, the peak densities reach the southeast corner (fig. 5.9(D)). This 4-year dynamic pattern, a traveling wave, will repeat itself over and over. The wave patterns emerging from spatial interactions appear to be nonunique, since several different moving patterns may be generated, depending on the initial values of the simulations. Different initial conditions in the spatial model may provide different traveling wave patterns in that the direction of the movement may change. However, all these patterns show the same qualitative properties: the 4-year periodicity always remains. In the homogeneous environment, the traveling waves may start from one corner and end up in another corner, or they may be born in the middle of the edge and finish at the opposite edge or the corners. Each set of initial values tested produces a traveling wave.

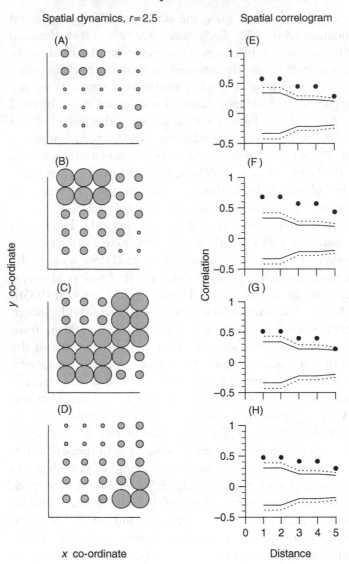

*Fig. 5.9.* Four-year cycle is common in spatial population dynamics of subpopulations interconnected by dispersal kernel I (p. 53). The relative size of the local populations is indicated by the size of the circles. The spatial population dynamics show a clear pattern of a traveling wave. (A) In the first sampling year, the local population levels are low in all patches. (B) At the next time step, six adjacent patches reach peak densities in the north-west corner of the space; (C) at the following time step, the peak densities have moved to south. (D) The peak densities reach the south-east corner. Spatial correlograms (E)–(H) indicate the presence of spatial self-organization such that the close-by patches tend to be more similar than those from remote areas (dots: spatial correlations from the data; confidence intervals, solid lines: 95%; dotted lines 99%, $r = 2.5$, $a_1 = a_2 = 0.05$, $m = 0.1$, and $c = 0.5$). Modified after Kaitala and Ranta (1998).

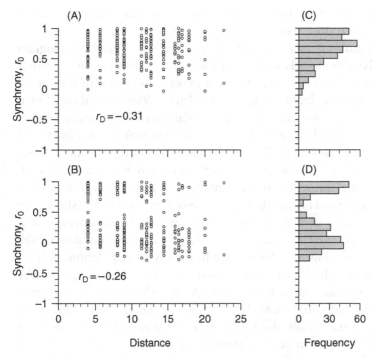

*Fig. 5.10.* A comparison of the synchrony among the local population dynamics indicates that the close-by populations tend to be in better synchrony than the remote patches. Thus, the degree of synchrony between the dynamics in local populations decreases with increasing distance. The cross correlation coefficients measure the degree (sign and magnitude) of the synchrony and range from $+1$ (fluctuations in step) to $-1$ (fluctuations in opposite phase). (A) Ricker dynamics with delayed density dependence, $r = 2.5$; (B) $r = 3.5$, the correlation coefficients between synchrony level and distance, $r_D$ are inserted. (C),(D) The corresponding frequency distributions of the synchrony measures. The time window used in the calculations was 200 generations. Modified after Kaitala and Ranta (1998).

At least two statistical measures are connected to the spatial-temporal self-organization. First, looking at the synchronicity patterns we may observe that the local dynamics among patches tend to be synchronized. The synchrony may not be perfect. Often, the synchrony levels among populations tend to fade away with increasing distance (Chapter 4; fig. 5.10). Second, the spatial snapshot patterns should show spatial auto-correlation (fig. 5.9(E)–(H)). Positive spatial autocorrelation indicates that, at a given moment, nearby subpopulations tend to be similar (Legendre and Fortin 1989). There are many different ways to compute spatial autocorrelation. Their usage depends on the type of data (Sokal and Oden 1978). Here the natural choice is to analyze interval (continuous) population data, which are regularly spaced.

For the simulated data (fig. 5.9), we compute an annual correlogram including five distance classes from 4.1 to 9.1 units with an interval of 1 unit (Kaitala and Ranta 1998). A correlogram is composed of the annual autocorrelation coefficients (Moran's I, Moran 1950b; Sokal and Oden 1978) for each distance class. For each spatial pattern we assumed binary joins. Here, two grid points (subpopulations) are connected only if their mutual distance is less than the distance class under consideration (Sokal and Oden 1978). The statistical significance can be tested for each individual spatial autocorrelation coefficient or for each correlogram at a time (Legendre and Fortin 1989). A correlogram as a whole is significant in statistical terms if at least one spatial autocorrelation coefficient is significant at $\alpha' = \alpha/w$, where $\alpha = 0.05$, and $w$ is the number of the points in the correlogram. Because we have only 25 population subunits we need to use the small sample correction for interval data in the statistical significance tests (Sokal and Oden 1978). Such a correction is advised for the sample size $10 < n < 50$. In figure 5.9(E)–(H), each correlogram is statistically significant. In other words, we can observe a positive autocorrelation at each time moment, each showing different spatial snapshot configurations. This suggests that the population sizes of the close-by areas are more similar than the population sizes of more remote areas.

In addition to spatial autocorrelation, we may also consider the patterns of synchrony. When population dynamics are spatially autocorrelated then the close-by populations tend to be similar in size to each other. Thus, if we look at the synchrony patterns as a function of distance we should see that close-by populations are more synchronous than remote patches. Indeed, the cross correlation coefficient indicates that the populations fluctuate in synchrony but that the degree of the synchrony declines, as expected, with increasing distance between subpopulations (fig. 5.10(A),(B)).

As in the single-population dynamics, increasing the growth rate complicates the spatially coupled dynamics. In the spatial population system the chaotic range of dynamics appears to emerge when $r > 2.7$. However, this value should be considered as an approximation since in the multidimensional patch configuration with nonunique dynamics, and with possibly different attractors co-existing, it is a painstaking task to identify an exact boundary for the chaotic range, if there is any. The general pattern is, however, that with increasing $r$ in the delayed Ricker equation the population densities tend to become more irregular, which can be seen in the peak densities in the first place, less in the period length. The 4-year period pattern is less sensitive to increasing values of the growth rate than the peak densities in the dynamics of single populations. In fig. 5.11(A)–(D) we illustrate a traveling wave with more complex dynamics for $r = 3.5$. We observe significant spatial autocorrelations (fig. 5.11(E),(G)). Interestingly, the spatial autocorrelation seems not to be present all the time. Instead, when it has appeared once, it will gradually weaken such that it cannot be detected, to reappear later on. The traveling wave pattern is visible also with $r = 3.5$, and it is also associated with a distance-dependent synchrony pattern (fig. 5.10(B)). In the present example, the complex dynamics with a clear four-year periodic component in the dynamics yield the similar patterns of traveling waves as in the more regular case with $r = 2.5$. Spatial autocorrelations and distance-dependent synchrony remain indications of the order. Kaitala and Ranta (1998) suggest that spatial order may be associated with spatial waves showing a certain degree of periodicity (Rohani and Miramontes 1995b; Bascompte *et al.* 1997).

### Evidence from Finland

We next take an opportunity to contrast and illustrate our theoretical developments with possible empirical evidence on traveling waves in vole dynamics (Ranta and Kaitala 1997). Several vole species annually cause severe damage in young sapling stands of pine, spruce, and birch. These are economically important forest trees in Finland. For this reason, the damage caused by voles has been recorded annually (fig 5.12(A)). These records, obtained from the different management districts in Finland, suggest an interannual fluctuating pattern in the geographical distribution of peak damage areas (Ranta and Kaitala 1997). The damage is locally characterized by 3- to 4-year cycles. The geographical area of the peak damage changes between years. Thus, the temporal change of the location of the peak damage creates a dynamic pattern, in which a certain degree of asynchrony among the observations from different areas can be observed. Locally, high peaks are usually preceded by a 2- or 3-year period of population increase, which then ends with a sudden decline of the local population to almost nonexistence (Hansson and Henttonen 1985).

The intensity of vole damage and geographical distance between the 19 forest management districts is negatively correlated (fig. 5.12(B),(C)). That is, nearby areas tend to have matching damage levels, but the

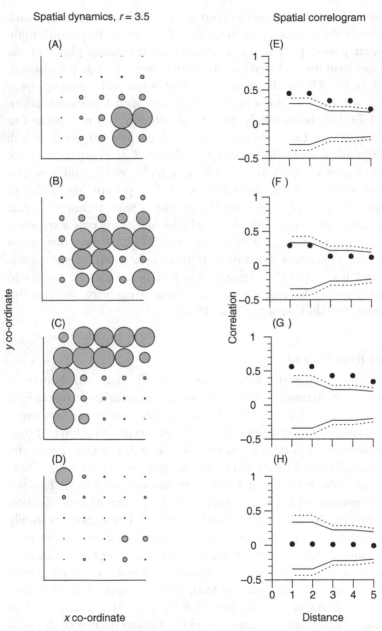

*Fig. 5.11.* Traveling wave (A)–(D) with complex population dynamics is obtained by increasing the growth rate ($r = 3.5$). Despite the increase in the growth rate the period length is not affected that much while the amplitude becomes irregular (c.f., fig. 5.9). This configuration also creates a significant spatial autocorrelation pattern. (E)–(H) Spatial correlograms indicate the presence and possible absence of spatial self-organization (confidence intervals, solid lines: 95%; dotted lines 99%, $r = 3.5$, $a_1 = a_2 = 0.05$, $m = 0.1$, and $c = 0.5$.). Modified after Kaitala and Ranta (1998).

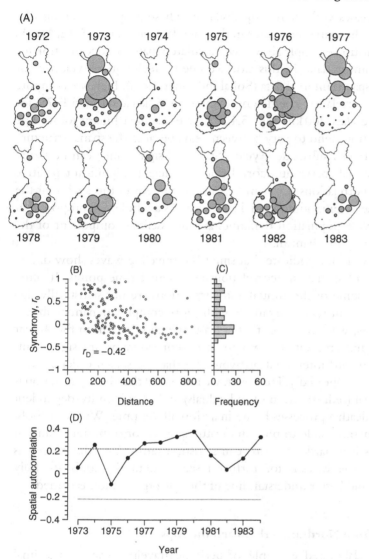

*Fig. 5.12.* (A) The vole damage in young plantations varies prominently over large areas between years (redrawn from Kaitala and Ranta 1998 using the original data). The data represent the number of seedlings destroyed, an indication of the annual local abundance of voles. The damage caused by voles to the tree stands is characterized by the three- to four-year periodicity, clustering of the damage, and the annual movement of the peak damage from one area to another creating a pattern of spatial asynchrony. (B) Synchrony at the level of annual vole damage graphed against the distance of the data sampling points (correlation coefficient inserted) and (C) the marginal distribution of the synchrony measures (cf., fig. 5.10). (D) Spatial autocorrelation with 2-km radius through time (95% confidence intervals are indicated). Modified after Kaitala and Ranta (1998).

match decreases with increasing distance. These properties can also be observed in the spatial structures in the vole damage data (fig 5.12(A)). As in our simulation example above, we calculated spatial autocorrelation for the vole damage data. In this case we need a technique developed for irregularly spaced interval data (Sokal and Oden 1978). Here we compute annual autocorrelation coefficients (Moran's I, Sokal and Oden 1978) for the distance class 200 km (fig. 5.12(D); we assumed binary joins. The autocorrelations tend to vary between years because the spatial structure changes between subsequent years. Statistical significance can be tested for each annual spatial autocorrelation coefficient. Significant positive spatial autocorrelations occur in 6 years. We interpret this such that the data include a nonrandom spatial structure. As a whole, we propose that the cyclic vole population dynamics are an essential component of the self-structuring mechanism.

The above results indicate that simulated traveling waves show different degrees of spatial autocorrelations at different time points. It often occurs that some of the annual autocorrelations are not statistically significant, as in our second example with more complex dynamics above. Nevertheless, we can expect that at least one correlogram in the 4-year cycle is significant. Often, two to three correlograms are significant. A straightforward interpretation would be that the spatial order appears and disappears repeatedly. However, we rather feel that the organization is due to global (redistribution of individuals) and local (density–dependent births and deaths) processes being in action all the time. With our tools, we are sometimes able to pinpoint features of self-organization, while at other times it is harder to detect in statistical terms. Yet, the process is there. This clearly calls for further research until the tools are fully developed and better understanding of the self-organization is gained.

### Evidence from Northumberland and the Alps

The most elaborated example of periodic traveling waves in animal population dynamics comes in cyclic populations of field voles, *Microtus agrestis*, in Northumberland, England (Lambin *et al.* 1998; Bjørnstad *et al.* 1999a; MacKinnon *et al.* 2001). Lambin and his associates have been studying vole population fluctuations since 1984 in the Kielder Forest and other surrounding forested areas in Northumberland. The Kielder forest data (1984–1998, 14 sampling sites) gave good evidence (Lambin *et al.* 1998) that in the small scale (approx. 10 × 15 km) field voles display a periodic traveling wave moving at the speed of 19 km annually from west

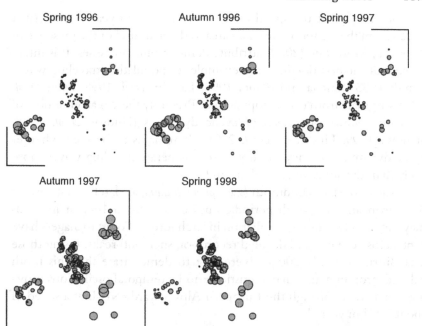

Spring 1996    Autumn 1996    Spring 1997

Autumn 1997    Spring 1998

*Fig. 5.13.* Spatial and temporal sequence of vole densities (size of the circle is in proportion to population density) for five sampling sessions in Kershope, Kielder, Wark, and Redesdale forests, Northumberland (redrawn from MacKinnon *et al.* 2001). The scale (lower left and upper right corners is 10 km, the lower left intersection is at 570 North, 344 East of the coordinates of Ordnance Survey British National Square NY).

to east. These estimates are based on very elaborate techniques (Sherratt *et al.* 2000; Sherratt 2001). Spatially more extensive, but temporally shorter data (fig. 5.13) confirmed these observations (MacKinnon *et al.* 2001). The new speed estimate was 14 km year$^{-1}$ traveling in a direction of 66° from north. Lambin and his associates are, at the time of writing this, somewhat uncertain of the causes of the vole traveling wave. However, they are suggesting (MacKinnon *et al.* 2001, p. 109) that nomadic predators and climatic factors are not responsible. Recently, an interest in spatial hetero-geneity in local population size has arisen as a possible driving force behind such population fluctuations (Sherratt *et al.* 2002).

It is also worth noting another example from northern parts of the British Isles, the long-term data on cyclic red grouse (*Lagopus lagopus scoticus*) in the Scottish Highlands, Kerloch moor (Moss *et al.* 2000). Using 18-year data from a 14-km$^2$ study area and similar analysis techniques as used with the Northumberland vole data above, Moss *et al.* (2000) came to the conclusion that the speed of the wave is 2–3 km year$^{-1}$ and that it travels from the center of the study area to the margin. For the presence of the wave, Moss et al. (2000) attribute demographic processes. It is interesting that the best documented examples of population traveling waves (Smith 1983; Kaitala and Ranta 1998; Lambin et al. 1998; Moss et al. 2000) originate from cyclic populations. Presently there is an ensemble of alternative views on causes of cyclic dynamics (Hudson et al. 1998; Stenseth 1999; Lindström et al. 2001). It remains to be seen whether cyclic dynamics, be they for any reason, generate traveling waves more easily than dynamics of any other kind.

Larvae of the larch budmoth (*Zeiraphera diniana*) are larch forest pests in the European Alps. With periodic outbreaks at 8- to 10-year intervals they cause wide-ranging defoliation in larch forests. Forest managers have kept a close eye on, and detailed records of, such outbreaks. Using those data Bjørnstad et al. (2002b) were able to demonstrate that this moth exhibits profound and characteristic periodic temporal oscillations moving like a wave through the European Alps towards NE-E at a speed of about 200 km year$^{-1}$.

### Evidence from measles in England and Wales

Further insight into repeated periodic waves, and alternative ways to understand them, has been gained in the studies of childhood microparasitic infections, in particular measles in England and Wales (Grenfell et al. 2001). An exhaustive analysis of exceptionally detailed spatial-temporal data of measles infections and host data reveals that traveling waves may hold and hide even more complicated structures than anticipated so far, based on data on animal populations. Grenfell and his colleagues (Grenfell and Harwood 1997; Grenfell and Bolker 1998; Earn et al. 2000) have studied childhood infections in order to uncover the characteristics of the dynamics of the epidemics and to predict the effects and efficiency of vaccination programs.

Grenfell et al. (2001) analyzed the measles dynamics in 354 administrative areas of England and Wales, ranging from large cities to small towns. The exceptionally long spatial weekly data cover the years 1944–1994. Wavelet power spectrum analysis (e.g., Nason and von Sachs 1999) allows us to detect and analyze the frequency structures of the time series even when they are nonstationary. Grenfell et al. (2001) found that the dynamics of measles epidemics show commonly seasonal cycles topped off with major long-term multi-annual cycles. The cycles are not regular, such that the relative strengths of the short- and long-term changes in incidence may vary considerably. Overall, the epidemics were characterized by biennial periods in the pre-vaccination era, which lasted until the late 1960s. During the vaccination era, the epidemics display longer temporal fluctuations. The dynamics of the epidemics in different areas are in synchrony such that the measure of the synchrony decreases with increasing distance between the locations.

The spatial-temporal patterns were investigated by calculating the phase difference between the measles epidemics at different data points. The analysis suggests that, in the pre-vaccination era, waves of infection moved from large cities to small, peripheral towns. Due to the fact that several large population centers act as sources for these waves the regional dynamics obtain a hierarchical structure. Vaccination caused a decreasing trend in the number of individual infections and changed the typical period length, but did not remove the traveling waves from the dynamics.

Grenfell *et al.* (2001) noted also that the analysis on spatial-temporal patterns in microparasitic infections may be jeopardized by the presence of nonstationary temporal variations in the dynamics. This means that the mean and variance of the time series may change with time. More interestingly, even the periods of oscillation may change, showing trends and sudden jumps, which makes the application of conventional frequency-domain analyses dubious as they assume stationarity. Obviously, we can expect that measles dynamics, and their kinds, continue to be important in developing our understanding of the spatial structures of populations, be it animal populations or microparasitic infections.

## Another view on self-organization

Features indicating self-organization in population data become increasingly important when we work on irregular grid structures. When the population network is irregular (as is the case with most natural systems), many of the visual patterns easily disappear and justification of the pattern loses its basis. For this reason, the need to develop new analytical and statistical approaches to distinguish spatial order from randomness has long been recognized. We shall now proceed to show that we can find signs of self-organization even when our visual guidance cannot observe any. In particular, we will now use spectral properties of the time series, in particular those of synchrony measures, as our aid. In the previous sections, we have become familiar with the fact that synchrony patterns vary with time. Here we will pay more attention to this problem. We will see that the dynamics in our system become self-organized such that the synchrony measures between population fluctuations in the subunits and their dependence on the inter-unit distance show a power law structure. This simply means that in irregular population networks the self-organization become quantitatively tractable by looking at the temporal synchronicity in the subunit dynamics. The synchrony is measured using a time windowing technique (Ranta *et al.* 1997a). We are able to show that such time-windowed synchrony measures show temporally scale-free, self-similar dynamics in time.

Consider again a system where the space is composed of dispersal-coupled population subunits, and each subunit hosts a local population renewal process. Let the spatial structure consist of $n$ randomly distributed units in a co-ordinate space. The dispersal follows kernel I (p. 53), where the distribution of the patches receiving dispersers depends on inter-patch distance. At each time step, a constant fraction $m = 0.1$ of individuals leaves their natal patch to reproduce elsewhere. To retain generality, we shall compare here several different models for local processes (Box 5.1): a Ricker model with delayed density dependence (Turchin 1990; Royama 1992) and linear autoregressive models (Box *et al.* 1994) of the order 1, AR(1), and 2, AR(2). The Ricker model with delayed density dependence (Turchin 1990; Ranta *et al.* 1997a, 1999b) allows us to generate cyclic population dynamics with 4-, 6-, and 10-year period lengths when the parameters are selected properly (Box 2.5). The temporal dynamics in each population subunit are affected by local noise and global Moran noise (Moran 1953b; Ranta *et al.* 1999a; Kaitala *et al.* 2001a). We are simply interested in how synchrony patterns may change in time (Ranta *et al.* 1997a, 1999b; Kaitala *et al.* 2001a). There are some indications that local populations may be in tight synchrony at times but not necessarily all the time (Ranta *et al.* 1997a). Thus, we may ask whether such a fluctuation in synchrony may be an outcome of chance, or whether there may be some other forces behind this phenomenon.

In the simulations, $n = 25$ local populations are randomly distributed on a $20 \times 20$ grid and initiated in random phase. The simulations are then run for $2^{10}$ time steps and the data for the next $2^{13}$ time steps are used in our analyses. We explore the simulated time-span with a moving time window technique (Ranta *et al.* 1997a). This is done in order to score temporal changes in the overall degree of synchronicity. Using this technique we also score the temporal changes between synchrony and its leveling off with distance, $r_D$ (Ranta *et al.* 1995a, 1999a). The temporal

---

**Box 5.1** · *Population renewal, dispersal and spatial self-organization*

A total of 25 population subunits are located randomly on a $20 \times 20$ grid. The local dynamics of the populations are affected by dispersal and local and global noise; assuming that a fraction $m$ of the population disperses annually we have

$$X_i(t+1) = (1-m)F[X_i(t), X_i(t-1), \mu(t)] + \sum_{s, s \neq i} M_{si}(t),$$

where $X_i(t)$ is the population size in patch $i$ at time $t$, $\mu(t)$ is the Moran effect and $M_{si}(t)$ is the number of dispersing individuals arriving at patch $i$ from patch $s$. The Moran effect is characterized by its annual probability of occurrence, $p(t)$, and its intensity, $\mu$, as follows

$$\mu(t) = \begin{cases} \mu & \text{if } \bar{p} \leq p(t) \leq 1 \\ 1 & \text{otherwise} \end{cases},$$

where $0 \leq \bar{p} \leq 1$ (here $p = 1/5$). Intensity $\mu$ is drawn from uniform random numbers between 0.5 and 1.5. We define the number of offspring alive after reproduction as follows

$$F = X_i(t)\mu(t)u_i(t)f[X_i(t), X_i(t-1)],$$

where local noise $u_i(t)$ is a random number drawn from a uniform distribution between 0.95 and 1.05. For the delayed Ricker dynamics we write

$$f[X_i(t), X_i(t-1)] = \exp\{r[1 + a_1 X_i(t) + a_2 X_i(t-1)]\},$$

where $r$ is the maximum per capita rate of increase, and $a_1$ and $a_2$ are parameters of density dependence. For the autoregressive model, we use the following

$$X_i(t+1) = \phi_1 X_i(t) + \phi_2 X_i(t-1) + \delta + \varepsilon(t),$$

where $\phi_1$ and $\phi_2$ are the autoregressive parameters, $\delta$ is a constant and $\varepsilon$ is normal random deviate (with mean 0 and variance 0.2). We choose to use $\delta = 2$. When we choose the parameter values such that $-1 < \phi_1 < 1$ and $\phi_2 = 0$ we have the AR(1) process, and for AR(2) the following inequalities should be satisfied simultaneously (Royama 1992): $\phi_2 + \phi_1 < 1$, $\phi_2 - \phi_1 < 1$ and $-1 < \phi_2 < 1$. In this system the dispersal obeys kernel I (p. 53) with parameters $m = 0.05$ and $c = 0.75$. For more details, see Kaitala *et al.* (2001a).

match in synchrony in pairs between the $n = 25$ (altogether 300 values) population subunits will be scored by using lag zero cross correlation, $r_0$ (Box *et al.* 1994). In order to analyze changing synchrony patterns we decided to build new time series using the synchrony patterns. In particular, we calculate here the difference between the upper and lower quartiles, $Q_{3,k} - Q_{1,k}$, in the $k$th time window for all pairwise correlation coefficients (Kaitala *et al.* 2001a). The upper and lower quartiles are indicated by the boundaries of the 25% highest and lowest scores of synchrony. For each time window we also measured the correlation $r_D$ between the level of synchrony and the distance among the populations compared. For the time window, we used $2^7$ time steps.

We analyzed the temporal behavior of the difference between the upper and lower quartiles, $Q_{3,k} - Q_{1,k}$, and the correlation between the level of synchrony and distance among the populations compared, $r_{D,k}$, where $k$ denotes the $k$th time window. Note that the narrower the quartile difference is, the higher the overall level of synchrony, which also makes the $r_{D,k}$ values close to zero. When the quartile differences are large there are some populations in tight synchrony, while there are also pairs of population subunits fluctuating out of phase or even randomly. In this way, $r_{D,k}$ may assume high negative values.

We first note that the fluctuations are periodic. Nevertheless, the temporal structure of the difference between the quartile boundaries of the synchrony measures shows noncyclic irregular variations over time (fig. 5.14(A)–(C)). This applies also to the correlation between the pairwise subunit synchrony against the distance between the units. However, when analyzed as time series using the power spectra of $Q_{3,k} - Q_{1,k}$, and $r_{D,k}$ there is a pattern in the frequency domain. The linear form of the power spectra of the time series analyzed (fig. 5.14(D)–(F)) suggests that the temporal dynamics of a spatially structured population are organized at the level of synchrony among population subunits. Such a form in the power spectra is a sign of the presence of power law in the spatial and temporal dynamics. Recall from Chapter 2 that the power law is of the $1/f^\alpha$ type, where $f$ is the frequency and $\alpha$ is a constant defining the slope of the power (on a log–log scale) and also the autocorrelation structure of the time series. The negative slope of the power spectra (positive $\alpha$, fig. 5.14(D)–(F)) indicates that long-term fluctuations dominate, giving the time series positive autocorrelation. More importantly, the linear structure of the power spectra is an indication of the presence of temporal self-similarity in the time series $Q_{3,k} - Q_{1,k}$, and $r_{D,k}$.

These results suggest that spatially structured populations contain a component of self-organization in their dynamics: the order is visible

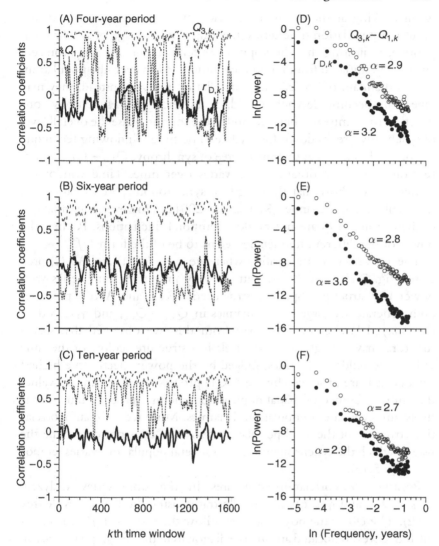

*Fig. 5.14.* The spatially structured population ($n = 25$) dynamics may become organized on the level of population synchrony. The population synchrony was measured using the upper and lower quartiles, $Q_{1,k}$ and $Q_{3,k}$, of the synchrony measures in pairs. When evaluated using time window techniques, we observe that the synchrony measures vary in time with local populations obeying (A) 4-year, (B) 6-year, and (C) 10-year period length in cyclic dynamics. Corresponding dynamics are also observed for correlation coefficients $r_{D,k}$ describing the relationship between the level of synchrony and the distance among the population subunits compared. (D)–(F) The power spectra of the time series $Q_{3,k} - Q_{1,k}$, and $r_{D,k}$. The linearity observed in the power spectra is a sign of the presence of power law in the spatial and temporal dynamics (the values of $\alpha$ are inserted). Modified after Kaitala *et al.* (2001a).

when looking at the dynamic behavior of synchrony patterns. In the simulated population time series, the long-term average of synchrony is positive, indicating that the populations tend to become synchronized, even when initiated from random conditions. However, the populations do not maintain the synchrony constellation forever. Instead, they may repeatedly become desynchronized, or, rather, a given synchrony constellation turns into another one, and in the shifting phase the overall level of synchrony breaks down for a while. The time windowing technique can be used to uncover that the degree of synchrony, $Q_{3,k} - Q_{1,k}$, and its relationship against distance, $r_{D,k}$, varies over time. Time windowing allows one to show that the length of synchronous periods varies in a temporal scale-free manner (Kaitala *et al.* 2001a). However, the variations of these time series are not random temporal fluctuations. Rather, the power of the different frequencies seems to be of the form $1/f^{\alpha}$.

The above results are verified when we used AR(1) and AR(2) processes (fig. 5.15). Synchrony patterns vary producing time series with power law structure. Spatially structured AR(1) and AR(2) processes both generate self-organized dynamics in $Q_{3,k} - Q_{1,k}$, and $r_{D,k}$, as does the delayed Ricker model. Apart from this, we observed that spatial structure may change the autocorrelation structure (color) of the time series. The reddest dynamics, judged by the power spectra (the highest values of $\alpha$), are produced by the AR(1) processes with parameter values that, in the absence of spatial population structure, produce blue (negatively autocorrelated) population dynamics. Moreover, the autocorrelation structure of the local population dynamics does not determine the way in which the synchronicity among several populations varies temporally (fig. 5.15).

Positive autocorrelation structures in the time series analyzed, $Q_{3,k} - Q_{1,k}$, and $r_{D,k}$, can also be illustrated using the IFS scores (Jeffrey 1992). The closer the power spectra follow the form of $1/f^{\alpha}$, the closer the IFS scores are located around the diagonals in the IFS graph (fig. 5.16). When the time series, $Q_3 - Q_1$, is represented by white noise, and the local patches are independent ($m = 0$), the IFS score fails to show any pattern (fig. 5.16(A)). When interactions among patches are added ($m > 0$) local dynamics are organized into spatial dynamics with self-organized structure. Several different kinds of local dynamics become organized due to global (redistribution of individuals) and local (density-dependent births and deaths) interactions. In the IFS graphs, points will aggregate around the diagonals which is observable for AR(1) and AR(2) processes (fig. 5.16(B)–(D)). Most distinct aggregations will be observed

AR(1) dynamics

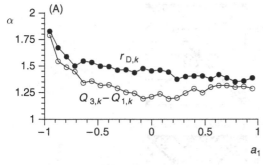

AR(2) dynamics

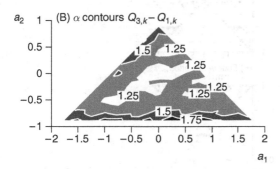

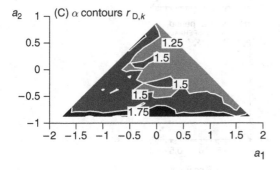

*Fig. 5.15.* The synchronicity degrees also fluctuate when the local dynamics obey AR(1) or AR(2) processes. The average values of the slope $\alpha$ of the power spectra of $Q_{3,k} - Q_{1,k}$ and $r_{D,k}$ for (A) AR(1) and ((B), (C)) AR(2) processes. The values are averages of ten replicated runs for each parameter value combination possible for the AR(1) and AR(2) processes. Modified after Kaitala *et al.* (2001a).

Quartile difference in synchrony, $Q_{3,k} - Q_{1,k}$

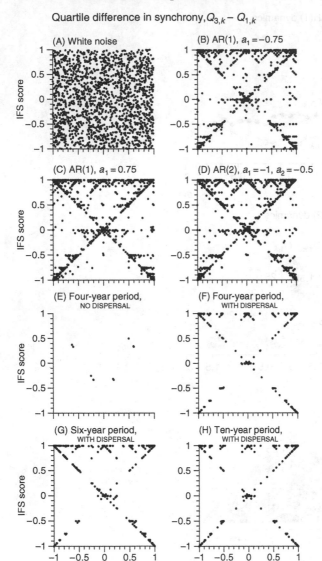

*Fig. 5.16.* The IFS scores can be used to detect power law structures in the time series, $Q_{3,k} - Q_{1,k}$. In the IFS graphs, temporal structures obeying power law are detected by aggregation of the dots along the diagonals. (A) White noise in a nonstructured population; (B) and (C) AR(1) processes; (D) AR(2) process; (F)–(H) spatially structured populations where local populations obey cyclic dynamics with 4-, 6-, and 10-year period lengths, respectively; (E) cyclic dynamics with period length of 4 years with no spatial interaction. Modified after Kaitala *et al.* (2001a).

in the cyclic population dynamics with different period lengths (fig. 5.16(F)–(H)). The cyclic fluctuations seem to be strongly self-organized, but only when dispersal is present (fig. 5.16(E)). It is the density-dependent feedback together with dispersal that "organizes" the dynamics.

### Evidence from Canada

The long-term time series from the whole of Canada, representing the comprehensive bookkeeping records of fur returns of the Canada lynx from Hudson Bay Company's trading posts (Elton and Nicholson 1942), are exceptionally rich in temporal and spatial patterns (e.g., Smith 1983; Ranta et al. 1997a,b; Stenseth et al. 1999). The data, which cover 96 years from eight provinces (Royama 1992), provide a good source for testing different population patterns. The first population ecologist to seriously analyze the large-scale temporal and spatial patterns of Canada lynx was Charles Elton who, in a joint effort with Mary Nicholson, compiled the data set. They pointed out that the population fluctuations are periodic which cover "the whole northern forest zone of Canada, from Labrador to British Columbia to Yukon." More recent studies have confirmed that the population dynamics of Canada lynx are highly variable (Stenseth et al. 1999), and that the dynamics show a pronounced 10-year cycle. Moreover, Elton and Nicholson (1942) explicitly stated that the fluctuations occur in synchrony such that "the most extraordinary feature of this cycle is that it operates sufficiently in line over several million square miles of country not to get seriously out of phase in any part of it" (fig. 5.17(A),(B)). Later on, Smith (1983) suggested that the population highs build up in the central parts of Canada, and then develop ("move") towards both the east and west. This is in agreement with the analyses by Ranta et al. (1997a,b).

Here, we set out to analyze the time consistency of the synchronicity patterns in the Canada lynx data (Ranta et al. 1997a). Using the sliding time window technique shown above, we see that the populations on average fluctuate in synchrony (fig. 5.17(C)). However, the degree of the synchrony varies through time (fig. 5.17(D)). A pair of populations may fluctuate in step for some time, then the synchronicity may disappear, to come back later. The power spectrum of the time series indicates that the results obtained for the Canada lynx are in good agreement with the theoretical results presented in the previous section (fig. 5.1(E),(F)). As in the simulations, the form of the power spectra suggests that the temporally varying degree of synchronicity is organized as a power law of $1/f^\alpha$ type

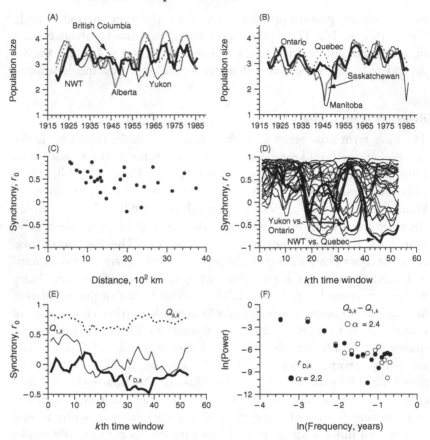

*Fig. 5.17.* The varying degrees of synchrony can be observed in the long-term dynamics of Canada lynx. ((A), (B)) The 1919–1985 data are annual lynx pelt harvest aggregated at the province level. (C) Also in the lynx data the level of synchrony (over the whole sampling period) levels off against distance of the province centers. (D) When a 15-year sliding time window is put through the data in (A) and (B) one finds that the level of synchrony in lynx population fluctuations between any two provinces is not time-invariant. (E) Both the lower $Q_{1,k}$ and upper $Q_{3,k}$ quartiles of the synchrony measures in pairs among the eight Canadian provinces, panel (D), and the correlation between synchrony level and distance among the provinces compared, $r_{D,k}$, fluctuate in an erratic manner over time. (F) Power spectra for $Q_{3,k} - Q_{1,k}$ (○) and $r_{D,k}$ (●). The data are from (D). Combined from Ranta *et al.* (1997a) and Kaitala *et al.* (2001a).

(Kaitala *et al.* 2001a). In fact, the analyses by Ranta *et al.* (1997a) support the idea that the Canada lynx dynamics is a traveling wave in space. This was first suggested by Smith (1983) and hinted by the graphs in Butler (1953). It remains to be seen whether the power law nature of the

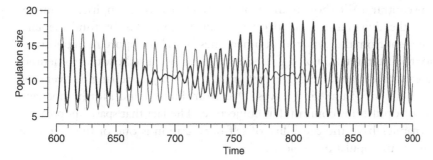

*Fig. 5.18.* Fluctuations in size of two randomly selected populations ($n = 25$), located in a 20 × 20 co-ordinate space. All populations obey locally 10-year periodic fluctuations; they are interconnected (kernel I, p. 53) by redistributing individuals (annually $m = 0.1$; more details in Ranta *et al.* 1997a). It turns out that global (dispersal) and local (density-dependent population) renewal will organize the spatial dynamics (fig. 5.17; Kaitala *et al.* 2001a). From the process, it also follows that the deterministic population cycle may wane, to reappear later on.

synchrony in dynamics, as discussed in this chapter, is a general sign of traveling waves in nature. When Ranta *et al.* (1997a, 1999b) simulated the Canada lynx dynamics in dispersal-coupled space, they observed that pairs of populations, initially in step in temporal fluctuations, might drift out of phase and then return to synchrony some time later on. A striking observation in this connection was that the deterministic local cycle often disappeared, to reappear later (fig. 5.18). The fluctuating synchrony and temporal loss of deterministic cycle are the results of the interplay between local (births and deaths) and global (immigration and emigration) processes jointly in action all the time in such systems. The differing processes are the factors organizing the spatial dynamics of populations.

## Summary

Introducing spatial structure may profoundly change the emerging dynamics, as we first saw in Chapter 3. In a trivial sense, spatial structure makes the dynamics more "complex." This is also true in the formal sense, i.e., that the emerging dynamics indeed are complex from a mathematical point of view. We here review a set of commonly used models that generate intriguing spatial dynamics, e.g., spiral chaos and traveling waves. Spatially complex behavior also shows that some emergent phenomena are in fact scale independent, i.e., self-organized at all levels of spatial resolution. This does not mean that the dynamics are time invariant – on the contrary. We show both theoretically and by some empirical examples that the dynamics observed over certain periods may be fundamentally different from the dynamics at other times. This has important implications for the interpretation of, e.g., time series data. We also show that sometimes the spatial dynamics, e.g., the patterns of synchrony, are changing over time in the form of traveling waves. Data from Finnish and Northumberland vole populations confirm that conjecture. The fact that spatial population dynamics have the capability of self-organization is one of the emergent "laws" in population ecology.

# 6 · Structured populations

Populations are not collections of identical individuals, albeit such approximations are often useful. In many cases, however, a higher resolution is needed to understand population and community processes. In this chapter, we introduce some more details by letting the population be divided into age or stage classes, all of them possibly with their specific vital rates. This can be achieved by rather small modifications of the simple models, yet qualitatively and quantitatively new phenomena will emerge.

The state of a population is usually thought of as its size or density. As we have seen, it is in fact the *density*-dependent feedback that is assumed to be important at the population level. The implicit assumption behind this is that all individuals are more or less identical when it comes to their demographic effects: their contributions to births and deaths (and immigration and emigration). This is, of course, not the case in most natural populations. Young individuals are often more susceptible to death than adults, and they often contribute less to reproduction. Individuals at some intermediate adult stage face less risk of dying and are the ones that reproduce successfully. Such differences are not necessarily only attributed to age, but also body size (often co-varying with age). For example, in most species with indeterminate growth, as in fish and reptiles, fecundity and survival are strongly dependent on size, rather than on age per se (Roff 1992; Stearns 1992).

The more general stage-structure problem has received considerable attention in the ecological and evolutionary literature. After some seminal contributions (e.g., Metz and Diekmann 1986; DeAngelis and Gross 1992; de Roos *et al.* 1992; Tuljapurkar and Caswell 1997; Caswell 2001), the field has become rich in ecological and evolutionary studies of structured populations. Quite clearly, the feedback environment of an individual organism, i.e., the target for natural selection, is in reality much more subtle than the simple one assumed in standard single-population models. As outlined in Chapter 1, the demography of a population is really the key to a more detailed and deeper understanding of population processes. Towards the end of this chapter, we are going to discuss some of the pros and cons of using simple and more detailed (consequently more complicated) models. This becomes an issue especially in some applied areas.

Many individual differences remain, however, even after having corrected for age and size. For example, in many birds and mammals, there is a strong reproductive skew such that only a small fraction of all the individuals in a population are the ones that get the opportunity to reproduce each breeding season (Clutton-Brock 1991; Sutherland 1996). In this chapter, we are going to have a closer look at populations that are structured one way or another. The focus will be on age-structured populations, but the approach taken here is relatively easily extended to other ways in which a population may be divided into functional groups.

## The projection matrix

The projection matrix, i.e., the mathematical device that changes a vector of population densities at one point in time into another, was briefly dealt with in Chapters 1 and 2. Let us recall eq. 2.3

$$
\begin{bmatrix} n_1(t+1) \\ n_2(t+1) \\ n_3(t+1) \\ \vdots \\ n_x(t+1) \end{bmatrix} = \begin{bmatrix} m_1 & m_2 & \cdots & m_x \\ l_1 & 0 & \cdots & 0 \\ 0 & l_2 & \cdots & 0 \\ \vdots & \vdots & \ddots & \vdots \\ 0 & 0 & \cdots & 0 \end{bmatrix} \begin{bmatrix} n_1(t) \\ n_2(t) \\ n_3(t) \\ \vdots \\ n_x(t) \end{bmatrix}, \tag{6.1}
$$

where the two vectors $\mathbf{n}(t)$ and $\mathbf{n}(t+1)$ indicate the population densities of age classes 1 to $x$ at time $t$ and $t+1$, respectively. The Leslie matrix $\mathbf{L}$ has as its elements the age-specific (net) fecundities, $m_x$, and survival rates, $l_x$. Caswell (2001) is a brilliant introduction to the details and extensions of this model. There are two important things to remember from this relatively simple equation. First, it is the dominant eigenvalue, $\lambda_{\text{dom}}$, of $\mathbf{L}$ that determines the growth rate of the population; second, that it is the (left) eigenvector (Caswell 2001) associated with this eigenvalue that informs us about the stable age distribution of the population. If the population is deterministically growing or shrinking, the age distribution remains the same (after some time of transients).

Here, we are going to highlight a couple of aspects of a structured population that have caught quite some attention in the ecological literature, although they have not been analyzed in much formal detail.

We will also undertake an analysis of a simplified version of this model with only two classes of individuals, as well as study the properties of structured populations in stochastic environments.

## Cohort and maternal effects

The environment experienced by individuals at an early age may have important effects on subsequent life history (Box 6.1; Lindström 1999; Metcalfe and Monaghan 2001), and cohorts of individuals may therefore differ considerably (e.g., Rose and Bradley 1998). Likewise, the environments experienced by parents can affect their allocation decisions. For example, propagule size variation may give rise to individuals of different

---

**Box 6.1** · *Early development and dynamics*

In many organisms, such as birds and mammals, the nutritional and other physiological and environmental factors experienced by individuals during early development, or any other key period of development may have a clear effect on survival and reproductive success later on during the life cycle. This means that the demography (the elements of the population transition matrix) is not constant but subject to year-to-year variability. Extreme years may then, e.g., change the future fecundity of a given cohort, but only temporarily. Such external factors may include maternal or parental (weaning) condition and behavior. As a consequence, this can have an important role in modifying life histories as well as the dynamics of populations (Benton *et al.* 2001). For example, young individuals often face nutritional deficit, which they may later compensate for by accelerated growth when the conditions get better (Metcalfe and Monaghan 2001).

The effects may be different between sexes. For example, weaning condition in grey seals may have a greater effect on survival for male pups than for females (Hall *et al.* 2001). This may have an impact on differential allocation between sexes among offspring. Hall *et al.* (2001), for example, suggested that females in good condition should invest favor in male pups because the marginal return in investing in the offspring is higher in males. These effects on early development can thus be called maternal (or parental effects) (Lindström 1999), or more generally cohort effects when a discernible cohort is thus affected, in turn affecting the dynamics of the entire population.

quality (Bernardo 1996). If all females experience the same environmental condition, then their effects can be expressed in the next cohort of offspring. So, the properties of the offspring due to the mother's experienced environment (maternal effects) may be passed on as an entire cohort effect (Beckerman et al. 2001). Cohort and maternal effects are a special variant of the more general problem of delayed density dependence, studied in a wide range of taxa (e.g., insects: Turchin 1990; Benado 1997; Bjørnstad et al. 1998, 2001; fish: Myers et al. 1997a,b; Fromentin et al. 2001; birds: Lindström et al. 1997b, 1999; Watson et al. 1998, 2000; mammals: Bjørnstad et al. 1995; Boonstra et al. 1998; Stenseth et al. 1998b; Erb et al. 2001; plants: Crone and Taylor 1996; Gillman and Dodd 2000).

The interaction between environmental variability and demography can also cause lags through cohort and maternal effects. A number of biological factors have been suggested to cause delays. Trophic interactions are regarded as one of the most important ones (e.g., Berryman 1996; Fryxell and Lundberg 1997; Turchin et al. 1999; Bjørnstad et al. 2001; Turchin 2001). Environmental stochasticity also has the potential to cause lagged dynamics, both actual and spurious (Chapter 2). Mousseau and Fox (1998) reviewed the empirical evidence for maternal effects and found examples across a wide range of environments and taxa. The life history consequences of early development are also well documented (Lindström 1999; Metcalfe and Monaghan 2001).

We can now consider eq. 6.1 as the starting point for our investigations of how demographic structure and environmental variability affect the dynamics of the entire populations. As noted earlier, the elements of the population transition matrix determine the population dynamics. Should they change over time, or be affected in a single pulse in a given year, the dynamics change. As far back as 1959, Leslie (1959) did, in fact, incorporate cohort effects into the first matrix models. Ever since, the dynamical consequences of maternal effects have been studied (Ginzburg and Taneyhill 1994; Crone 1997; Inchausti and Ginzburg 1998; Benton et al. 2001). As may be expected, these studies suggest that such intrinsic lags can destabilize dynamics much in the same way as do extrinsic lags.

If individual resource acquisition is density dependent, then changes in age- or stage-specific densities will have effects on population growth rate. Thus modeling cohort effects requires an understanding of the distribution of competitors across the life cycle. In many species, age or size may not be strong indicators of how and where resources are acquired. In such species, density-dependent processes act as in nonstructured populations: the number of competitors is the total population size. However, there are several cases where the age classes of the population may be ecologically very different, sometimes to the extent that the direct interaction between them is very weak or nonexistent.

## Ontogenetic niche shifts

In many species diet changes with stage or age, and it is quite common that different age groups of a single species live in different habitats. Such changes in diet, or shifts in habitat, with age are known as ontogenetic niche shifts (Wilbur 1980; Werner and Gilliam 1984; Werner and Anholt 1993; Arendt and Wilson 1997; Plaistow and Siva-Jothy 1999). Hence, the feedback environment of an individual is again not only contingent on the demography and the densities of the different stage classes in the population, but potentially also on how the demography is spatially located.

### A flexible demographic model

The model we are developing here (Lundberg *et al.* unpublished) is a rather straightforward extension of Leslie's original ones (Leslie 1945, 1948, 1959). In those models, the density-dependent feedback was experienced as the sum of all individuals in the population, regardless of age. In our modification of Leslie's models, we allow an ontogenetic niche shift to reduce the number of individuals (from adjoining age groups) affecting the density-dependent feedback of the target age group.

Let the population consist of individuals belonging to $x$ age groups. Their numbers at any time $t$ can be given in a column vector $\mathbf{n}$ of length $x$. Each age group $i$ has its specific fecundity $m_i$ and probability $l_i$ that a female of age $i$ at $t$ will survive to $i + 1$ at time $t + 1$. In the Leslie matrix $\mathbf{L}$, the values of $m_i$ are the elements of the first row and the values of $l_i$ are the elements of the first subdiagonal. The other elements of $\mathbf{L}$ are all zero, eq. 6.1. The number of individuals in the different age groups at $t$ can now be mapped to $t + 1$ as

$$\mathbf{n}(t + 1) = \mathbf{L}\mathbf{n}(t), \tag{6.2}$$

and for the population size we have $N(t) = \sum n_i(t)$. Note that eq. 6.2 is just a compact form of eq. 6.1.

Leslie (1948) incorporated density-dependent feedback by postulating that population density at each time affects survival of the different age groups. He defined the quantity

$$q(t) = 1 + aN(t), \tag{6.3}$$

where

$$a = \frac{\lambda - 1}{K}. \tag{6.4}$$

The constant parameter $a$ is the coefficient of density dependence and $\lambda$, the dominant eigenvalue of $\mathbf{L}$, is the population growth rate. The parameter $K$ is the carrying capacity of the environment, hence $a$ is the strength of the density dependence. Note also that $q$ is time dependent because $N$ is. The $q(t)$ values are the diagonal elements of a matrix $\mathbf{Q}$, which can now be combined with eq. 6.2 to achieve growth of an age-structured population in a resource-limited environment

$$\mathbf{n}(t+1) = \mathbf{L}\mathbf{Q}^{-1}\mathbf{n}(t). \tag{6.5}$$

Leslie (1959) also introduced time lags into eq. 6.3. If the lags are taken to be the number of relevant age groups in the population for each age group $i$ one has

$$q_i(t) = 1 + bN(t - i - 1) + aN(t), \tag{6.6}$$

where $b$ is the effect of density at birth on the probability of survival at some later time $t + x$. The constants $b$ [age group number, $x$, lagged density dependence, lag($x$)] and $a$ (direct density dependence) are both $> 0$ and their relative magnitude, $b/(b + a)$, determines how strong the impact of the cohort effect is on the density dependence feedback in eq. 6.6. For clarity we shall write the elements of $\mathbf{Q}(t)$ for a population with four age groups

$$\mathbf{Q}(t) = \begin{bmatrix} q_0(t) & 0 & 0 & 0 \\ 0 & q_1(t) & 0 & 0 \\ 0 & 0 & q_2(t) & 0 \\ 0 & 0 & 0 & q_3(t) \end{bmatrix}. \tag{6.7}$$

Equation 6.5 now characterizes the dynamics with the lagged effects taken into consideration.

### Ontogenetic niche shift

A straightforward way to incorporate the effect of an ontogenetic niche shift into the Leslie matrix model is to assume that the different age groups either completely overlap in their relevant niche dimensions or that the different age groups have nothing in common except that they belong to the same population. The former case indicates that all age groups live in the same environment (and have identical diets), while in the latter different age groups live in entirely different environments (or their diets do not overlap). The degree to which those cases apply can be handled as follows.

Let $\mathbf{A}$ be an $x$-by-$x$ matrix, where the elements $\alpha_{ij}$ indicate how much the niche of the $i$th age group overlaps with that of the $j$th age group. The diagonal elements $\alpha_{ii}$ are standardized to unity. If all $\alpha_{ij} = 1$ we have the original Leslie model (with different age groups competing fully on common resources), while with $0 \leq \alpha_{ij} \leq 1$ we have relaxed competition among age classes. In this case, it is natural to assume some hierarchy in the niche overlap values, e.g., $\alpha_{12} > \alpha_{13} > \alpha_{14}$. When the off-diagonal elements of $\mathbf{A}$ are all zero there is a complete ontogenetic niche shift among ages. The upper and lower off-diagonals in $\mathbf{A}$ can be either symmetric (sequential niche shift) or asymmetric (Wilson 1975).

A population model of an age-structured population with the cohort effect and with the ontogenetic niche shift needs the following modification to eq. 6.6.

$$q_i(t) = 1 + b \sum_{i=1}^{x} \alpha_{ij} n(t - i - 1) + a \sum_{i=1}^{x} \alpha_{ij} n(t). \qquad (6.8)$$

The summations give the overlap-specified weights for age-group-specific densities. Note that when $\mathbf{A}$ is a matrix of 1s we are back to Leslie (1959), and when the diagonal elements that make $\mathbf{Q}(t)$ are all matching, we are back to Leslie (1948).

## Dynamics of cohort effects and niche shifts

In the explorations of the model, we used $K = 100$ for all age groups regardless of the values in the $\mathbf{A}$ matrix. Using Leslie's original values for the four age groups in $\mathbf{L}$ we had four different $\mathbf{A}$ matrices. The first one, composed of 1s as elements, corresponds to that of Leslie's (1948, 1959) approach, where all individuals are equal competitors. The entries of the second $\mathbf{A}$ matrix were 1s along the diagonal and 0s elsewhere. This corresponds to perfect ontogenetic niche segregation. The third (all elements below diagonal and diagonal values $= 1$, others being 0) and fourth (diagonal and above diagonal elements equal to 1, others equal to 0) $\mathbf{A}$ matrices were asymmetric. These correspond to the scenarios where the adult environment/resources can be used by adults and

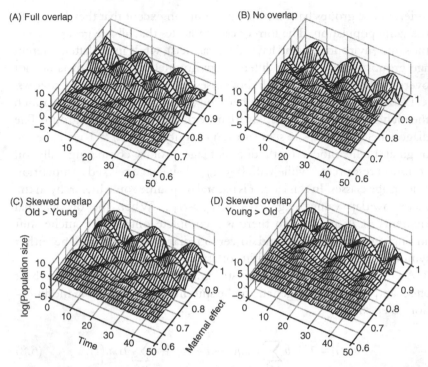

(A) Full overlap

(B) No overlap

(C) Skewed overlap Old > Young

(D) Skewed overlap Young > Old

*Fig. 6.1.* Fluctuations in population size [log[$N(t)$]] over a 50-year period (*x* axis) after a neutral period of 5000 generations has elapsed. Maternal effect, *z* axis (range from 0.6 to 1, from modest effect to maternal effect dominating the density dependence feedback) depicts the relative contribution of the maternal effect [$b/(b + a)$, where *b* is the density-dependent component lagged by *x* time steps, and *a* is the direct density dependence]. The four panels give four differing scenarios in the ontogenetic niche shift. In panel (A), **A** is a matrix of 1s, in (B) there is no niche overlap among the different age groups (i.e., all elements in **A** are 0s, exception the diagonal of 1s). In panel (C), older age groups completely overlap with younger age groups, while individuals of the age group *x* share 50% in common in their diet with the subsequent age group *x* + 1. In panel (D), the case of (C) is reversed. The Leslie matrix elements are taken from Leslie (1959, p. 154).

juveniles, but the juvenile environment or resources are only accessible to juveniles (third case) and vice versa (fourth case). We initiated the system with Leslie's (1959) values for **n**(0) and let the system stabilize for 10 000 generations and sampled the next 50 generations of population dynamics. The simulation was repeated for a range of values from 0.6 (with a step of 0.02) to 1 for the impact of the cohort effect, $b/(b + a)$.

Our results suggest that the presence of ontogenetic niche shifts stabilizes the population dynamics (fig. 6.1). With **A** deviating from full-niche overlap induces the periodic phase of population fluctuations to start with larger values of $b/(b + a)$ (around 0.85) than with Leslie's (1959) original formation. To substantiate the first conclusion, we used another set of simulations. We let the number of age groups range from $x = 3$ to $x = 6$, we also let the parameter $\lambda$ range from 1 to 20 [Leslie's (1959) $\lambda$ for four age groups was 3.0]. However, only two kinds of $\mathbf{A}$ were tested: the full overlap and the complete ontogenetic niche differentiation. The $b/(b + a)$ ratio was set to 0.7 throughout. The system was initialized with random numbers for $\mathbf{n}(0)$. For these data we calculated the coefficient of variation for the log-transformed population sizes. The system was repeated 100 times for each combination of $\lambda$, $x$ and $\mathbf{A}$, and the averages of these values are reported.

For this system we find persistent (smooth cyclic; see fig. 6.1) fluctuations over all values of $x$ and $\mathbf{A}$ used only when $\lambda$ is large enough, the value depending on the number of age classes, $x$ (fig. 6.2). Second, and most importantly, population variability with the cohort effect but without an ontogenetic niche shift is substantially higher than that with a complete ontogenetic niche shift (fig. 6.2). Experimentation shows that the conclusion holds for $\mathbf{A}$ matrices between the two extremes used. The closer the values of $\mathbf{A}$ are to the full ontogenetic niche shift values the smaller is the population variability under the cohort effect.

## A few comments

The approach we have taken is highly flexible. For example, a full range of differential resource use can be modeled from complete niche overlap amongst all age classes, through asymmetric resource partitioning to complete niche separation. Numerous studies have indicated that incorporating biological detail (such as age structure, noise, different functions of density dependence, and so on) may radically affect the conclusions from a model. This particular modeling framework can be extended to incorporate noise, and may therefore be useful in predicting population dynamics in applied problems. Furthermore, it is now well established that behavior acting on demography (e.g., nonlethal effects) is responsible for a large array of trophic interactions (direct and indirect), patterns of habitat use and subsequent demography (Lima and Zollner 1996; Wellborn et al. 1996; Schmitz et al. 1997). As our model is based on a matrix describing the distribution of competitive interactions over a life history, it is quite capable of accommodating both direct and indirect

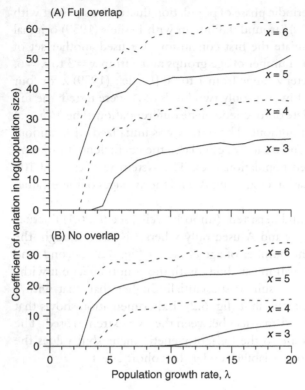

*Fig. 6.2.* Population variability (measured as coefficient of variation over time in log-transformed population size) against different population growth rates $\lambda$ in age-structured populations with the number of age groups $x$ varying from 3 to 6. In panel (A) matrix **A** is a matrix of 1s, while in (B) there is no food overlap among the different age groups (i.e., all elements in **A** are 0s, exception the diagonal of 1s).

impacts on a life history through density dependence (Yodzis 1989; Brommer *et al.* 2000).

Delayed effects often tend to destabilize the dynamics of populations. The general conclusion from our model is that adding niche differentiation between age classes reduces the impact of the delay due to the cohort effect. This is in line with the general belief that age structure together with differentiation among population members tends to stabilize the dynamics (Fryxell and Lundberg 1997). It is clear that the competition between age classes has a major effect on the stability properties of the system. The impact of the cohort effect therefore depends on the number of competitors faced by the youngest individuals, rather than the existence of ontogenetic niche shifts per se.

# Doers and viewers

The model presented above is a very general solution to the age/stage problem. In this section, we are going to have a closer look at an application and special case of the previous theory, namely a population divided into only two classes. We will not, however, make any detailed assumptions about the exact background of the class division. That is, they may not necessarily represent, for example, ages or sizes. We will only assume that a fraction of the population is responsible for the recruitment to the population next year (the *doers*), whereas the other fraction is only present and using resources, but otherwise not engaged in reproduction (the *viewers*). The doers and viewers are, e.g., representing the breeding and the floater segments of a bird population. Moreover, this fraction is very likely not constant from one generation to the next (see, e.g., Rohner 1996). We will instead assume that this fraction is environmentally determined such that, in a good year, a large fraction of the population is going to have the opportunity to reproduce, whereas a bad year means that only a few of the individuals will have that possibility.

There is good evidence that in many natural populations only a fraction of individuals of mature age are capable of reproduction. The reasons for this are manifold; sometimes it is due to lack of food, sometimes to lack of breeding territories, or even the presence of predators might make some individuals postpone their reproduction. Most obvious examples are helpers in birds where mature offspring may assist their parents to breed (Brown 1987). Further, in social species performing co-operative breeding the degree of reproductive skew may be related to environmental conditions. In dwarf mongoose (*Helogale parvula*) and meerkats (*Suricata suricatta*), for example, reproductive performance in subordinates depends on age and food abundance (Creel and Macdonald 1995; Waser *et al.* 1995; O'Riain *et al.* 2000). Blumenstein and Armitage (1999) suggested that the degree of co-operative breeding in marmots could result from environmental constraints. Brood-adjusting strategies as a response to fluctuations in prey populations, e.g., in the vole-eating Tengmalm's owl, *Aegolius funereus*, (Korpimäki 1987; Korpimäki and Hakkarainen 1991), may cause the fraction of mature and reproducing females to fluctuate. Finally, in four allopatric species of mockingbirds (*Nesomimus* sp.) co-operative breeding is maintained where limited availability of preferred habitat constrains dispersal (Curry 1989).

## A population model for doers and viewers

Again, we will use the Ricker dynamics as the skeleton of the population renewal process for the doers and viewers. Each year $t$, a proportion $p(t)$ of all individuals $N_T(t)$ in the population is not able to reproduce for one reason or another. They are the viewers, $N_V(t)$, while the rest are the doers $N_D(t)$ who take care of the reproduction. For the renewal process under these circumstances we have

$$N_T(t+1) = N_D(t) \exp\{r[1 - N_T(t)]\}$$
$$N_V(t+1) = p(t+1)N_T(t+1) \qquad (6.9)$$
$$N_D(t+1) = N_T(t+1) - N_V(t+1).$$

Parameter $r$ is the population growth rate. We used here $r = 1.5$ (yielding stable dynamics). We let $p(t)$ be a random variable, but assumed that it is temporally autocorrelated. We therefore made $p(t)$ an autoregressive process of order 1, AR(1), according to Ripa and Lundberg (1996) with the parameter $\kappa$ assuming three values, $-0.7$ (blue), 0 (white), and 0.7 (red) so that the $p(t)$ assumed values from 0 to $p_{MAX}$ (see eq. 2.12, p. 27).

The proportion of viewers in the population affects overall population variability (fig. 6.3). When the proportion of viewers is small (0.1–0.3, fig. 6.3), the viewers are the most variable segment of the population. As the environment goes from blue to red, the coefficient of variation in population size, CV(viewers), decreases, whereas the CV(doers) increases. If the proportion of viewers is high (0.6–0.8, fig. 6.3), the CV(viewers) now increases as we go from blue to red environments and so does the CV for both the doers and the total population. However, in red (and white) environments, the CV(doers) now exceeds the CV(viewers). Population variability is hence much dependent on the composition of the population (the proportion of doers versus viewers) and the properties of the environmental variability (blue versus red). When looking at the variability of the total population it follows the same pattern. The CV(total) decreases as we go from blue to red environments if the proportion of viewers is low (left column, fig. 6.3), but increases as we go from blue to red if this proportion is large (right column, fig. 6.3). For any one color of the environmental variability, the CV(total) invariably increases with increasing proportion of viewers. This highlights the fact that it might be important to know the population structure and not just basic vital rates before conclusions about population stability conditions can be drawn.

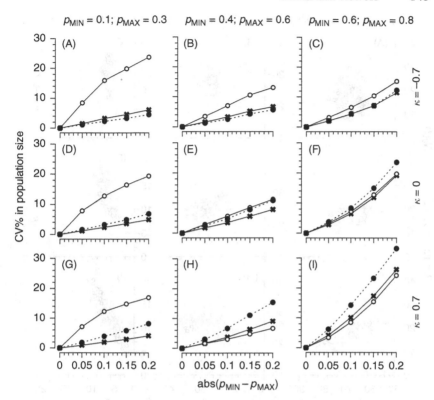

*Fig. 6.3.* The nine panels show how population variability (measured as the coefficient of variation in population size of the last 1000 generations in the simulations extending over 5000 time steps) varies with both the average proportion of viewers in the population (columns) and with the color of the variability of the proportion of viewers (rows; upper: blue ($\kappa = -0.7$), middle: white ($\kappa = 0$), bottom: red ($\kappa = 0.7$)). As the variability of the proportion of viewer goes up along the *x* axis in each panel, so does naturally the CV of the populations.

Another interesting feature of this system is illustrated in fig. 6.4. With two segments of the population inversely related by the parameter *p*, we would perhaps expect them also to be inversely related under most circumstances when measured from data. The data we produced here are simulated trajectories of the doers and viewers according to eq. 6.9. The two population segments were plotted against each other for small to large proportions of viewers (columns in fig. 6.4) and under different environmental variability (from blue to red; rows in fig. 6.4). When the environment was blue, there was no relationship at all between doers and viewers, regardless of the proportion of viewers. When the environment was red, on the other hand, the expected negative relationship between

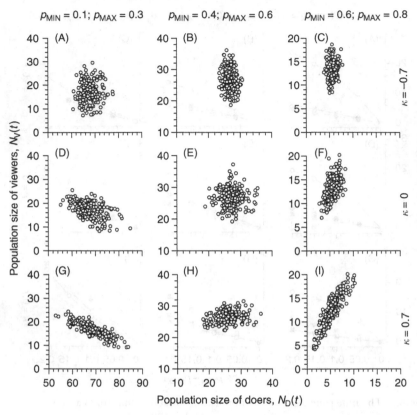

*Fig. 6.4.* The relationship between the number of doers and viewers in the population (final 200 generations of simulations lasting 5000 generations). The top row of panels shows that when the environment is blue ($\kappa = -0.7$), the relationship between doers and viewers is very weak and that the proportion of viewers in the population (columns) has very little influence on this relationship. When the environment is either white ($\kappa = 0$) or red ($\kappa = 0.7$), the relationship changes from being negative (left column) to positive (right column). See also fig. 6.3.

the two groups was revealed, but only if the proportion of viewers was small. For high proportions of viewers, there was a positive relationship between the groups. When the viewers constituted roughly half of the total population, there was no discernible relationship no matter what the environmental variability was like. This again underscores the fact that the interaction between the environmental driver (here the variability of the proportion of nonreproductive viewers in the population), demography, and population structure is an intricate determinant of the emergent dynamics.

# Stage structure and the *Tribolium* example

One particularly nice example of the analysis of stage-structured population comes from the experiments on the dynamics of the flour beetle, *Tribolium* (Box 6.2) carried out in Park's (1948) tradition by Costantino, Cushing, Dennis and colleagues (e.g., Costantino *et al.* 1995, 1997, 1998; Cushing *et al.* 1998a,b; Dennis *et al.* 1997, 2001). In many invertebrates (and also many plants), the subclasses of a population are both easily discernible and naturally grouped, unlike, for example, the size classes used when studying fish or reptiles. In the *Tribolium* populations studied (actually the two species

---

**Box 6.2** · *The* Tribolium *cannibalism experiment*

The three-stage life cycle of *Tribolium*, including cannibalism of eggs by larvae and adults and pupa by adults, may produce highly complicated population dynamics patterns. The model of the development and interaction of larval, pupae, and adult *Tribolium* numbers is given as (Costantino *et al.* 1997)

$$L(t+1) = bA(t)\exp[-c_{el}L(t) - c_{ea}A(t)]$$
$$P(t+1) = L(t)(1 - \mu_l)$$
$$A(t+1) = P(t)\exp[-c_{pa}A(t)] + A(t)(1 - \mu_a).$$

$L$, $P$, and $A$ are the numbers of larvae, pupae, and adults, respectively. The constant fractions $\mu_l$ and $\mu_a$ are larval and adult mortality rates. The density-dependent fractions $\exp[-c_{el}L(t)]$ and $\exp[-c_{ea}A(t)]$ describe egg survival in the presence of larvae and adults, and $\exp[-c_{pa}A(t)]$ is the survival of pupae in the presence of adults.

We may predict the behavior of the dynamics by simulating the age-structured model with cannibalism on pupae, $c_{pa}$, as a bifurcation parameter. The other estimated parameters are as follows: $b = 6.598$, $c_{el} = 0.01209$, $c_{ea} = 0.01155$, $\mu_l = 0.2055$, $\mu_a = 0.96$. Figure B5 illustrates the change of the dynamics between different types depending on the value of the bifurcation parameter, cannibalism of pupae by adults. The bifurcation diagram, with a further study of the attractors, indicates that as the cannibalism rate increases, the dynamics range from stable equilibrium to different periodic, quasiperiodic, and chaotic population trajectories (Costantino *et al.* 1997). Interestingly enough, the range $0.423 < c_{pa} < 0.677$ spawns multiple attractors.

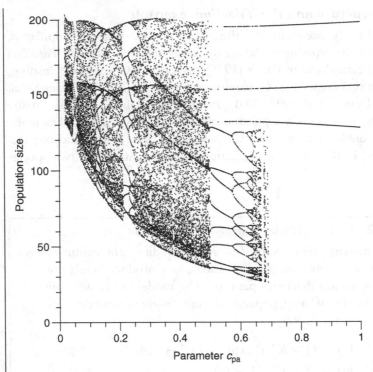

*Fig. B5.* Bifurcation diagram of the total population size (adults + pupae + larvae) in the three-stage *Tribolium* model (Box 6.2; Costantino *et al.* 1997) as a function of adult mortality $c_{pa}$. The initial population sizes are random numbers and different for each value for the bifurcation parameter $c_{pa}$.

Here a stable period-three cycle coexists with stable cycles of period eight or higher and with chaotic dynamics. Which attractor is reached in this parameter range depends on the initial population sizes.

Cannibalism can be controlled in the laboratory experiment on *Tribolium*. Thus, Costantino *et al.* (1997) posed the problem of whether such complicated dynamics, as predicted by the theoretical modeling, can be repeated in an experimental laboratory. The experimental population paths indeed were a good match with those predicted by the model. This experiment gave a lot of hope for complexity studies in population ecology: "fluctuations in natural populations might often be complex, low-dimensional dynamics produced by nonlinear feedback" (Costantino *et al.* 1997).

*T. confusum* and *T. castaneum*) the obvious classes are larvae, pupae, and adults. The three-stage model can hence be written as

$$
\begin{bmatrix} n_1(t+1) \\ n_2(t+1) \\ n_3(t+1) \end{bmatrix} = \begin{bmatrix} 0 & 0 & m_3 \\ l_1 & 0 & 0 \\ 0 & l_2 & l_3 \end{bmatrix} \begin{bmatrix} n_1(t) \\ n_2(t) \\ n_3(t) \end{bmatrix}, \tag{6.10}
$$

where $n_1$ refers to larvae, $n_2$ and $n_3$ to pupae and adults, with their respective net fecundities (only adults reproduce) and survival rates. Just as in the cohort effect and ontogenetic niche shift in the preceding sections, some of the elements of the transition matrix are density dependent. In the *Tribolium* case, it is the adult fecundity ($m_3$, sensitive to both adult and larvae density), and the survival of pupae into the adult stage ($l_2$). The density effect is due to cannibalism. The eggs are eaten by both larvae and adults (hence the effect on net fecundity), and pupae are eaten by adults.

This model has very rich dynamics (Box 6.2), which are also seen in the experimental populations. Costantino *et al.* (1997) were able to experimentally manipulate the cannibalism, i.e., adults predating on pupae, and also to estimate the population dynamics from the experimental results. The agreement between experimental results and predictions from the iterated model with a varied cannibalism coefficient is very good (fig. 6.5).

The *Tribolium* experiments are a nice example of how data and theory can meet. The details and sophistication of such experiments, however, require strict control of the experimental situation. Apart from the fact that this is only possible for certain model systems there is another interesting challenge when interpreting the results. The model for the population, eq. 6.12, includes the population parameters for a strictly isolated population only, and that is exactly how the experiment is also set up. Hence, the feedback environment has been reduced to its bare bones. This is exactly how it has to be done in order to reveal the detailed dynamics in relation to the model. The rest of the environment of the population under natural circumstances is then, of course, lacking. (We note in passing, though, that the *Tribolium* project has also dealt with critical aspects of the abiotic environment, e.g., Costantino *et al.* 1998.) Revealing the underlying mechanisms of a population or community process is one thing, translating it into a wider context is another challenge. Conversely, inferring from observable patterns the underlying mechanisms is definitely not a trivial task. This transition from the mechanistic processes to large-context manifestations, and the reverse, is one of the major challenges in ecology, but unfortunately too rarely taken seriously.

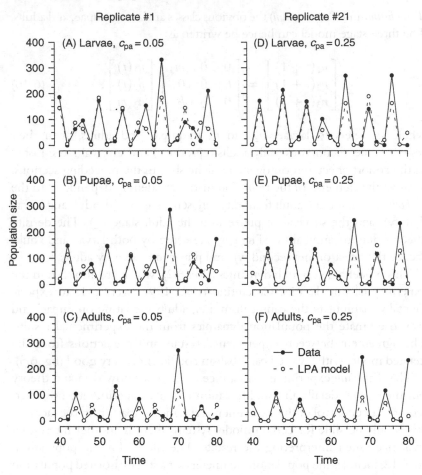

*Fig. 6.5.* Fluctuations in numbers of larvae, pupae, and adults in the *Tribolium* experiments by Dennis *et al.* (2001). The data (solid line, closed dots) are displayed for their replicates 1 (A)–(C) and 21 (D)–(F) with different values of the parameter $c_{pa}$. The model (Box. 6.2) prediction is indicated with dashed line and open dots.

## Finnish grouse dynamics

As another illustration of the problem of generalizing larger pictures from smaller, sometimes fragmented or isolated data, here we are going to have a closer look at the data from Finnish woodland grouse population dynamics collected by the Finnish Game and Fisheries Research Institute (FGFRI). The data include time series of counts of both young (first year) and adult birds in late summer across the entire country. The details of the data collection and their limitations are found in, e.g., Lindén (1981) and Lindén and Rajala (1981).

The life history of grouse is relatively straightforward. The population can essentially be divided into two groups, young birds or first-time breeders and older adults. Both categories of the birds reproduce. We can then use the following model for the dynamics

$$\begin{bmatrix} n_1(t+1) \\ n_2(t+1) \end{bmatrix} = \begin{bmatrix} m_1 & m_2 \\ l_1 & l_2 \end{bmatrix} \begin{bmatrix} n_1(t) \\ n_2(t) \end{bmatrix}, \tag{6.11}$$

where $n_1$ and $n_2$ are the densities of young first-time breeders and older adult birds, respectively (this is a revised model from Lindström *et al.* 1997b). The first-time breeders survive with probability $l_1$, thus moving to the group of older adults. The older adult birds also remain as adults, i.e., they stay alive from one year to another and reproduce successfully next year, with probability $l_2$.

Before analyzing this model, we introduce two more assumptions. First, we let the net per capita reproduction be density dependent. This density dependence is set to be a second-order process, i.e., implying an unspecified cohort effect such that it is the time at birth that partly determines the fecundity in young birds during the next year, and that the investment of older birds is discounted by the reproductive investment during the previous year. This is indeed a crude way of implementing the more straightforward cohort effects modeled in the previous sections. With grouse, however, there is another factor that likely plays an important role, and that is predation (e.g., Lindström *et al.* 1995, 1999). As shown earlier, predation generally enters as a second-order time lag. Second, we assume that the first age class is slightly less fecund than the older birds. In fact, we simply scale their respective fecundities thus; $m_1 = \nu m_2$, where $\nu$ is [0, 1].

The above model fits excellently to the data from the grouse in Finland across the geographic provinces. Figure 6.6 shows an example of the dynamics of young and old birds in one of the provinces, together with the simulation results. Here, the model parameters ($l_1 = 0.4$, $lm_2 = 0.5$, $m_2 = 2.1$) yield dynamics with six-year periodicity. The outcome of the model is compared with the 21-year data on the dynamics of hazel grouse population in the province of Mikkeli, Finland. Not only do the dynamics match very well at a more superficial level, but the more detailed cross correlation pattern (see Chapter 4) is also consistent with data. To achieve this fit, one does not have to use exceptional parameter values. Instead, the

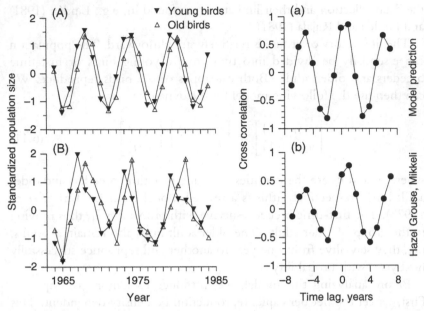

*Fig. 6.6.* Population dynamics of young and adult game birds. (A) Predictions by a stage-structured model where the fecundity is subject to delayed density dependence. (B) The dynamics of hazel grouse in the province of Mikkeli for 1964–1984. The right-hand panels provide cross correlation coefficients between adult and juvenile birds. Both the data and the simulation results are detrended (drawn after Lindström *et al.* 1997b).

parameter values used in the model are in fact also easily derived from the data. Hence, the simple model exercise shown here indicates that, indeed, rather crude stage structuring of the population is flexible enough qualitatively and quantitatively to account for processes at large geographical scales and for systems embedded in real complexity, such as the grouse populations in Finland are (see also Chapter 4).

## Summary

This chapter has dealt with the more detailed and refined aspects of population ecology. The often-used assumption that the vital rates of a population can be averaged over all individuals may not often be true. We therefore need a more detailed resolution of the birth and death processes. Births and deaths are often stage (e.g., age, sex, size) dependent and differences in those stage-specific rates may be critically important for population dynamics, abundance, and distribution. We review the basic building blocks for stage-structured population modeling. We also show that a rather minor extension of Leslie's original matrix models of population growth can capture a wide range of features, including maternal effects, cohort effects, and ontogenetic niche shifts. The former two effects are related to temporally variable environments, the former implicitly potentially (but not necessarily) assuming spatial structure. We are hence equipped with more sophisticated tools for understanding not only population dynamics under various temporal and spatial circumstances, but also the very basis of life history evolution. We also give two brief examples (a classic insect laboratory study and one on Finnish woodland grouse) of how stage-structured modeling can elucidate the mechanisms behind observed dynamics. We also introduce the simple structuring of the population into two parts: the doers and the viewers, i.e., individuals that contribute to population renewal and those who do not. Also, such a simple subdivision has notable influences on the resulting population dynamics and the interpretation of population data without consideration of population structure.

# 7 · Biodiversity and community structure

This chapter is devoted to biodiversity, viz. the local and global number of coexisting species. We first show how community assembly critically depends on the interaction terms $\alpha_{ij}$ of the community matrix **A**. In an isolated community, elimination of a single species easily leads to cascading extinctions. Attempts to reintroduce the species lost may not always succeed, and may even lead to further extinctions. Extending community assembly into space enhances local and global diversity but too much dispersal among communities may considerably reduce maximum achievable species richness. We also suggest that harvesting species from a community with strong interactions may result in unexpected extinction cascades.

## Community assembly

The issue of species richness ultimately translates into the concept of the ecological niche, defining the resource utilization profile of any single species either when alone or in a network of interactions with other species with closely matching profiles (e.g., Levins 1968; Emlen 1984; Lundberg *et al.* 2000b). The classic question now becomes how many and how similar species can (locally) coexist. This problem of local species richness crystallizes into a simple set of questions (Diamond 1975):

- To what extent are the component species in a given locality or community mutually selected from a larger species pool to fit with each other?
- Does the resulting constellation resist invasion?
- If so, how?
- To what extent is the final species composition of a community uniquely specified by the properties of the physical environment, and to what extent does it depend on chance events?

These questions outline much of the skeleton in research of community structure, the study on the pattern of species richness and species abundance relationships (Pielou 1974; Cody and Diamond 1975; Brown

1984; Diamond and Case 1986; Lawton *et al.* 1994; Gaston 1996; Gaston *et al.* 1997; Johnson 1998; Tokeshi 1999; Weiher and Keddy 1999). Ecologists in the 1970s were very optimistic when suggesting that it is likely that competition between species plays a key role in the integration of species communities (a view pioneered by MacArthur 1972; for a critical tune, see Strong *et al.* 1984). Diamond (1975, pp. 347–348) put it very explicitly by stating that, "Real or potential utilization of some of the same resources could be an obvious explanation why similar species do not occur in the same community, unless their resource utilizations are somewhat co-adjusted." The research tradition on biodiversity and species assembly was born. In this chapter, we shall trace the principal achievements of this tradition. We first focus on biodiversity in an isolated community context. Then we extend our explorations of community assembly into spatially structured communities. In Hubbell's (2001) recent terminology, we build on the "niche-assembly" approach, but will also explore, at the end of this chapter (p. 172), the "dispersal-assembly" approach. However, at the time of writing, our attitude towards Hubbell's new theory on biodiversity is neutral.

Ecologists have long attempted to understand the rules for how ecological communities are assembled (Diamond 1975; Pimm 1982; Post and Pimm 1983; Drake 1991; Anderson *et al.* 1992; Law 1999). That work has taught us about the limits to species diversity, expected relative abundance of species in natural and undisturbed communities, and the tempo and mode of ecological succession processes, both in genuinely virgin habitats (e.g., newborn volcanic island) and in secondary succession (e.g., reforestation after logging). The theories of community assembly have always struggled with the null model problem. For example, defining the species pool from which potential members of the community are drawn is often problematic (Law 1999). Both theoretical (Pimm 1980; Post and Pimm 1983; Law and Morton 1996; Morton *et al.* 1996; Law 1999; Tilman 1999) and empirical (Drake 1991; Jenkins and Buikema 1998; Law *et al.* 2000) work has shown that once a community has been assembled, there are certain limitations to the addition of further species.

## Community matrix

In an attempt to answer the four questions above, we shall take a pragmatic approach and build on the concept of the community matrix (Levins 1968; May 1974). The elements of the community matrix **A** are the pairwise interspecific interaction terms, specifying how the different species influence each other. Suppose there are $S$ species sharing (at least to some degree) a common resource. We can now write $S$ simultaneous equations, each one specifying the dynamics of species $i$. Let species $i$ have density $N_i(t)$ at time $t$, and let the population dynamics be described by the Ricker model, and we have

$$N_i(t+1) = N_i(t) \exp\left\{ r_i \left[ 1 - \frac{N_i(t) + \sum\limits_{j \neq i}^{S} \alpha_{ij} N_j(t)}{K_i} \right] \right\},$$

$$i, j = 1, \ldots, S.$$

(7.1)

This is a discrete time analogue of the Lotka–Volterra competition equation where densities are mapped from time $t$ to $t+1$, $r_i$ is species-specific maximum per capita growth rate, and $K_i$ is the species-specific carrying capacity of the system. The coefficients $\alpha_{ij}$ tell us how much species $j$ (per capita) contributes to the density-dependent feedback of species $i$. If all the $S$ equations are solved for the respective $K$ values, we have for species $i$ (Levins 1968)

$$K_i = N_i^* + \sum_{j \neq i} \alpha_{ij} N_j^*.$$

(7.2)

This is a system of linear equations with coefficients $\alpha_{ij}$. This linear system can be expressed in matrix form, and we have

$$\mathbf{K} = \mathbf{A} \mathbf{N}^*,$$

(7.3)

where $\mathbf{K}$ is a column vector of species-specific carrying capacities, and $\mathbf{N}^*$ is a column vector of the equilibrium population densities for the $S$ species in the community. The community matrix $\mathbf{A}$ with the interaction terms $\alpha_{ij}$ is given as

$$\mathbf{A} = \begin{bmatrix} 1 & \alpha_{12} & \alpha_{13} & \cdots & \alpha_{1S} \\ \alpha_{21} & 1 & \alpha_{23} & \cdots & \alpha_{2S} \\ \alpha_{31} & \alpha_{32} & 1 & \cdots & \alpha_{3S} \\ \vdots & \vdots & \vdots & \ddots & \vdots \\ \alpha_{S1} & \alpha_{S2} & \alpha_{S3} & \cdots & 1 \end{bmatrix}.$$

(7.4)

Note that all the diagonal elements ($\alpha_{ii}$) are normalized to 1s. Going back to eq. 7.1, we see that if an interaction term $\alpha_{ij}$ is positive, then species $j$ has a negative effect on the equilibrium density, $K_i$, of species $i$. Should $\alpha_{ij}$

be negative, then species $j$ has a positive effect on $i$. Hence, if $\alpha_{ij}$ and $\alpha_{ji}$ are both positive, then species $i$ and $j$ are competing (perfectly symmetrically if $\alpha_{ij} = \alpha_{ji}$). Should the interaction terms have opposite signs, then in effect species $i$ and $j$ are a resource–consumer pair.

## Assembling communities

Equation 7.3 can be used to make predictions of the species assembly (Emlen 1984). That requires, of course, that we know the $\alpha_{ij}$ values of $\mathbf{A}$. By knowing the equilibrium population densities, $N_i^*$, it is possible to calculate $\mathbf{K}$. Let us assume that we have the relevant information for an $S$ species community, as specified by eq. 7.3. We can now ask questions about the effects of deleting and adding species. The effect of removing species $s$ from the community is found by solving eq. 7.3 for $\mathbf{N}^*$ after deleting the $s$th row and the $s$th column from $\mathbf{A}$ and the $s$th element from $\mathbf{K}$. We rewrite eq. 7.3, now with one species less, and we have

$$\mathbf{N}^* = (\mathbf{A}')^{-1}\mathbf{K}'. \tag{7.5}$$

The vector $\mathbf{N}^*$ gives the new population equilibrium values (Emlen 1984). $\mathbf{A}'$ is the new community matrix with species $s$ removed.

For the community to persist then, obviously all species must have equilibrium population size greater than zero (i.e., all the elements of the vector $\mathbf{N}^*$ have to be positive). To find out which one of the $S$ species can be eliminated, we have to calculate the expected equilibrium densities after eliminating one species at a time. By doing this, we first find all feasible communities of $S - 1$ species, then $S - 2$ species, ..., and finally we are left with some persistent two-species combinations. The rules of combinatorics tell us that the number of unique combinations of $s$ species out of the total of $S$ species is: $_S C_s = S!/[s!(S-s)!]$, where $S!$ is the factorial of $S$ (e.g., the factorial of 6 is $1 \times 2 \times 3 \times 4 \times 5 \times 6 = 720$). In a three-species system, there are three one-species combinations ($a$, $b$, and $c$), three two-species combinations ($ab$, $ac$, and $bc$), and one three-species combination ($abc$). In a system of six species there are naturally six different ways for one-species communities, 15 unique two-species combinations, 20 for three species, 15 for four species, 6 for five species and, of course, one six-species assemblage. It is easy to see that with increasing $S$ the number of $S - s$ species combinations increases rapidly (e.g., from a species pool of $S = 20$, unique ten-species combinations can be taken no less than 184 756 times). This naturally creates problems if one aims to compare realized species assemblages with those expected purely by random causes. The sample size of repeated assemblages that are observed in nature has to be big enough before one can say with certainty which of the species combinations are forbidden ones. Before going further, let us remind ourselves that the method to calculate the vector $\mathbf{N}^*$ (with all values positive) does not tell us whether the equilibrium community is stable or unstable (Box 7.1). It only lets us decide that the given species constellation is possible.

The technique outlined above can, however, be used for reconstruction to address problems of species assembly. To demonstrate this we have selected a species pool of six species. These species all have $r = 1.5$ and $K = 1$ in eq. 7.1 and the $\mathbf{A}$ matrix (off-diagonal elements generated by drawing random numbers between 0.5 and 1.0) is displayed in Table 7.1.

With the pool of six species, the potential number of assemblages is 63. However, in this example we have only 37 realized ones. Naturally, all the one-species assemblages are found, as were all of the potential two-species assemblages. However, only 11 (out of 20 feasible) three-species assemblages are possible, while no more than four (out of 15) of the four-species constellations are possible. None of the five-species systems is possible with the $\mathbf{A}$ matrix used (Table 7.1(B)). The "forbidden combinations" for each assemblage size are those with some values in the $\mathbf{N}^*$ vector that are $\leq 0$. Table 7.1(B) also demonstrates the "Humpty Dumpty" effect in community assembly (Pimm 1991): once a larger

---

**Box 7.1** · *Local stability analysis of the Ricker community*

Consider community dynamics defined by the Ricker equation, eq. 7.1, for which the population equilibria are given in eq. 7.2. The local stability properties of the equilibrium population size are defined by the eigenvalues of the linearized matrix (see Chapter 2)

$$\mathbf{B} = \begin{bmatrix} 1 - \frac{r_1}{K_1} N_1^* & -\frac{\alpha_{12} r_1}{K_1} N_1^* & \cdots & -\frac{\alpha_{1n} r_1}{K_1} N_1^* \\ -\frac{\alpha_{21} r_2}{K_2} N_2^* & 1 - \frac{r_2}{K_2} N_2^* & \cdots & -\frac{\alpha_{2n} r_2}{K_2} N_2^* \\ \vdots & \vdots & \ddots & \vdots \\ -\frac{\alpha_{n1} r_n}{K_n} N_n^* & -\frac{\alpha_{n2} r_2}{K_n} N_n^* & \cdots & 1 - \frac{r_n}{K_n} N_n^* \end{bmatrix}.$$

If the absolute values of all the eigenvalues are $<1$ then the community equilibrium is locally stable. Note that $r$ and $K$ both influence the entries in the $\mathbf{B}$ matrix. See also Box 2.2 and 2.4.

Table 7.1. *A community matrix for six species (A) and feasible assemblages of* $S < 6$ *(B).*

(A) A six-species ( $a-f$ ) community matrix **A**.

|   | $a$ | $b$ | $c$ | $d$ | $e$ | $f$ |
|---|-------|-------|-------|-------|-------|-------|
| $a$ | 1.000 | 0.999 | 0.808 | 0.853 | 0.944 | 0.552 |
| $b$ | 0.772 | 1.000 | 0.858 | 0.645 | 0.830 | 0.740 |
| $c$ | 0.610 | 0.998 | 1.000 | 0.786 | 0.837 | 0.657 |
| $d$ | 0.698 | 0.865 | 0.869 | 1.000 | 0.536 | 0.892 |
| $e$ | 0.814 | 0.761 | 0.911 | 0.752 | 1.000 | 0.688 |
| $f$ | 0.584 | 0.841 | 0.718 | 0.962 | 0.710 | 1.000 |

(B) Assemblage size number and {Possible assemblages} from a species pool of six species with all $r = 1.5$ and $K = 1$ based on the community matrix displayed above. (Note that none of the five-species assemblages was possible.)

No. 1: {a} {b} {c} {d} {e} {f}

No. 2: {a,b} {a,c} {a,d} {a,e} {a,f} {b,c} {b,d} {b,e} {b,f} {c,d} {c,e} {c,f} {d,e} {d,f} {e,f}

No. 3: {a,b,c} {a,b,f} {a,c,d} {a,c,f} {a,e,f} {b,c,f} {b,d,e} {b,d,f} {b,e,f} {c,d,e} {c,e,f}

No. 4: {a,b,e,f} {b,c,d,f} {b,c,e,f} {c,d,e,f}

No. 6: {a,b,c,d,e,f}

community (in terms of $S$) is broken down, it cannot be easily reassembled. With our example, the reassembly will result in a community of four species (see also the Community closure section below).

We can now extend the above results to analyze in more detail the problem with "forbidden combinations" in assemblage sizes close to the size of the entire species pool $S_{POOL}$. We begin by letting the species pool be small, intermediate, or large ($S_{POOL} = 6$, 11, and 16). All species in all species pools have $r = 1$ and $K = 1$. We also let the communities be characterized by two sets of interaction terms. In the potentially more competitive one the off-diagonal elements of the **A** matrix were randomly drawn from a uniform distribution between 0 and 1. In the less competitive scenario, the elements were randomly selected between 0 and 0.5. Using eq. 7.5 we ensured that all entries in the $\mathbf{N}^*_{POOL} > 0$. We first rarefied the assemblage size by removing one species at a time from the pool, resulting in a community $S_{ASS}$ (equal to $S_{POOL} - s$), and scored whether all elements of $\mathbf{N}^*_{ASS} > 0$ for all combinations of $S_{POOL} - s$. Second, we took any $S_{POOL}$ from 16 to 2 and reduced each by one and again scored whether $N^*_{i,ASS} > 0$ for all $i$ and for all combinations of $S_{POOL} - 1$. We repeated the two procedures 100 times (each time drawing a new **A** for the species pool) for all combinations to get

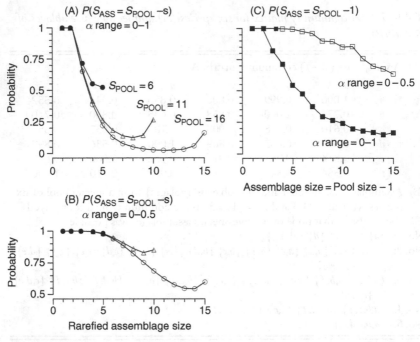

*Fig. 7.1.* Probability of observing an assemblage of a give size, $S_{ASS}$, as a function of the rarefied assemblage size, $S_{ASS} = S_{POOL} - s$, when the rarefaction is done species by species. In (A), the values of the interaction terms range from 0 to 1, representing a more competitive situation compared to (B), where the interaction term values are smaller (0–0.5). Trajectories for three differing $S_{POOL}$ sizes (6, 11, and 16) are shown. (C) Probability of observing an assemblage reduced in size by one species from the species pool. The interaction terms are either drawn between 0 and 1 (filled symbols) or between 0 and 0.5 (open symbols). The rightmost data points are for assemblages reduced from $S_{POOL} = 16$ to $S_{ASS} = 15$, the next data points to the left are for $S_{POOL} = 15$ reduced to $S_{ASS} = 14$, and so on. The results are averages of 100 replicates for each parameter combination.

an estimate of the probabilities that a rarefied or a reduced assemblage is feasible. The results (fig. 7.1) unambiguously support the preliminary finding presented in Table 7.1. Smaller communities are far more likely than larger communities to assemble from the species pool, to values closer to the original species number in the pool. Assembling subsets of any size from the $S_{POOL}$ becomes easier if the pool is less competitive, i.e., the interaction terms in **A** are closer to 0 than 1. This echoes the old sagacity: the closer the species are packed together the harder it is to get them to coexist (Levins 1968). This is the principle of competitive exclusion (Gause 1934).

These results have interesting consequences for the maintenance of biodiversity. Despite that fact that the species in the "pool" can happily coexist, it is far harder to get subsets of the species to coexist. Hence, if one species is lost, more are likely to join it in the fall. Our model communities are, however, not entirely general in one important sense. The interaction coefficients defined in the **A** matrix imply communities that are strictly competitive (all the interaction terms are positive). To what extent species assemblages in nature are build on competition rather than on other kinds of interaction, is a matter of argument (Hubbell 2001). It has been (Strong *et al.* 1984), and still is (Tokeshi 1999), a focus of observational and experimental research as well as theoretical explorations. One could argue, of course, that the model communities we have used here are in fact representing at least the subset of communities called guilds (e.g., MacArthur 1972). A guild is a set of species making use of common resources. Good examples of guild-based communities are found among insects, e.g., nectar and pollen-feeding bumblebees, freshwater zooplankton (*Bosmina, Ceriodaphnia, Daphnia*) feeding on algae, and tits (Paridae), small passerine birds in boreal forests, and dabbling ducks (*Anas*). Another restriction of this initial exploration is that there is no spatial structure in the community organization. The results in this section implicitly suggest that species-rich communities are easier to build up on a spatially structured basis than on a local basis. That is, having several units of suitable habitats for species in the focal guild enhances the species diversity pooled over all local units. For example, the three-species assemblages in Table 7.1(B), when pooled together, maintain the six species of the community (the same conclusion is, of course, true for the one-, two-, and four-species assemblages). We will extend the analysis of spatial structure and community organization below.

## Species abundances

Equation 7.3 suggests a relationship between the species abundance at the equilibrium and the interaction terms $\alpha_{ij}$. However, the relative importance of intra- and interspecific interactions, or some combination of them, is not directly discernible. To examine this we calculated the equilibrium population densities $\mathbf{N}^*$ from eq. 7.5. We first did this for $S = 15$ and the equilibrium population sizes were graphed against $\Sigma\alpha_{ji}$ (the influence of the focal species on others), $\Sigma\alpha_{ij}$ (the influence of other species on the focal species) and the net effect $\Sigma\alpha_{ji} - \Sigma\alpha_{ij}$. Figure 7.2(A)–(C) suggests that the relationship between the equilibrium

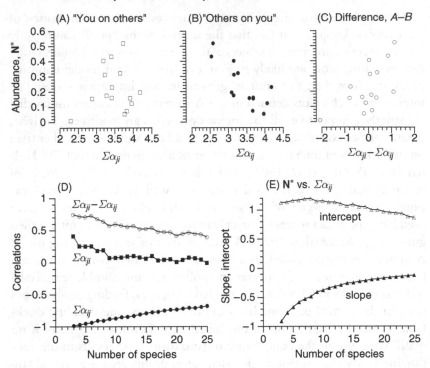

*Fig. 7.2.* (A)–(C) Scatter plots of population equilibrium density, $N^*$, versus measures of overall interactions characterized by the community matrix **A**. Panel (A) indicates how the focal species influences the other species $\sum \alpha_{ji}$, while (B) indicates how other species influence the focal species $\sum \alpha_{ij}$, and (C) gives the difference between the two $\sum \alpha_{ji} - \sum \alpha_{ij}$ (intra-specific interactions excluded). Panel (D) shows the corresponding correlations against the number of species in the community. Panel (E) shows the regression slopes and intercepts between $N^*$ versus $\alpha_{ij}$ with increasing numbers of species. For the results in (E) and (F) the mean of 100 replicated runs are indicated for each $S$, $\min(\alpha_{ij}) = 0$, $\max(\alpha_{ij}) = 0.5$, $r = 1$, $K = 1$.

density of the focal species and how strongly the other species interact with it (specified by the **A** matrix) is tighter than the relationships between the two other measures. This was also confirmed in an analysis where $S$ ranged from 3 to 25. The correlation between $N^*$ and $\sum \alpha_{ij}$ was negative and the tightest (fig. 7.2(D)). However, the correlation gradually reduces with an increasing number of species in the community. Linear regression between the two variables indicates that with an increasing number of species the slope of $N^*$ versus $\sum \alpha_{ij}$ gets shallower (fig. 7.2(E)). The regression line between $N^*$ and $\sum \alpha_{ij}$ can be called the "assembly line." It is the line in abundance–interaction space that defines the feasible

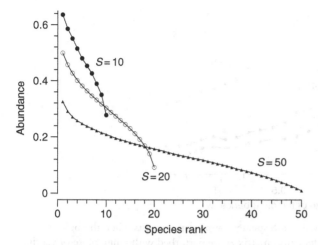

*Fig. 7.3.* Species abundance rank graphs for equilibrium communities, eq. 7.5, of three different sizes [mean of 100 replicated runs are indicated for each $S$, $\min(\alpha_{ij}) = 0$, $\max(\alpha_{ij}) = 0.5$, $r = 1$, $K = 1$].

community. Wilson *et al.* (2003) used a mean-field approximation of generalized Lotka–Volterra models like the one we have used to show that the results shown in fig. 7.2 are due to the covariance between interaction coefficients and population densities.

A more conventional way to summarize abundance patterns in a community is to rank the species in descending order and to graph these ranked abundances. As seen in fig. 7.3, the rank versus abundance graphs generated from the model communities resemble those from real communities (Tokeshi 1999; Hubbell 2001). Several alternative models have been put forward to explain such patterns (Tokeshi 1999). However, for the highly linked communities (no 0s in **A**) the "assembly line" suffices here. It efficiently summarizes the relevant characters of **A** and **N**$^*$ of the equilibrium community.

### Stochasticity and biodiversity

The analyses in the preceding sections have all assumed that the environment is constant and that only the deterministic properties of the community matter. Here, the role of environmental stochasticity for community structure will be briefly dealt with. External perturbations are attributed to enable the coexistence of several species with similar ecological requirements (Hutchinson 1961). Environmental fluctuations

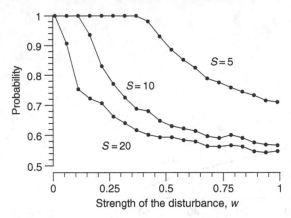

*Fig. 7.4.* The probability that a $S$ species community persists when the off-diagonal values of the interaction matrix **A** are perturbed with white noise of varying strength, $w$. For each parameter value, the system was replicated 1000 times, and the number of cases with all elements of $\mathbf{N}^* > 0$ was scored.

are an unavoidable part of real systems. Here, we will let the interaction terms in the community matrix be subject to random perturbations. We shall modulate the interaction terms in the **A** matrix by multiplying the off-diagonal elements with $\mu = 1 + \varepsilon$, where $\varepsilon$ is a uniformly distributed random number between $-w$ and $+w$, where $0 < w < 1$. Increasing the value of $w$ (here $w$ ranges from 0 to 0.9) increases the intensity of the noise but does not affect its expected value. For each $w$ we generated 1000 **A** matrices and calculated the probability that all species from the species pool ($S_{POOL}$ being 5, 10 or 20 species) were persistent. As shown in fig. 7.4, the probability of having a full community declines with increasing $w$.

The results suggest that large communities are more vulnerable to species losses when disturbed. To explore in more detail the impact of stochasticity on community structure, we let two sources of stochasticity affect the community. We let either all off-diagonal elements of **A** be multiplied by a random number $\mu(t)$ ranging between $1 - w$ and $1 + w$, i.e., the perturbation is global for all species, or by $\mu_i(t)$, when the perturbation is species-specific. As before, the off-diagonal elements of **A** were random numbers from a uniform distribution between 0 and 1. We used $S = 5$ and $S = 10$, with three different values of $r$ (1.5 = stable dynamics, 2.25 = two-point fluctuations, 3.25 = chaotic fluctuations). The initial population vector was also a set of random numbers from a uniform distribution between 0 and 1. For each $w$, we iterated the community dynamics for 300 time steps after which we scored the number of species present. A species was

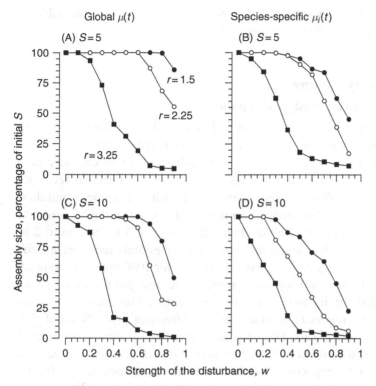

*Fig. 7.5.* Assemblages of $S_{INIT} = 5$ and $S_{INIT} = 10$ were disturbed with white external noise of varying strength $w$. After 300 iterations, we scored the number of species remaining, $S_{FINAL}$. If the $S_{FINAL} = S_{INIT}$, the index of assembly persistence is 100% ($y$ axis). The external perturbation was either matching (global) for all species, or species specific. The system was explored using three different values of $r$. Averages of 100 replicated runs for each parameter combination are shown.

said to be present if at $t = 300$ its population size was $> 1 \times 10^{-5}$. This procedure was repeated 100 times for all parameter combinations.

The results shown in fig. 7.5 confirm the previous observation that larger assemblages are more sensitive to environmental stochasticity (the lines in fig. 7.5 for $S_{10}$ drop earlier from 100% than for $S_5$). Another interesting feature is that global disturbance [matching modulation, $\mu(t)$, for all species], i.e., all species fluctuate in perfect synchrony, yields fewer extinctions than species-specific perturbations [with $\mu_i(t)$ and with increasing $w$ the different species will begin to fluctuate out of synchrony]. Thus, across-species synchrony enhances local diversity, while asynchrony reduces it. This is in contrast to the observations of single-species systems, where synchrony increases local extinctions (e.g., Heino *et al.* 1997a). Finally, not so unexpectedly, chaotic systems ($r = 3.25$) are more vulnerable to species extinctions than stable ones (fig. 7.5).

## Community closure

### Invasions from a closed species pool

Assembling model communities is technically rather straightforward, but the rules for doing it are far from trivial. Lundberg *et al.* (2000c) used eq. 7.1 as the basis for their explorations. For each community size they kept $r = 1.5$ and $K = 1$ for all $S$ species. As above, the interaction terms $\alpha_{ij}$ were randomly drawn from uniformly distributed random numbers between 0 and 1. For each community size $S$, they sought persistent communities and created 50 of them, all of sizes of $S$ ranging from 2 to 10. From each such community a species was randomly removed and the remaining community was allowed to interact for 1000 time steps to reach a new equilibrium. In that process, more species than just the target species were weeded out due to interspecific competition. The resulting community, after deleting one random member, was frequently (31.6% out of 450) less than $S - 1$ (fig. 7.6). The frequency of cascading extinctions differed among original community sizes. For small $S$, the new community size was very close to the expected $S - 1$ but with increasing community size the probability of cascading extinctions increased steeply (fig. 7.6(B)).

Lundberg *et al.* (2000c) kept track of the identity of the species removed and attempted to reintroduce it back to the community resulting after its disappearance. The post-extinction community was kept at its equilibrium and the previous member was reintroduced with a population density of $10^{-2}$ (i.e., two orders of magnitude below carrying capacity). The reinvasion was mostly successful. However, in 8.4% of cases the former member of the community could not invade. The failure also caused extinction cascades (fig. 7.6(C)). Reinvasions were more successful into initially small communities (fig. 7.6(D)). Thus, local extinction loss of a species potentially has cascading effects such that the community erodes, but also that community restoration may be impossible. This was termed "community closure" (Lundberg *et al.* 2000c).

### Extinction and invasion cascade versus stability

There are some interesting extensions of the above analysis. Our toolbox (eq. 7.5, Box 7.1) enables us to not only distinguish between

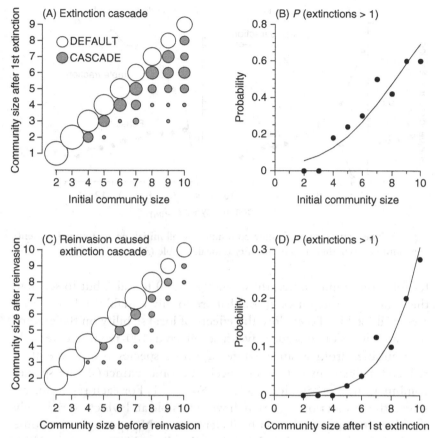

*Fig. 7.6.* (A) The number of species remaining after one randomly selected species is eliminated from the initial community. The dots indicate the size of the remaining communities. (B) The probability that more than one species becomes extinct with the loss of a randomly selected species increases with increasing community size. A logistic regression model $P(k) = \exp(\alpha + \beta k)/[1 + \exp(\alpha + \beta k)]$ was fitted to the data ($\alpha = 0.46$ and $\beta = -3.8$) to give the probability that at least one extra species was eliminated with the extinction of the target species (the extinction cascade). (C) After allowing the weeded out community to stabilize for 1000 time steps the target species was reintroduced to the community with a density of $10^{-2}$. Most of the time, the introduction was successful. However, sometimes the introduction caused new extinctions. The dots indicate the size of the resulting communities after reintroduction. (D) The probability of the reinvasion extinctions increased with increasing community size ($\alpha = 0.6$, $\beta = -6.9$; modified after Lundberg *et al.* 2000c).

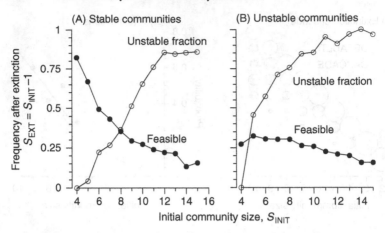

*Fig.* 7.7. The proportion of feasible communities (all initial species minus 1 present) depending on whether the community is locally stable (A) or unstable (B).

feasible communities, i.e., those with $N_i^* > 0$ for all $i$, but to separate these into two categories: those that are locally unstable and those that are locally stable. To explore the effects of local stability on the extinction and invasion cascades, we first selected 100 replicated feasible communities from a small range of initial species $S_{INIT}$. We then reduced their size by letting one species become extinct (all one-species combinations of extinctions) $S_{EXT} = S_{INIT} - 1$. For each replicate, we scored the number of $S_{EXT}$ which were feasible and which were locally unstable. It turns out that with all elements in $\mathbf{N}_{INIT}^* > 0$ the outcome much depends on whether the $S_{INIT}$ is locally stable or unstable. With stable $S_{INIT}$ the proportion of feasible communities (all elements in $\mathbf{N}_{EXT}^* > 0$) is rather high if the initial community size is small (low $S_{INIT}$), but levels off soon at about 25% when the number of species approaches ten (fig. 7.7(A)). With initially unstable communities, the proportion of feasible communities is much lower (25% to 15%) and more stable throughout the range of $S_{INIT}$ examined (fig. 7.7(B)). In the feasible $S_{EXT}$ communities, the proportion of unstable ones sharply rises with increasing number of $S_{INIT}$ in the community. The data also show that with large-enough initial communities the outcome is roughly matching regardless of whether the initial community was locally stable or unstable (fig. 7.7).

These results suggest that the locally stable and unstable communities do differ in terms of both extinction cascade and invasion cascade. We addressed this in more detail by using 100 initially stable and unstable communities of eight species. For each replicate, we eliminated one species (target) and after 1000 iterations of eq. 7.1 we scored the number of remaining species (a species was considered extinct if its density dropped below $1 \times 10^{-5}$). The expectation is that the number of species remaining in the community should be seven. However, due to extinction cascades $S_{EXT} = 7$ was observed only in four cases with locally stable initial communities and once in locally unstable initial communities. The average number of species after the extinction cascades was 5.1 with the initially stable communities and only 3.8 with the initially unstable communities (fig. 7.8(A)). The target species was returned back to the community (with an initial density of $1 \times 10^{-4}$) and again after 1000 iterations of eq. 7.1 we scored the number of remaining species in the reinvasion communities. In our simulations, we sought for cases where the extinction cascade resulted into 100 replicates of each $S_{EXT}$ from 2 to 7. Out of the 600 replicates with initially stable communities of $S_{INIT} = 8$, the invasion from $S_{EXT}$ to $S_{EXT} + 1$ took place in 63% of the cases; with the initially unstable communities the corresponding figure is 49% (a two-by-two test yields $\chi^2 = 19.6$, $p < 0.001$). Thus, the two initial community categories locally stable and unstable, differ in terms of both extinction and invasion cascade. When a species leaves a community via extinction, the doors of the community are often closed and thus reinvasion becomes impossible.

It should be kept in mind that these model communities are far from realistic representations of natural species assemblages. The results, nevertheless, highlight several important community processes. Previous studies have shown that cascading extinctions are positively related to species abundance and connectance (Pimm 1980; Law 1999). Here connectance was held constant and all species had equal weight excepting entries in the **A** matrix. Species diversity was the only variable property of the communities. Yet, we observe that communities that are more diverse were not more resistant in relative terms to species loss. That locally unstable communities have more extinction cascades is understandable perhaps just due to their unstable character. We do not have any good explanation as to why the locally stable and unstable communities differ in terms of invasion cascade. This observation clearly calls for more research.

Cascading extinctions are only one of the problems with species loss (Borrvall et al. 2000). Should the extinction be local and the original species pool still extant, then reinvasion is far from certain. Experimental studies also indicate that species richness in the resident community has only weak

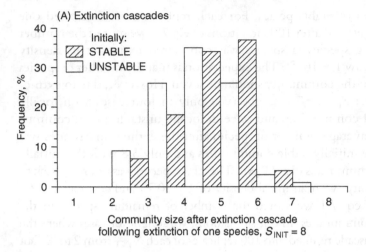

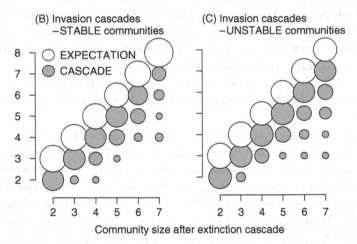

*Fig. 7.8.* Extinction and invasion cascades in a community initially having eight species. (A) The extinction cascades are shown for 100 replicated runs with stable and unstable communities (Box 7.1). In each run a randomly selected species was removed and the number of remaining species was scored after 1000 time steps (eq. 7.1, $r = 1.5$, $K = 1$, $\min(\alpha_{ij}) = 0$, $\max(\alpha_{ij}) = 1$). The difference in the remaining number of species in the initially stable and unstable communities is significant in statistical terms: $t_{198} = 8.95$, $p < 0.001$. ((B), (C)) The extinct species was returned (with an initial density of $1 \times 10^{-4}$) at $t = 1001$ and again after 1000 time steps the number of species present was scored. The invasion analyses were replicated 100 times for each $S_{EXT}$ after the extinction cascade. The size of the circle refers to the number of cases (e.g., in (B) the number of observations in {2,2} is 34, in {2,3} 64).

effects on invisibility (Law *et al.* 2000). Community closure can be a problem in ecological time. For example, cod (*Gadus morrhua*) has been driven to commercial extinction (Northern cod off Newfoundland, Canada; Myers *et al.* 1997c) or is very close to being so (Baltic cod; Kuikka *et al.* 1999). Despite moratoria or dramatically reduced catches, the two cod populations seem to have severe problems recovering. In the Baltic Sea, the loss of cod has been followed by a decrease in the numbers of one (herring, *Clupea harengus*), and an increase in the numbers of the other (sprat, *Sprattus sprattus*) dominating species in fish community (ICES 1999). These examples from fisheries are by no means evidence for community closure, but might give us a hint for where to start looking.

Community closure can also be a problem in evolutionary time. An evolutionarily stable (ESS) community defined such that all the (current) adaptive peaks are occupied by the potential members (Brown and Vincent 1992) may lose species that are not able to reinvade. Similarly, ESS communities may not even be reached in the first place. Sudden losses from the community will alter the adaptive landscape such that the peak where a former member resided no longer exists and, although it would have all the ecological and evolutionary characteristics for a successful return, the route (and thereby the continual change of the adaptive landscape) to that potential peak is closed by the resident community members.

This study (Lundberg *et al.* 2000c) has shown that the irreversibility of species loss from communities adds to the dangers of severe distortions of natural ecosystems and that the problem may have both ecological and evolutionary dimensions. Community closure may have far-reaching consequences for conservation biology, harvesting, and ecosystem restoration.

## Invasion from an open species pool

In the previous section, the initial number of the species was fixed, and the only change in the community structure was through extinctions. We now modify the model, using eq. 7.1, by allowing invasion of new species into the community. We begin by having two species in the community. At a random point of time an invading species, with known interaction terms $\alpha_{ij}$ with the extant members, attempts to enter the community. The interaction terms are drawn from uniformly distributed random numbers between 0 and 1 (still, $\alpha_{ii} = 1$). The invasion, with a density of $1 \times 10^{-4}$, is either successful or not. The success is scored after a randomly selected point of time (uniform random numbers between 5 and 50 time units). All species with an abundance $<1 \times 10^{-5}$ are considered extinct. The system is repeated for 200 invasion cycles. To score an established presence, a species has to remain for at least two cycles in the community.

As we learned on p. 168, the invasion might not always increase the number of species in the community. It may also happen that an extinction cascade follows the invasion (fig. 7.8). The succession of species is rapid, and presence time for species in the system is characterized by many short stays and a few long ones. The presence time is longer for chaotic population dynamics than for stable ones. Yet, species composition between any subsequent time cycle is about 90% matching. A typical feature for the interactive open-pool invasion system is that the number of species present seldom increases above 10 (figs. 7.9, 7.10). Systems with stable dynamics have more species compared to systems with chaotic dynamics. These features owe much to the fact that the interaction terms in the **A** matrix are random draws from a uniform distribution between 0 and 1. Narrowing the range of $\alpha_{ij}$ terms will increase species numbers.

An assembly of species in time is vulnerable to invasions by a species with average $\alpha_{ij}$ larger than in the extant assembly. For example, in the succession data (fig. 7.9(A)), invading species which established themselves had mean $\alpha_{ij} = 0.503$, while the corresponding figure for unsuccessful invaders was 0.478, and in the target assembly it was 0.386. These invasions cause extinction cascades in about 10–20% of cases when extinctions take place. We generally find the "assembly line"; that is, there is a strong negative relationship between species abundances $N_i$ and how strongly the other species influence the target species, $\Sigma \alpha_{ij}$ (see fig. 7.2(B),(D), on p. 160). The overall correlation between $N_i$ and $\Sigma \alpha_{ij}$ for the 200 cycles of time averages was approximately $-0.7$ for the stable dynamics and a bit less, approximately $-0.6$, for the chaotic dynamics. The "assembly line" correlation weakens a bit after extinctions but strengthens again at events of extinction cascades.

## Spatial extension

The logical next step is spatially extended community dynamics and here we ask how spatial heterogeneity influences biodiversity. By space we, of course, refer to the spatially coupled system of population subunits. This coupling is due to either dispersal or shared environmental fluctuations (Chapter 4).

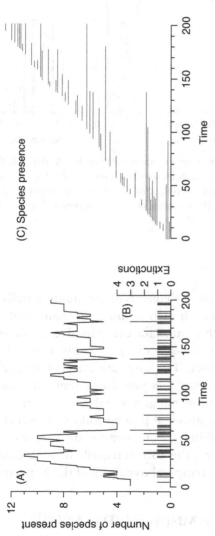

*Fig. 7.9.* A community was initiated according to eq. 7.1. At random points of time an additional species attempts to invade the system. The invasion may be successful, or it may fail. Whether it is successful or not, the invasion might also lead to extinctions of one or more species. Panel (A) gives the succession of species numbers for a system with stable dynamics ($r = 1.5$), while (B) gives the number and timing of extinctions and in (C) the temporal presence of the 105 species in the system is shown.

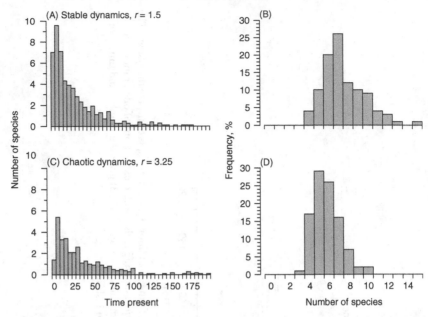

*Fig. 7.10.* The species invasion system is run for 200 cycles. At the end, the length of the time present is scored for all species involved (see Fig. 7.9(C)). For stable and chaotic dynamics the frequency distributions of persistence times (averages of 100 replicated runs) is given (A), (B), together with the frequency distributions of the terminal species numbers, (C), (D).

## Dispersal

We first assume that a landscape consists of $n$ subunits initially inhabited by a total of $S$ species. In each such subunit, the community dynamics are given by eq. 7.1. Each subunit for every local species is sufficiently independent that the species-specific population renewal process is governed by local processes. The units are linked together with dispersing individuals. Each year, after population renewal, a fraction $m$ of each species in each location emigrates and will redistribute into the other subunits. Here, the dispersal process follows kernel II (Box 3.2, p. 54) with $d_{MAX}$ specifying the maximum distance the dispersing individuals can reach (here, the 25% percentile of all distances among the $n$ sites). Thus, for the temporal dynamics of the $S$ species community we write

$$\mathbf{N}(t+1) = \mathbf{MN}(t) \exp\{r[\mathbf{D} - \mathbf{AN}(t)]\}. \qquad (7.6)$$

Here $\mathbf{N}$ is a matrix with $S$, $n$, and time $t$ as dimensions. $\mathbf{D}$ is a column vector with all $S$ elements equal to 1. The elements of $\mathbf{N}$ are species-specific population densities in each locality. The matrix $\mathbf{M}$ is the dispersal matrix

$$
\mathbf{M} = \begin{bmatrix}
1-m & m/(n-1) & m/(n-1) & \cdots & m/(n-1) \\
m/(n-1) & 1-m & m/(n-1) & \cdots & m/(n-1) \\
m/(n-1) & m/(n-1)1-m & \cdots & & m/(n-1) \\
\vdots & \vdots & \vdots & \ddots & \vdots \\
m/(n-1) & m/(n-1) & m/(n-1) & \cdots & 1-m
\end{bmatrix}.
$$

(7.7)

The dispersing fraction of individuals $m$ (0.0, 0.05, 0.1, or 0.5) is the same for all species. The initial number of the species was $S = 15$, and the off-diagonal elements of the community matrix $\mathbf{A}$ are random numbers from uniform distribution between 0 and 1 (we ensured that for all elements in $\mathbf{N}^* > 0$, and of course $\alpha_{ii} = 1.0$). For each subunit, we first determined by uniform random numbers how many species (from 1 to $S$) and which ones will be present, and their populations were initialized by using uniformly distributed random numbers between 0 and 1. For all the $S$ species we set $r = 1.5$, and the presence (with density $> 1 \times 10^{-4}$) of species was checked after 1000 iterations of eq. 7.6. The number of local units $n$ was 5, 10, 20, or 50. For each parameter combination, we replicated the simulations 100 times.

In this system with $m = 0$ we have the situation indicating the sole effect of space. As there is no dispersal, local species richness per subunit will settle down to matching levels despite the number of subunits in the global system (fig. 7.11). When the number of local units is low, the global species richness will remain much below the maximum of $S = 15$ possible. However, when $m > 0$ local species richness starts to increase, but for a substantial effect in local diversity the number of subunits $n$ has to be sufficiently high. In dispersal-coupled systems, global and local species richness will show opposite patterns. With increasing dispersal, global species diversity goes down with increasing $m$, while local diversity is enhanced with increasing $m$ (fig. 7.11).

## Interactions and distribution

In the dispersal-coupled community (eq. 7.6) with $S$ species and $n$ locations we have for each time step $S \times n$ distribution matrices. Such matrices are often used to score ecological distance among the $S$ species (e.g., Gauch 1982). There are several differing ways to calculate the entries, $e_{ij}$ (similarity

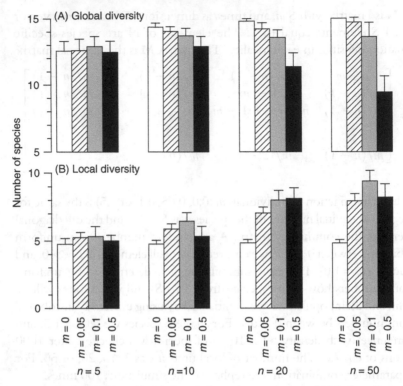

*Fig. 7.11.* Global (A) and local (B) number of species at $t = 1000$ in iteration of eqs. 7.6 and 7.7. Initially, $S = 15$ using a randomly drawn **A** matrix. Results are presented for four values of $m$. When $m = 0$ (open bar), the spatial effect is alone in operation, with $m > 0$ the $n$ units are coupled with redistribution of individuals; the spatial effect now interacts with redistribution. The results are averages (+95% confidence limits) of 100 replications for each combination of $m$ and $n$.

in habitat use between species $i$ and $j$), of the ecological distance matrix **E**. With the proportional similarity index $e_{ij} = \Sigma \min(p_{ij}, p_{jh})$, $p_{ij}$ and $p_{jh}$ are the fractions of individuals in the locality $h$, and the summation is over all the $n$ habitat subunits. We now have two different measures of similarity among the $S$ species, the community matrix **A** and the ecological similarity matrix **E**. By taking $S = 15$ in the initialization phase, and by using $r = 1.5$ and 3.25, **A** from random draws between 0 and 1 (except 1s for diagonal elements), and $m = 0.05$ and $d_{\text{MAX}}$ of the dispersal kernel II (p. 54) $= 25\%$ percentile of the interpatch distances, we scored the match between the two similarity measures. With $n = 50$ the system was left running for 500 iterations of eq. 7.6, and the final time step was used to score the correlation between elements of the two matrices **A** and **E**

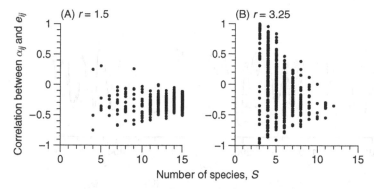

*Fig. 7.12.* Correlation between elements in the **A** (interspecific interactions, $\alpha_{ij}$) and **E** (similarity of species distribution $e_{ij}$ over $n = 50$ sites). The expectation is for a negative relationship between the two variables. Both stable (A) and chaotic (B) Ricker dynamics (*r* values indicated) are explored (in both graphs results of 500 replicated runs are displayed).

(diagonal values of both matrices excluded). Due to the cascading extinctions we did not control for the weeded-out species number, instead we repeated the system 1000 times for both values of *r* used.

The expectation is that interspecific interactions (the **A** matrix) when strong enough should also affect the distribution and abundance of species over the *n* sites. If so, the correlation between the elements of the **A** and **E** matrices are expected to be negative. This is what we also tend to find (fig. 7.12). However, the association between the elements of **A** and **E** is by no means always very strong. In fact, with chaotic dynamics one quite often also scores high positive correlations between the elements of the two matrices (fig. 7.12(B)). Thus, the results warn against drawing conclusions too hastily on the strength and direction of ecological interactions based on the terms in the **E** matrix (Connell 1980). Our experimentation further shows that the relationship between **A** and **E** (fig. 7.12) is rather robust to changes in *m* and $d_{MAX}$.

An interesting feature emerges when the temporal dynamics of the relationship are examined: the relationship between **A** and **E** is by no means temporally stable. Periods of tighter association are intervened with periods of lower association (fig. 7.13). In addition, one finds rather high levels of synchrony both within species ($r_0 \approx 0.9$ for all species in this example) and more interestingly also across species (fig. 7.13(B)). Temporal fluctuations in the correlation between entries in **A** and **E** fit well to the power law (fig. 7.13(C)). This pattern suggests that dispersal and the density-dependent feedback system in a multispecies community

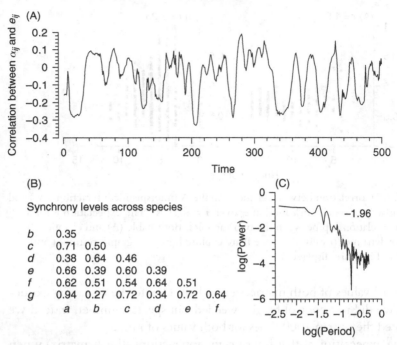

*Fig. 7.13.* (A) An example of temporal dynamics of correlation between the entries in the interaction matrix **A** and the ecological similarity matrix **E** of seven species' distribution over 50 subunits. (B) The table gives synchrony values, $r_0$, in population fluctuations calculated across the species. (C) The power spectrum of the fluctuations in (A). The final 500 time units of a simulation lasting 1500 time steps are used for this graph.

act in a similar manner as in spatially coupled single-species systems (see Chapter 5, p. 104; Kaitala *et al.* 2001a).

## Management implications

We shall close this chapter by considering some of the implications for population and community management that may result from the analyses above. The starting point will be the **B** matrix given in Box 7.1.

The community stability criterion is that the absolute values of all eigenvalues of the **B** matrix are less than 1 (which means that perturbations away from the community equilibrium are diminishing). An important feature of the **B** matrix is that the entries include both $r_i$ and $K_i$. This has implications for population management both in terms of harvesting and conservation biology. Consider, e.g., a fisheries system where one of

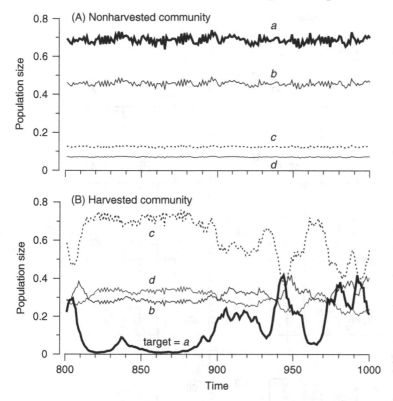

*Fig. 7.14.* Population fluctuations in a four-species community without (A) and with (B) 10% harvesting on the most abundant species *a*. Note the periods of transition (around $t = 810$ and $t = 940$) in species abundance relationships.

the species in the community is harvested. We simulated this scenario with a four-species community. We assembled a community using matrix **A** with $r_i = 1.75$, and with initial values $\mathbf{N}^* > 0$. The system was left running for 1000 generations. In the first scenario, one system was disturbed only with external noise (the Moran effect, $w = 0.1$, $\kappa = 0.8$), while the numerically dominant species in the other one was subject to harvesting. This system was also subject to external disturbance. The annual harvesting rate was selected to be a colored random variable between 0 and 20% with $\kappa = 0.8$. The nonharvested community displayed only minor fluctuations around $\mathbf{N}^*$, while population fluctuations in the harvested community were much wilder (fig. 7.14). Interestingly, the annual harvesting rate (with long-term average of 10%) drastically changes species abundance relationships in the community. The formerly most abundant species now becomes the least abundant, and one of the

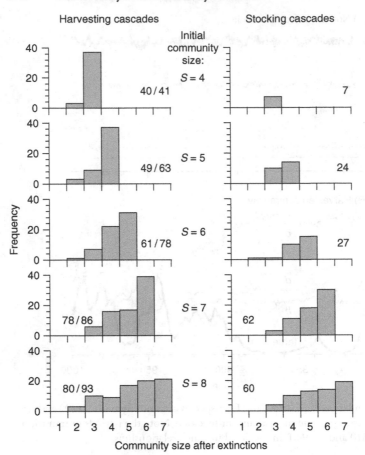

*Fig. 7.15.* Harvesting (left-hand column) and stocking (right-hand column) extinction cascades in communities of $S = 4$ to $S = 8$. The inserted numbers give the number of cases (out of 100 replicated runs) when species extinctions were recorded. In the harvesting cascades, the first number gives the number of cases (graphed in the histograms) when some species other than the harvested one became extinct. The second figure gives the number of cases when the target species became extinct.

rare species becomes the top-abundant species. There are also interesting transition periods in the community structure (fig. 7.14(B)).

We extended the above analysis by initiating 100 random communities with **A** so that $N_i^* > 0$ for all *i* for $S = 4$ to $S = 8$. Harvesting was as above, and at $t = 1000$ we scored the number of species present. The harvesting caused quite a number of species extinctions, and often the species becoming extinct did not include the target (randomly selected from the $S$) species (fig. 7.15). The number of cascading extinctions increased with increasing number of species in the community. Our next step was to reverse the system. Instead of harvesting, we enhanced the population size of a randomly selected target species. The annual stocking rate was a colored random variable with a range from 1 to 1.05 ($\kappa = 0.8$) by which the $N_{TARGET}(t)$ was multiplied. After enhancing the population of the target species with the average annual rate of 2.5% of the population size, we observed stocking cascades (fig. 7.15). They were not as many as with harvesting, but their numbers did increase with the increasing community size.

These results extend the extinction/invasion cascade to the domain of population management. Harvesting a (randomly selected) target species in a multispecies community may easily lead to extinction of the target species. We also find a rather high number of cascading extinctions and single-species extinctions that do not include the target species.

## Summary

Species co-existence is an intriguing problem in ecological systems. Likewise, the assembly process towards a more or less coherent community is no less challenging. In this chapter, we extended our models to include entire communities, and we revisit a number of classic problems in community ecology. For this purpose, we use very simple models to highlight a number of fundamental problems without attempting to recreate realistic representations of extant natural communities. Community assembly is equivalent to what has been termed succession – the sequential inclusion of new species interacting with the ones already present. We show that both community assembly and the resulting final community are nonequilibrium systems under very modest stochasticity. This chapter takes a closer look at one aspect of that stochasticity, namely the loss and reintroduction of species from putative stable ones. The loss of species is often followed by extinction cascades, the relative size of which depends on the initial size of the community. This may often also lead to "community closure," i.e., that former members of a persistent community cannot reinvade the community to which they once belonged. Such cascades are not only confined into actual extinction, but also significant reductions in population size, e.g., when species are harvested. When these processes are taken to a spatial context, a number of other phenomena emerge. For example, we show that there is an intricate interaction between expected local and global diversity depending on the rate at which species can redistribute between landscape elements. With increasing dispersal ability of the community members, global diversity decreases whereas average local diversity increases. This chapter thus concludes that spatial structure and temporal stochasticity are indeed major components not only in single-species dynamics, but in community dynamics as well.

# 8 · *Habitat loss*

Our focus is on population fluctuations, extinction risks, and on species coexistence in a fragmented landscape. As this is an entirely theoretical enterprise, we can control for various aspects (constant carrying capacity, fixed density of fragments per unit area, increasing isolation, etc.) while altering others. This is often impossible to achieve with natural systems. The concept of refugee-arrival (former inhabitants of habitats lost) caused perturbations in local populations will be introduced. In addition, we shall explore population fluctuations in the center and border of a species' distribution range. Finally, we shall provide a few explanations as to why species with periodic multi-annual dynamics may lose the cycle.

## What is meant by habitat loss?

The issue of habitat loss brings into mind various aspects of changes over time in pristine habitats in nature. Often we tend to associate habitat loss with human-caused consequences: intensifying agriculture initially, timber logging for sawmills and pulp mills of the paper industry, not forgetting urban development (fig. 8.1). The picture that we tend to have in mind when somebody mentions "habitat loss" is that in the beginning there was large widespread homogeneous coverage of uniform habitat areas all over a given biome. Ever since those times everything has deteriorated, fragmented, and we have lost habitats. Areas suitable for breeding of a given species have become isolated from each other. Many species have become extinct and numerous others have become threatened. This has all happened in a very short historical time. Let us take a more analytical approach when attempting to articulate an answer to the question: what is meant by habitat loss? The answer is, naturally, very much dependent on the target species, the target habitat, and the target context. In what follows, we shall first aim for a clarification of what habitat loss might mean in its various different views. Then we shall explore what

Espoo, Finland
Forested areas
in 1752 and 1976

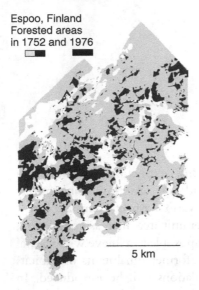

5 km

*Fig. 8.1.* Loss of forested area in Espoo, southern Finland during the past 200 years (modified from Halme and Niemelä 1993).

consequences the different meanings of habitat loss may have on population and community dynamics in space and time.

The most obvious answer to our question is that there used to be homogeneous coverage of a given habitat and now only splinters of it are left (fig. 8.1). Other views are that the suitable habitat for our target species or community has always occurred in various forms of spatially scattered areas suitable for breeding, surrounded by areas that are more hostile to the population renewal process. Over time some of these areas have been lost, seen as a reduced number of suitable areas (fig. 8.2(A)), or reduced resource availability in some areas, while in others it remains at the original level (fig. 8.2(B)), or the profitability reduction is global (fig. 8.2(C)). This is not all: habitable areas may become lost in a given spot to appear elsewhere later on. Inhabitants of the lost areas, *refugees*, have to move elsewhere in search of habitable areas (fig. 8.2(D)). During ontogeny of a given species, juveniles and adults may live in different habitats perhaps not exactly matching in per capita profitability for the different age groups (fig. 8.2(E)). Finally, for many species still living in their natural habitats, the availability of habitable areas is different throughout their distribution range. Often at the center of the distribution area habitat availability is higher, while towards the edges of the range habitable areas are more scanty (fig. 8.2(F)). The different scenarios of habitat loss (fig. 8.2) are further

What is habitat loss?

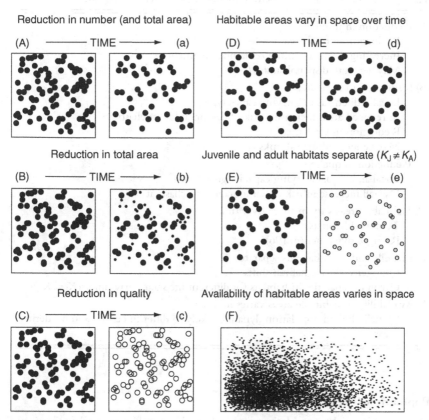

*Fig. 8.2.* Examples of different varieties of habitat loss (Table 8.1) In panel (B)–(b) symbol size refers to habitat area, in (C)–(c) ● is a habitat patch of higher quality than ○, while in (E)–(e) ● refers to juvenile habitats and ○ to adult habitats (likely differing in their carrying capacities, too).

commented on in Table 8.1. The exploration of various kinds of habitat loss can be summarized by stating that, ultimately, it is resource loss.

## Consequences of habitat loss – a primer

It is worth noting that many of the different routes to habitat loss not only yield habitat loss, but are often associated with other changes in the ambient environment (Table 8.1). We shall now explore the consequences of habitat loss under controlled conditions, where just one aspect of habitat loss is in action at a time.

Table 8.1. *Different ways of perceiving habitat loss (see also fig. 8.2)*

| |
|---|
| (A) Reduction in number |
|     Reduction in total area |
|     Increasing distance among remaining fragments |
|     May generate dispersal sinks, $I_i < E_i$ |
| (B) Reduction in local area |
|     Local reductions in carrying capacity, $K_i$ |
|     In reduced areas increasing risk of demographic stochasticity |
|     Reduction in total area |
|     May generate dispersal sinks |
| (C) Reduction in quality |
|     Local and global carrying capacity goes down |
|     Overall increase in risk of demographic stochasticity |
| (D) Habitable areas vary in space over time |
|     Temporal habitats |
|     Refugee dynamics (p. 196) |
| (E) Juvenile and adult habitats separate |
|     Ontogeny-caused habitat shifts |
|     Carrying capacities in habitats for different life stages may differ ($K_J \neq K_A$) |
| (F) Availability of habitable areas varies in space |
|     Spatially linked population dynamics likely to differ in central and bordering areas |

## Population-level consequences

Our focus is on the population size and on the number of local extinctions (when the number of local sites $n > 1$). The tool for our purposes is described in Box 8.1. Losing resources – a reduction in carrying capacity – is the most obvious method of habitat loss (fig. 8.3). Reduction in habitable area, e.g., due to timber logging, is one way this can take place. It is feasible to think that at least some residents of the clear-cut take refuge in the intact habitat area. This feature – refugee forged dynamics (p. 196) – can potentially cause, via the density-dependent feedback, a drastic drop in the population size of the intact area (fig. 8.3). The kickback may also cause the resident population to become extinct even though the remaining habitat would easily maintain a viable population.

Let us take a uniform single habitat ($n = 1$) with $K = 200$ as a reference. As expected, the mode of the population size at $t = 1001$ is 200. However, with demographic stochasticity (p. 185) there is some variation around this value (fig. 8.4(A)). We now split the habitat into $n = 20$

**Box 8.1**  ·  *Demographic stochasticity and Ricker dynamics*

Ricker dynamics will be taken as the skeleton of the population renewal process. It will be set into a spatial frame, so that we are able to address a few general questions of habitat loss. The renewal process is enriched with demographic stochasticity to make addressing small population dynamics more realistic. Thus, we write

$$X_i^*(t+1) = \text{Poisson}\left\{ X_i(t) \exp\left[ r\left( 1 - \frac{X_i(t)}{K_i} \right) \right] \right\}.$$

The number of population subunits ranges from 1 to $n$, depending on the question we are addressing. $K_i$ is the carrying capacity of the $i$th subunit, $r$ is the population growth rate (here $r = 2$, i.e., stable dynamics) and $X^*$ refers to population size before any dispersal adjustments have taken place. Dispersal follows kernel II (p. 54), where the mortality of dispersing individuals increases with distance, and none will survive distances longer than $d_{MAX}$. The system is initiated with $X_i(0)$ drawn from i.i.d. random numbers ±10% around the carrying capacity $K_i$, and is left running for 1000 generations. The data of ensemble population size (summed over all $n$) and on the number of extinct units out of the $n$ are collected at $t_{1001}$. We replicated the simulation 1000 times for each parameter combination, and shall display results of these frequency distributions.

fragments (randomly allocated into a $10 \times 10$ co-ordinate space, 0.2 per unit area), with $K_i = 10$, thus retaining the global $K_E = 200$. The change from uniform habitat into 20 fragments generates a problem for the dispersing individuals ($m = 0.05$) to disperse across the hostile environment separating the fragments (fig. 8.4(D)). In our particular example, the median distance is six units among the $n = 20$ local units. Assuming that the best dispersers of the focal species can barely manage this distance, i.e., in kernel II $d_{MAX} = 6$ (and that local population sizes less than 0.01 become extinct due to various stochastic reasons), the ensemble population size $X_E = \Sigma X_i$ (the summation is over all $n$ at $t = 1001$) has the mode of 100 (fig. 8.4(B)). This is all due to habitat loss increasing fragmentation, viz. the distances among the remaining habitable units. Retaining all other things, but increasing the maximum distance that individuals are capable of dispersing to and surviving, eq. 8.2, to $d_{MAX} = 9$ increases the mode of the $X_E$ back to 200 (fig. 8.4(C)). Notably, the ensemble

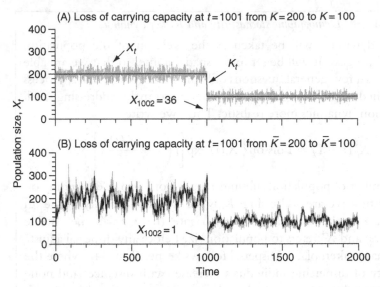

*Fig. 8.3.* Population fluctuations (grey line) in two systems, where the initial carrying capacity (black line) $K = 200$ is reduced to $K = 100$ at $t = 1001$. In (A) the carrying capacity is constant before and after the habitat loss, while in (B) carrying capacities in the two phases are stochastic (obeying AR(1) process with $\kappa = 0.9$). The populations obey Ricker dynamics with $r = 2.0$ and demographic stochasticity (p. 185), accounting for the variation in population size in (A).

population size does not peak at such high values as observed in the uniform nonfragmented habitat (fig. 8.4(A)). These findings underline the significance of occasionally successful long-distance movements (e.g., Turchin 1998) in affecting the sustainable population size in a fragmented landscape. However, it is not only $d_{MAX}$ that is significant – parameter $m$ also affects the outcome. The bending isoclines (fig. 8.4(E)) of (the median of) the $X_E$ contour graph show that dispersal agility and maximum distances dispersed are not related linearly: to reach matching population sizes as in the pristine nonfragmented habitat, dispersal distances should be long, but the agility to disperse should not be that high (fig. 8.4(E)). High dispersal rates and long maximum distances dispersed keep the incidence of local extinctions low (fig. 8.4(F)).

One element of habitat fragmentation is that while the habitat thins out the fragments become isolated. We simulated the effects of increasing isolation by gradually increasing the co-ordinate space where the $n = 20$ subunits were randomly located. Understandably, isolation due to habitat loss leads to reduced population size, and increased risks of local (and

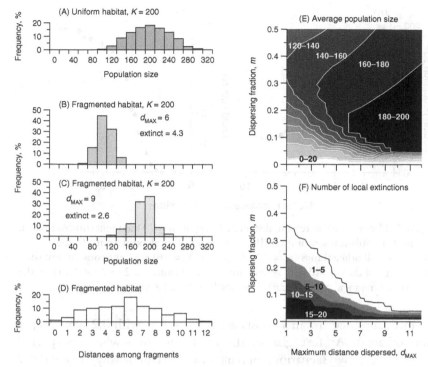

*Fig. 8.4.* (A) Frequency distribution of population sizes (Ricker dynamics with demographic stochasticity and $r = 2$) of 100 replicated runs at $t = 1001$ in a single homogeneous habitat with $K = 200$. The habitat is split into 20 fragments, $K_i = 10$ in each (frequency distribution distances among the fragments is shown in (D)), thus retaining the total carrying capacity of 200. The dispersing fraction is $m = 0.1$ of the local population size. In (B) the maximum distance at which the dispersing individuals survive is $d_{MAX} = 6$ (matches the median distance among the fragments (D)). The median ensemble population size is 90, much less than the 200 in the nonfragmented habitat (A). Increasing $d_{MAX} = 9$ recovers the median population size of the ensemble to 200 (C). Note, however, that in the fragmented system the tail of high population sizes is truncated as compared to the nonfragmented habitat (A). Experimentation in (B) and (C) suggests that the rate of local extinctions depends on dispersal distance. The effect of dispersal rate $m$ and maximum distance dispersed $d_{MAX}$ (a parameter of the dispersal kernel II, p. 54) on the ensemble population size and the number of local extinctions is explored in the contour diagrams in panels (E) and (F), respectively. The results for the fragmented system are for 100 replicated runs (for each parameter combination) at $t = 1001$.

eventually also regional) extinctions (fig. 8.5). The next step is to keep the carrying capacity of the ensemble constant at $K = 200$, but to increase the number of subunits. This was done in two different environments (with four combinations of dispersal parameters $m$ and $d_{MAX}$), in the $10 \times 10$

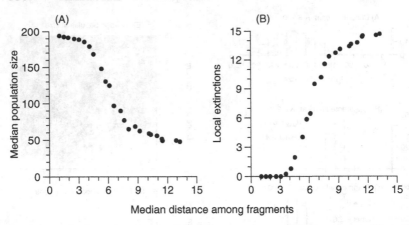

*Fig. 8.5.* The effect of increasing distance among the $n = 20$ fragments on (A) median ensemble population size and on (B) the number of local extinctions. In the simulations all other things ($m = 0.1$, $d_{MAX} = 5$, $K_i = 10$) are kept constant but the dimensions of the co-ordinate system are increased from $2 \times 2$ to $25 \times 25$. Here the measure of fragmentation ($x$ axis) is the median distance among the fragments.

co-ordinate space and in a constant-density space (with 0.2 subunits per unit of area). At first glance, the results are somewhat unexpected (fig. 8.6). The two scenarios (constant area, constant density) do not differ much in terms of $X_E$ and risk of extinction; rather, the dispersal parameters seem to be the discrimination parameters here. The more one disperses and the further one goes, the more "safe" (in terms of larger population size, smaller extinction risk) is the outcome.

One way to summarize the impact of habitat loss is to explore its significance on population extinction risk. A straightforward assumption is that loss of habitat will affect the carrying capacity, $K$. The relationship between carrying capacity and extinction risk can be simply exemplified by applying the techniques described, e.g., in Burgman *et al.* (1993). Let us assume the Ricker dynamics with demographic stochasticity (Box 8.1) with different growth rates. We can now calculated the terminal population size at $t = 1001$ for 1000 independent replicates for various $r$ and $K$ values. From these distributions, one can estimate (Burgman *et al.* 1993) the risk of population extinction. The outcome of this exercise is straightforward (fig. 8.7). Populations with periodic and chaotic fluctuations have a much higher risk than stable populations of becoming extinct. More important is the observation that the smaller the carrying capacity, the larger the extinction risk (fig. 8.7), even in stable populations. Thus, a habitat-loss-caused reduction in carrying capacity increases the risk of population extinction.

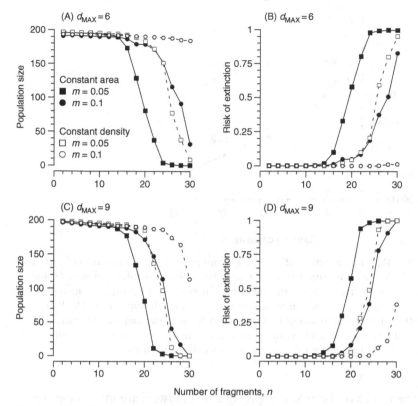

*Fig. 8.6.* The effect of fragmentation on ((A), (C)) the population size (pooled over all $n$) and on ((B), (D)) the risk of extinction of the $n$ fragments. The effect of constant area (10 × 10 grid) and constant density of 0.2 fragments per unit area is explored using two differing dispersal rates, $m = 0.05$ and 0.1. In (A) and (B) the maximum dispersal distance is $d_{MAX} = 6$ (the median distance among 20 fragments in a 10 × 10 co-ordinate space), while in (C) and (D) it is $d_{MAX} = 9$. See also fig. 8.4.

## Community-level consequences

It is natural to think that populations of any species are not entirely isolated from the web of ecological interactions in which they live. For example, a given pair of species may often encounter each other – more often than at random – in using the same resources that limit the population growth of other species. Alternatively, one of the species benefits from the presence of the other one by consuming some of them. We have either a community of two competing species or a genuine food web. We shall first explore the consequences of habitat fragmentation on the co-existence of two competing species by the system described in Box 8.2.

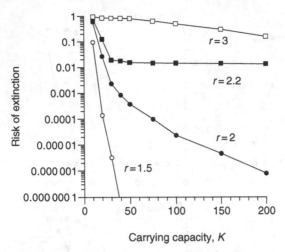

*Fig. 8.7.* Risk of extinction of a homogeneous population as a function of carrying capacity *K*. The population renewal obeys the Ricker dynamics (different *r* values indicated) with demographic stochasticity. The risk of extinction is estimated based on the frequency distribution of population sizes (e.g., Burgman *et al.* 1993) at *t* = 1001 of 1000 independent replicates of each parameter value combination. In these calculations, we assume that demographic stochasticity takes care of effects of inbreeding and problems of mate finding in small populations.

The results of the two competing species system are interesting. First, with the parameter selection used, when *n* = 1, species 2 became extinct in 63% of the cases (fig. 8.8(A)). However, with increasing *n* the probability of the outcome that both species persist in the system also increases. In this scenario (Box 8.2), with *n* = 25, both species will persist in the network of the *n* population subunits (fig. 8.8). It also turns out that population sizes in the two-species community of competitors are reasonable for both species with these parameter selections (fig. 8.8(B)).

The next step is to analyze a resource–consumer interaction (Box 8.3). The parameter values for the resource–consumer, *R* and *C*, dynamics are taken from Leslie and Gower (1960) when they simulated the long-term persistence of *R* and *C* in the system. We selected the Leslie–Gower "second" set (fig. 8.9), as this keeps the equilibrium densities of the two species low, "making chance extinction of either the predator or the prey population a very real possibility." Indeed, with *n* = 1, the probability of extinction of *C* was as high as 0.83 (for *R* it was 0.12; fig. 8.9). When increasing the degree of fragmentation an interesting diversification, in terms of extinction risk, took place depending on whether the noise was

**Box 8.2** · *Two-species competition*

Leslie and Gower (1958) have a discrete-time version of a two–species competition model, which we shall extend here to cover dispersal-linked dynamics in an $n$ patch system

$$N_{1,i}^*(t+1) = \frac{\lambda_1 N_{1,i}(t)}{1 + \alpha_1 N_{1,i}(t) + \beta_1 N_{2,i}(t)}$$

$$N_{2,i}^*(t+1) = \frac{\lambda_2 N_{2,i}(t)}{1 + \alpha_2 N_{2,i}(t) + \beta_2 N_{1,i}(t)}.$$

Here $\lambda$ is the growth rate, $\alpha$ refers to the density-dependent effect that the species has on itself, while $\beta$ is the interspecific competition term $(\alpha_1, \alpha_2, \beta_1, \beta_2 > 0)$. The system is in equilibrium when $\lambda$ equals the divisor. Leslie and Gower (1958) derived the following relations

$$y = \frac{\beta_1(\lambda_2 - 1)}{\alpha_2(\lambda_1 - 1)} \text{ and } x = \frac{\beta_1 \beta_2}{\alpha_1 \alpha_2}$$

to indicate the outcome of the competition (when $n = 1$). Here we are particularly interested in $1 < y < x$, which says that the stationary state is unstable, and either species 1 or 2 will become extinct depending on the initial number of the two species. $N^*$ refers to population size before any redistribution of individuals has taken place. The dispersal is after kernel II (p. 54). We initiated the system with i.i.d. random numbers ±10% around the equilibrium densities for the two species. The system was left running for 1000 generations. We scored the presence of the two species checking whether $N_1(1001) > 0.01$ and $N_2(1001) > 0.01$ in any of the $n$ fragments. We varied $n$ from 1 to 50. Each parameter combination was replicated 1000 times.

local only or global only (fig. 8.9). With solely local noise, the risk of extinction for both $R$ (already at $n = 3$) and $C$ went down (at $n = 12$). However, no such reduction of extinction risk was observed with only global noise (fig. 8.9). This finding has a natural explanation. With only local noise, populations of $C_i$ and $R_i$ keep fluctuating independently. At times some $C_i$ (rarely $R_i$) may be extinct in some of the local fragments while in others they thrive well. In contrast, when the only external modulator of population dynamics is the global noise [the Moran effect (Chapter 4)], synchrony of the population fluctuations is the outcome. When the AR(1) noise with $\kappa = 0.7$ generates a sequence of bad years in

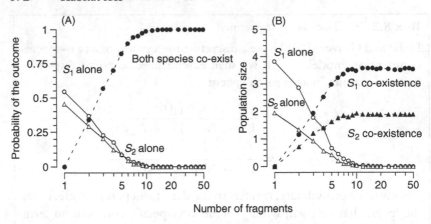

*Fig. 8.8.* A system of two competing species that is unstable (i.e., either $S_1$ or $S_2$ wins, depending on initial densities) when the number of fragments is small [when $n=1$, $P(S_1 \text{ wins}) = 0.63$] becomes a stable co-existence system for the two species when the number of fragments exceeds 10. The parameter values used are: $\lambda_1 = 1.5$, $\lambda_2 = 2.2$, $\alpha_1 = 0.01$, $\alpha_2 = 0.23$, $\beta_1 = 0.77$, $\beta_2 = 0.95$, $m_1 = m_2 = 0.1$, $d_{MAX,1} = d_{MAX,2} = 6$ (see also Leslie and Gower 1958). The outcome is based on 1000 replicated simulations for each parameter value combination.

one place, it generates a spell of bad years all over. With a high degree of synchrony, extinction of one local population means the simultaneous extinction of all other populations (Allen *et al.* 1993; Heino *et al.* 1997a). We also experimented with both global noise and local noises simultaneously. As expected, the results depend very much on the strength of the two processes relative to each other. For example, with $\mu(t)$ and $\varepsilon(i, t)$ drawn from the same range, the number of fragments reducing the risk of extinction to zero stabilizes after $n \approx 30$ around 0.62 for $C$ and to 0.03 for $R$ with the parameter values as in fig. 8.9. When global disturbance dominates the system, synchrony prevails and the extinction risk, especially of $C$, is high. When local disturbances dominate, asynchronous, local-unit-independent fluctuations prevail. This leads to low risks of extinction (Allen *et al.* 1993; Heino *et al.* 1997a).

## Habitat-loss-generated population fluctuations

We are here examining the extent to which the dynamics of resource availability affect the population dynamics of the resource users. We imagine a system where the population renewal process is indirectly affected by a variable environment by explicit changes in resource supply.

---

**Box 8.3** · *Resource–consumer dynamics*

Leslie and Gower (1960) studied a discrete-time version of predator–prey dynamics. For the dispersal-linked dynamics in an $n$ patch system (here $i$ refers to the $i$th patch) we write

$$R_i^*(t+1) = \frac{\lambda_R R_i(t)}{1 + \alpha_R R_i(t) + \gamma C_i(t)} \mu(t)\varepsilon(i, t)$$

$$C_i^*(t+1) = \frac{\lambda_C C_i(t)}{1 + \alpha_C \frac{C_i(t)}{R_i(t)}} \mu(t)\varepsilon(i, t).$$

Here $\lambda$ is the growth rate, $\alpha$ is the density-dependent effect that the species has on itself, and $\gamma$ is the voracity of the predator ($\alpha_1, \alpha_2, \gamma > 0$). $R^*$ and $C^*$ refer to population sizes before any redistribution of individuals has taken place. The dispersal is after kernel II (p. 54). The Leslie and Gower (1960) resource–consumer dynamics calls for stochastic elements. We have implemented the stochasticity here by using global external disturbance $\mu(t)$ and local noise $\varepsilon(i, t)$. For the global noise, when in action, we used AR(1) random numbers between 0.5 and 1.5 with $\kappa = 0.7$. When the global disturbance was off we had all $\mu(t) = 1$. For the local noise at each $t$ we selected $n$ by i.i.d. random numbers between 0.5 and 1.5. In runs with no local noise, all $\varepsilon(i, t) = 1$. We experimented $n$ from 1 to 15, with 1000 replicates for each parameter combination. Populations were initiated with i.i.d. random numbers $\pm 10\%$ of the equilibrium values (Leslie and Gower 1960). Our measure of extinction of either $R$ or $C$ is simply that all the $n$ populations are extinct at $t_{1001}$. When $n > 1$ $m_R = m_C = 0.1$ and $d_{\text{MAX,R}} = d_{\text{MAX,C}} = 6$, the dimension of the co-ordinate space, into which the $n$ fragments were allocated, was always $10 \times 10$. The threshold for extinction was set to 0.5 (the 5% percentile of the long-term frequency distribution of the noise-disturbed dynamics of $C$).

---

Models of population dynamics in patchy landscapes have been much studied (Gilpin and Hanski 1991; Bascompte and Solé 1997; Fryxell and Lundberg 1997; Tilman and Kareiva 1997). However, with a single exception (Hanski 1999b), the patch availability has invariably been assumed to be constant, i.e., there has been population dynamics and dispersal, but no dynamics of the landscape itself. Consider a system where the resources are patchily distributed but the void in between them is not entirely hostile to the consumer population. Individuals

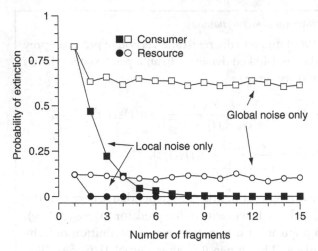

*Fig. 8.9.* Probability of extinction in resource–consumer dynamics as a function of the number of fragments. Two differing noise scenarios exist: local noise (i.e., each unit has its own noise) and global noise (all units are disturbed in a matching way). The parameter values used are: $\lambda_R = \lambda_C = 1.25$, $\alpha_R = 0.001$, $\alpha_C = 0.5$, $\gamma = 0.025$ (see also Leslie and Gower 1960). The probabilities are based on 1000 replicated simulations for each parameter value combination.

leaving one habitable patch are likely to locate to another patch. The patches are assumed to be sufficiently distant that the local dynamics within a patch do not directly influence the dynamics in the other patches. These systems are characterized by having some localities that are persistent over time, and some that are occasionally unable to support a subpopulation of the focal species. Some sites are more vulnerable than others to periodic catastrophes or spells of adverse weather. For species with a large latitudinal range, it may be that local populations near the distribution center are always suitable, but those towards the distribution margin are only extant at most benign times.

Imagine a landscape consisting of two patches with identical carrying capacities, $K_1 = K_2$. The focal species inhabits both resource patches whenever they are available. We let density-dependent renewal occur before dispersal between the patches takes place. We will now assume that one patch has a certain probability of disappearing before the next breeding season. Thus, the former residents, the refugees, of the patch lost have to redistribute themselves. The refugees, either all of them or a given fraction of them – depending on the mortality rate while in transit – subsequently arrive at the remaining patch. Because of the arrival of the refugees, the patch becomes crowded. However, local density-dependent renewal takes care of over-crowding. The resource dynamics can also be seen from the other side of the coin. A single patch exists, but at good times the number of inhabitable patches, where reproduction can take place, doubles. After colonization of the newly emerged patch, there is a certain probability that the extra patch disappears and the residents have to settle back into the remaining patch. A model for such dynamics is outlined in Box 8.4.

## A two-patch refugee system

In the Ricker model with $r = 1$ not much happens, but a dynamic land-scape with refugee arrival (Box 8.4) changes the picture dramatically: the local dynamics are destabilized (fig. 8.10(A)). Typical for the fluctuations is that once one of the resource patches disappears, the refugees will increase the population size in the remaining patch. Due to density dependence, the population quickly returns towards the carrying capacity and even well below it. Increasing refugee mortality slightly dampens the fluctuations (fig. 8.10(C), 8.10(D)). However, even with 50% mortality, substantial fluctuations of the population size remain.

We conclude that the dynamic appearance and disappearance of resource patches, and the subsequent arrival of refugees, cause stable local dynamics to turn into complex dynamics. Here, we have used the Ricker model but experimentation with many other population renewal models gives similar results. This suggests that refugee-caused complex dynamics are not the anomaly of the Ricker model. It is true that the strong over-compensatory dynamics of the Ricker model easily destabilize population dynamics, but in our study it is environmental stochasticity at the landscape level, and not exactly local overcompensatory population responses, that causes complex dynamics. It is tempting to suggest that a substantial part of the observed complexity in the dynamics of many natural populations might simply be due to underlying resource dynamics and the episodic arrival of refugees. Refugee arrival and subsequent population crashes due to density dependence can thus operate as the key elements modulating a stable population equilibrium into more complex dynamics.

## Local dynamics and ensemble dynamics

We now extend the above scenario to a population inhabiting several patches ($n = 40$), all carrying the same amount of resources. In the local

**Box 8.4** · *Refugee-forced population dynamics*

We shall assume that the renewal of our focal population in discrete time $t$ follows the Ricker equation

$$X_i(t+1) = X_i(t) \exp\{r[1 - X_i(t)]\},$$

where $X$ refers to the population size of the $i$th population, and $r$ is the maximum per capita population growth rate. In this model, the equilibrium population density (without migration) is unity. Assume further that the population subunits, or patches, occur in implicit space (kernel 0, p. 53) and are coupled via dispersing individuals. Following, e.g., Allen *et al.* (1993), this can be modelled as

$$\tilde{X}_i(t) = (1 - m)X_i(t) + m\overline{X}(t),$$

where $m$ is the dispersing proportion of individuals and $\overline{X}(t)$ is the mean density of individuals taken over all populations at time $t$. The population size after the dispersal is $\tilde{X}_i(t)$.

We shall first deal with a two-patch system. Of course, in a two-patch case with equal population renewals the dispersal is symmetric. Once the other patch is gone, the residents will disperse to the remaining patch. In this system, the dispersing fraction $m$ allows for colonization of the patch that emerges with benign times to come. While in transit, a given proportion $d$ of the refugees may die before they reach the resource patch of destination. One more parameter, $p$, the probability of the disappearance and reappearance of the resource patch, is needed.

The next step is to make the number of subunits much larger, $N = 40$. On top of this system, we shall overlay the temporal dynamics of the resource patches. For the annual habitat loss (and reappearance of resources) we write

$$k(t) = h\left[\sin\left(\frac{2\pi t}{P}\right) + 1\right],$$

where $k(t)$ is rounded up to integers, $P$ is the period length, $h$ is a constant affecting the maximum $k(t)$ can achieve [we used $h = 10$; making $k(t)$ to range from 0 to 20, i.e., 0% to 50% of all available patches]. Thus, after the annual loss of habitat patches the remaining patches are simply $n_t = N - k(t)$. The patches were initiated in phase by setting $X_i(1) = 0.5$, the system was left running for 500 generations, and we used the final 100 generations for our analyses.

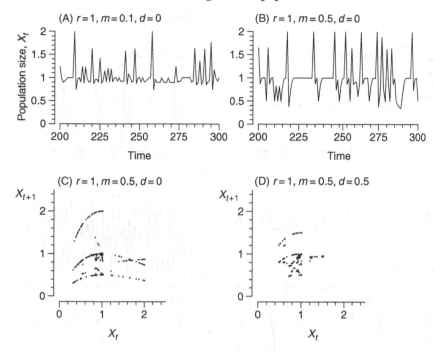

*Fig. 8.10.* Examples of the complexity of refugee-arrival-generated dynamics in a patch remaining after its sister patch is lost and gained (the probability of patch disappearance/reappearance, $p = 0.2$). The uppermost panels give dynamics for two differing dispersal values without dispersal mortality ($d = 0$). In the phase portrait (C) the data are the same as in (B), and they should be compared with (D) where everything else is equal, but dispersal mortality is $d = 0.5$. Throughout $r = 1$, $m$ is the dispersing fraction.

units, population dynamics are assumed stable. The resource patches have their own dynamics such that the patch numbers fluctuate in a cyclic manner with 10-year periodicity (Box 8.4). At times, a fraction of the units is lost. The former residents of the lost patches disperse to locate to a new patch. In what follows, we shall deal with the emerging dynamics at a single-unit level and in the ensemble dynamics, i.e., population fluctuations in time summed over all the subunits. The single-population fluctuations match a system where a local population is assessed independently of others. While the ensemble dynamics match a system such as a migrating species breeding in several subareas, e.g., in the arctic tundra, that are annually counted (at a bird observatory) on their way back to the over-wintering grounds. We consider a scenario where the patches are ranked according to their persistence ability. Patch loss follows the rank order. That is, the

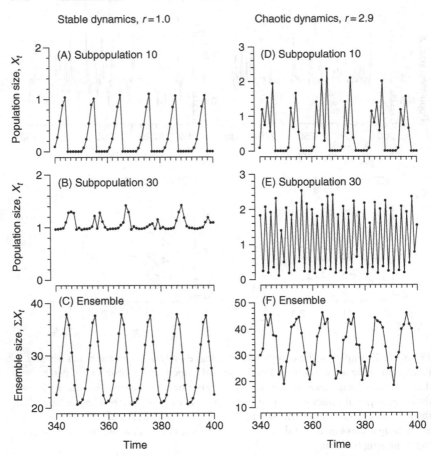

*Fig. 8.11.* An example of local population dynamics ((A), (B), (D), (E)) and ensemble dynamics ((C), (F)) with systematic loss of habitat units. Two subpopulations are graphed, ((A), (D)) is for a patch which is absent ≈ 50% of its time, ((B), (E)) for a patch which is always extant. The left-hand panels are for the stable skeleton of the dynamics, while in the right-hand panels the skeleton is chaotic. Note that the periodicity of the habitat loss is differently pronounced in different local populations (and varies after *r*), while in both cases the ensemble dynamics display clear 10-year periodicity.

different breeding sites are differently susceptible, e.g., for climatic forcing that comes in cycles (e.g., Burroughs 1992).

As in the two-population case, the stochastic arrival of refugees breaks down the stable local dynamics, creating local population trajectories with complex features (fig. 8.11). Dynamics in the most heavily tolled patches are pulse-periodic with high amplitude (irrespective of the population growth rate *r*). Most importantly, however, at the local-patch level, the forcing periodicity of the habitat loss disappears, while at the ensemble level of the 40 patches the 10-year periodicity of habitat loss is clearly visible over the range of stable to chaotic population dynamics. An interesting detail is that habitat-loss-caused refugee disturbance does not synchronize the dynamics of the local populations (in stable dynamics synchrony averages $r_0 = 0.3$, in chaotic dynamics $r_0 = 0.07$).

### Habitat loss and refugees

The inflow and outflow of individuals to and from a local population is of course an example of environmental disturbances and it may have profound implications for the local dynamics (Fryxell and Lundberg 1993, 1997; Ranta *et al.* 1997a,b,c; Chapter 4). We have shown here that variability in the underlying resource supply (through the refugee effect) can alter very stable dynamics to complex ones *indirectly*. This has consequences for how we interpret time series data from natural populations. If we fail to recognize that not only can direct external stochasticity change the dynamics, but that the indirect effects of fluctuating resources can do so as well, then it will be difficult to disentangle the deterministic and stochastic components of the observed dynamics. Our results also emphasize the importance of the underlying spatial structure. It is both the temporal variability of resources and other factors, as well as how that variability manifests itself in space, which matter.

Another important result is that any periodicity in the structural perturbation affecting the population system interacts in a complex manner with both spatial scale and the fundamental frequencies of the within-patch population processes. This has important implications for defining an appropriate spatial scale for empirical study of the phenomena described here. With spatially structured loss, some frequencies of perturbation do show through strongly in local population time series whereas others do not. This phenomenon is likely to be caused by some frequencies of perturbation being strongly amplified because they resonate with "natural" frequencies of the local population dynamics.

We argue that population networks, where the number of habitable areas fluctuates from year to year, are likely to be common in nature. Of particular interest is the strong effect of vulnerable patches on the population dynamics of other patches within the system. This has important potential consequences; for example, for the management of conservation areas. It is usually assumed that the more habitable areas that are added to a population network, the better this would be for persistence of the target species, but this may not be true. If the added fragments are of low quality, such that they are often unavailable, then adding them to a population network may have a strongly negative effect by amplifying the range of population fluctuations in other habitat fragments. Conversely, however, it may have a beneficial effect, in breaking down synchronization of population dynamics between sites. Hence, further work is required to fully understand the management importance of these effects in specific situations.

## Life on the edge

In this section, we take a closer look at the dynamics in different parts of a species distribution. Local extinctions can be assumed to be more frequent in the areas near the border of the distribution than in the distribution center. Habitable patches at the border will receive fewer dispersing individuals than areas in the center because there is nobody coming beyond the border. Thus, simply by demographic stochasticity bordering areas might become locally extinct for a longer period than areas surrounded by habitable patches. An alternative explanation for the distribution border is that the rate of loss of habitable patches is higher there than in the center. The bordering areas might, e.g., be located in climatically more harsh environments than in the central areas. Thus, in bad years the availability of habitable areas along the distribution borders might be much less than in benign years. With a loss of a patch, one of two things can happen for the resident individuals. Either they die out with the patch, or at least some of them, the refugees, may manage to redistribute themselves into the remaining habitable patches (as above, p. 196).

### Dynamics in the center and at the border of the distribution range

Imagine a population living in a landscape of habitable patches matching in resource availability. For our argument, we shall let the $n = 40$ patches form a one-dimensional array, the spacing between adjacent subunits being one distance unit. Local population renewal follows the Ricker model (Box 8.4). In each generation, a given fraction $m$ of individuals in the local subunits disperse. Their dispersal success is assumed negatively distance-dependent. Every time step a fraction $v$ of the $n$ available habitat units will become hostile to reproduction. The distribution center is the least disturbed, while patches in the distribution margins at both ends of the landscape vector will often be unavailable for colonization and hence for successful reproduction. After reproduction, the annual distance-dependent dispersal takes place. Each patch sends out $mX_i(t)$ immigrants and the number of emigrants received in patch $i$ follows the dispersal kernel I (p. 53). The loss of habitable patches (takes place after reproduction) from each end of the vector follows a negative binomial distribution with parameters $k = 4$ and $p = 0.5$. The refugees follow the same distance-dependent rule as the dispersing individuals so that the patches next to the destroyed area receive more refugees than other redistributing individuals. The reverse takes place in the central patches. The simulations of the system (populations initiated with random numbers between 0 and 1) were run for 1124 generations, from which the first 100 were omitted. From the census after habitat loss, we recorded the coefficient of variation in population size and the color of the population time series.

The extent of annual habitat loss had a median of 10% of the habitable patches being unavailable for reproduction each year. The frequency of years when no habitat loss took place was 6%, while only rarely close to 50% of the patches were eliminated. The habitat loss created a clear gradient in population variability. It increased from relatively low values in the distribution center towards much larger values in the distribution margins (fig. 8.12(A)). This effect is dependent on the spatial extent of the dispersal: with localized dispersal $c = 1$ (dispersal kernel I, p. 53) the overall variability was higher than when the distance among the habitable patches was not that important, $c = 0.05$. The color of the dynamics of local populations was blue in the least-disturbed central areas. This matches to the blue color of Ricker dynamics of a single population. In the area where the coefficient of variation reached its highest values, the population fluctuations were dominated by long-term variations (fig. 8.12(B)). The bluish nature of the Ricker dynamics turned to red due to the stochastic temporal loss of habitable units in bordering areas of a species distribution.

The result that long-term fluctuations are redder at the distribution border than in the central areas inspires further explorations of this pattern. We now take $n = 41$ subunits, arranged in a one-dimensional array as above, and we modified the renewal process slightly to have one additional lag

$$X_i^*(t+1) = X_i(t)\exp\left\{r\left[1 - \frac{X_i(t-1)}{K_i}\right]\right\}. \qquad (8.1)$$

Lag to $(t-1)$ makes the populations fluctuate cyclically (fig. 8.13(A)). Population size before dispersal is $X_i^*$, and each generation a fraction $m_i$ of

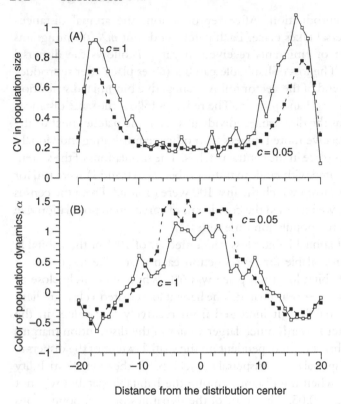

*Fig. 8.12.* Coefficient of variation (A) in population size and (B) color of the long-term population fluctuations along the distribution range. Habitable patches close to the 0 co-ordinate in the $x$ axis are in the distribution center, while getting away from these areas in both directions leads towards the margins of the distribution. Two different dispersal-distance scenarios (long distance $c = 0.05$, and short distance $c = 1$) are displayed; the Ricker parameter $r = 2.2$.

the residents of the $i$th subunit take leave and redistribute themselves after the dispersal kernel I (p. 53).

The landscape structure within a distribution range of a species resembles that in fig. 8.2(F), i.e., habitable areas are more numerous per unit area in the center of the distribution than at the borders. Thus, we took unit 21 as the most central one; we placed 20 units on both sides of this unit, to achieve symmetrical distribution. In one set of simulations, the units were of increasing distance from each other (fig. 8.13(B)) to mimic the distribution of habitable units when moving from the center towards the border in fig. 8.2. Thinning out of a population at the margins of its distribution range can also be due to carrying capacity $K_i$ being lower

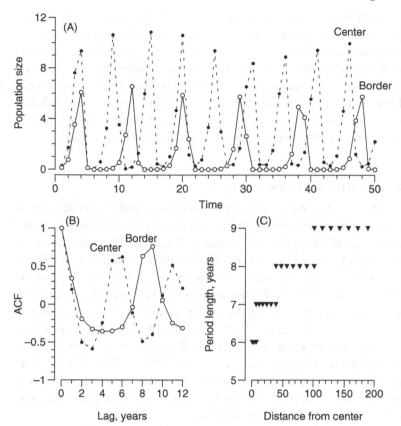

*Fig. 8.13.* (A) Population fluctuations in the distribution center and at the border of distribution. The autocorrelation function, ACF, for the data in (A) is given in (B). Note that the center population has a period length of 6 years, while the border population peaks at 9-year intervals. (C) Cycle period lengths graphed for all populations against distance from the distribution center. In this particular simulation, the parameter values were as follows: $r = 1.75$, $m = 0.1$, $c = 0.5$, $K = 1$ for all patches (interpatch distance increases to the power of 1.75 of the patch rank from the center).

there than at the center. In what follows, interpatch distances are either constant or they gradually increase towards the distribution border, and local carrying capacity is either constant or it decreases towards the border. In our simulations, populations are initiated with random numbers $\pm 10\%$ of $K_i$ and $r = 1.75$. Our focus is on the length of the cycle in various parts along the gradient from center to the distribution border.

Experimentation with the system shows that nothing very interesting happens with constant spacing of the $n$ units, regardless of $K_i$ being constant or a decreasing function of distance from the border. However, changing the spacing between adjacent units to increase with the distance from the center (thinning out of the habitat) brings an extra feature: the cycle length increases towards the distribution border, i.e., the dynamics redden (fig. 8.13). In our particular setting, populations at the center fluctuate with 6- to 7-year periodicity, while towards the distribution border cycle period gradually increases to 9 years (fig. 8.13(C)). This feature remains (though not as prominent) even with constant carrying capacities in the local units.

## Cyclic dynamics – a special feature

Elton (1924) was the first to draw the attention of population ecologists to the peculiar feature that many animal populations in northern latitudes tend to display rather regular self-repeating dynamics. This phenomenon is especially pronounced in the Canada lynx and snowshoe hare in North America, both species showing 10-year cycles (Elton and Nicholson 1942). Fennoscandian voles (*Clethrionomus* spp., *Microtus* spp.) are another good example of species with cyclic dynamics, here with a period of 3–5 years (Hansson and Henttonen 1985; Stenseth 1999). In addition, many forest grouse species show regular periodicity (6- to 7-year) in their dynamics (Lindén 1981; Lindström *et al.* 1995, 1999). Since Elton (1924), research on cyclic populations has been the center of population ecology. Therefore, we shall dwell a bit on the possible consequences of habitat loss on cyclic population dynamics.

A common feature of these animal groups is, perhaps apart from the Canada lynx, that in some parts of the distribution range the populations show regular cyclic dynamics, and in other parts long-term dynamics are either stable or fluctuate with no clear periodicity (e.g., Smith 1983; Hansson and Henttonen 1985; Lindström 1994; Stenseth 1999). Finding explanations for the cyclic population dynamics has been a preoccupation of population ecologists for the past three-quarters of a century (Elton 1924; Stenseth 1999; Lindström *et al.* 2001). Several explanations have been suggested for this observation (Batzli 1992). Therefore, the enigmatic observation that a given group of species displays cyclic dynamics in some parts of their distribution range but not everywhere has caused much trouble and confusion (Lindström *et al.* 2001). We shall now provide a model to explain why cyclic dynamics may exist in certain parts of the distribution range of cycle-prone species, but not everywhere. The model is an extension of the models used in the previous section.

## Population cycles, but not everywhere

The landscape of a cycle-prone species consists of a network of habitable patches suitable for reproduction, and of less hospitable areas. The patches are connected via dispersing individuals. For a dispersing individual it is less costly to reach nearby habitable sites than favorable localities far away. We assume that the patches can be arranged in a gradient of increasing vulnerability to stochasticity, e.g., due to weather. The south-to-north (lowland-to-highland, temperate-to-alpine) gradient in stability of the areas may suffice as a metaphor, but it may also be the increasingly fragmented border area between the pristine taiga forest biome and agricultural landscape. We assume that the habitable patches are located in one-dimensional space and that the patches are coupled by dispersal.

We used the delayed Ricker model for the simulations, and dispersal according to kernel I (p. 53)

$$X_i(t+1) = (1 - m)X_i(t)\exp\{r[1 + a_1 X_i(t) + a_2 X_i(t - 1)]\} + \sum_{s,s \neq i} M_{si}(t). \tag{8.2}$$

Here $X_i$ is population size in the $i$th subunit at time $t$, $m$ is the dispersing proportion of the population, and $M_{si}$ is the number of immigrants arriving to patch $i$ from patch $s$. The parameters $a_1$ and $a_2$ specify the direct and delayed density dependence, respectively. They were selected so that the resulting cycle period length was 4, 6, and 10 years. With the parameter values (fig. 8.14) the resulting population fluctuations are damped. There is good empirical support for both direct, $a_1$, and delayed, $a_2$, density dependence in the systems we are interested in (e.g., Hörnfeldt 1994; Berryman and Turchin 1997). In the best-documented cases, the delayed Ricker equation, when fitted to data, indicates that the skeleton of the population dynamics are damped (Canada lynx: Stenseth et al. 1998b, 1999; voles: Stenseth et al. 1998b,c; Stenseth 1999; grouse: Lindström et al. 1999).

To make a patch vulnerable to temporal loss – or at least unsuitable for population growth – we did the following. The $n$ habitable patches were arranged into a one-dimensional vector. The local populations in the habitable patches were initialized by drawing random numbers between 0 and 1. The populations were left to renew for 500 time units (to get rid of transients). From this time on at any given year, $\nu$ of the $n$ patches were temporarily lost, to reappear some time later on. This was achieved by setting $n = 50$, and drawing $\nu$ for each year from a uniform random distribution

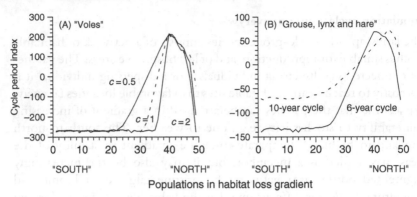

Populations in habitat loss gradient

*Fig. 8.14.* Cycle period index values for the delayed Ricker dynamics with parameter values yielding a damped 4-year period ($r = 2.25$, $a_1 = -0.075$, $a_2 = -0.05$; eq. 8.4) in cyclic dynamics ("voles"), 6-year ($r = 1.2$, $a_1 = -0.035$, $a_2 = -0.074$) cycles ("grouse") and 10-year ($r = 0.46$, $a_1 = 0.05$, $a_2 = -0.095$) cycles ("lynx and hare"). Large positive values of the index indicate that the corresponding period length is recognizable in the population dynamics. The $x$ axis gives the position of the habitable patches relative to the habitat loss gradient. Close to the origin, there is no habitat loss, while habitable patches 41–50 are vulnerable to temporary loss. In (A) the effect of three different values of the parameter $c$ (of the dispersal kernel I, p. 53) are shown separately, in (B) $c = 0.5$, and $m = 0.1$ in both cases.

between 0 and 10. The lost habitable patches were always eliminated from the "northernmost" (locations 41–50) cells in the vector. Thus, the patch indexed $i = 50$ was always lost whenever a patch was lost, while patch $i = 41$ was lost only when $n = 10$. Individuals from the patches lost, refugees, were left to redistribute themselves into the remaining patches. Note that patch loss does not imply loss of individuals. The redistribution also followed dispersal kernel I (p. 53). Thus, both the individuals dispersing annually among the habitable patches and refugees attempting to locate to extant patches obeyed the same distance-related dispersal pattern.

With power spectral analysis, we scored the contribution of the cycle period length to population fluctuations by calculating the periodicity index. For the 4-year period we calculated the sum of the log(power) over period length ranging 3–5 years; for the 6-year period the range is 5–7 years and for the 10-year period it is 8–12 years. The value of this cycle period index is larger the more the cycle period dominates the power spectrum (fig. 8.14). Our results suggest that cyclic dynamics can be found in one part of the distribution range of a species but not in other parts. In this particular example, the areas with cyclic dynamics have to be close enough to areas where habitat loss occurs now and then and residents of the habitable areas lost have to redistribute themselves to the remaining patches. The extent of the distribution range of a given species with prevailing cyclic dynamics and the region of the stable dynamics both depend on the overall extent of the area where habitat loss regularly takes place. It also depends, to some extent, on the dispersal parameter $c$ (fig. 8.14(A)).

Several hypotheses have been suggested to explain the geographical gradient of cyclic population dynamics in small mammals in Fennoscandia (Stenseth 1999; Lindström et al. 2001). The current dominating idea is that the predator communities exploiting the grouse and rodents differ from north to south. The north is dominated by specialist predators tending to destabilize the dynamics, whereas generalist predators dominate in the south, implying more stable predator–prey dynamics (e.g., Hansson and Henttonen 1985, 1988; Hanski et al. 1991, 1993). This effect is thought to be reinforced by differences in the winter climate such that a thick and hard to penetrate snow cover in the north makes generalist predation less favorable. In the south, a thin or absent snow cover in winter puts no limits to generalist predators. Our findings here do not exclude the predation possibility, but could merely reinforce it. Here, we have chosen to be parsimonious with the only critical assumption that environmental stochasticity is more severe towards the edge of the population's range than in the center. That, together with well-substantiated assumptions about local population dynamics and dispersal between landscape elements, is enough to create a gradient of local dynamics from nonperiodic to periodic one. It is perhaps not surprising that environmental stochasticity, here modeled as random loss of landscape fragments and subsequent redistribution of refugee individuals to remaining patches, causes deterministically damped oscillations to become persistent. This finding highlights the fact that landscape structure (or spatial structure in general) may have a profound influence on local population dynamics. The properties of the landscape together with dispersal mechanisms and local dynamics can explain a suite of large-scale population phenomena ranging from region-wide synchrony of fluctuations to spatial differences in local dynamics (Ranta et al. 1995a, 1997a,b,c, 1998, 1999a,b). Here, we have reinforced that general conclusion, but also added an explanation for the geography of Fennoscandian vole and grouse cycles.

## Cycle lost

The drive to crack the secrets of cyclic dynamics has obsessed empiricists (Kalela 1962; Hansson and Henttonen 1985; Krebs et al. 1995; Korpimäki and Norrdahl 1998) as well as theoreticians (Stenseth 1977; Stenseth *et al.* 1998a,b; Hanski *et al.* 1993; Turchin and Hanski 1997). Despite many efforts, there is no consensus of the causes that keep the clockwork ticking (Stenseth 1999). Ironically, just as the mysteries of the cyclic dynamics began to yield to scientists (Stenseth 1999), the voles and woodland grouse ceased to cycle in vast areas of Fennoscandia. Prior to the mid 1980s voles in Fennoscandia fluctuated with clear 4-year periodicity over decades (Kalela 1962; Henttonen 1985). Since 1986, the regular cycle waned to mere irregular variation in a forest-covered study area (fig. 8.15(A)). The cycles have ceased also in Finnish woodland grouse (fig. 8.15(B)), which have been showing 6-year periodicity from at least the beginning of the last century (Lindström *et al.* 1995). Where have all the cycles gone?

Naturally, a number of explanations are possible. One of the most obvious is that the environment in which the populations live has changed. Apart from a slight but noticeable recent change in climate, we know that the Finnish boreal landscape has changed such that it has become more fragmented and the fraction of old forest has declined drastically due to forestry.

Here, we explore the hypothesis that habitat fragmentation may be responsible for the changes in dynamics. We assume that the population dynamics obey eq. 8.4. By selecting the parameters properly, one can easily reproduce cyclic dynamics with the period of 4, 6, and 10 years (fig. 8.16, see also previous section). The landscape is built up of habitable areas coupled by dispersing individuals. We also assume that dispersal among patches close to each other is far easier than dispersing long distances. In the increase phase of a cycle, a substantial proportion of individuals in the population is assumed to be young individuals prone to dispersal, while during the decline the proportion of young individuals is lower, resulting in a lower migration rate. We used $n$ habitable units arranged in one-dimensional space. We modeled habitat fragmentation by modifying the interpatch distances (fig. 8.16). In one end of the array, they were short, and increased towards the other end. The distance dependence of dispersal was selected so that migrating individuals were able successfully to cover short interpatch distances, but the probability of reaching the most distant ones was low. Regardless of the position of a subunit on the gradient, all units keep sending dispersing individuals. The dispersing fraction of individuals, $m$, leaving population, $i$, is taken to be a function of change in population size between subsequent time intervals. In particular, when the difference $X_t - X_{t-1}$ is positive (increasing

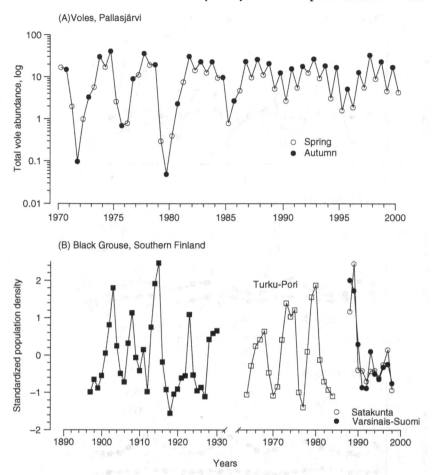

*Fig. 8.15.* (A) Long-term data on vole dynamics at Pallasjärvi (boreal forest), Northern Finland (by courtesy of Heikki Henttonen). (B) Long-term data on fluctuations of black grouse in South Western Finland (by courtesy of FGFRI). For heterogeneity in sampling methods, the time series for each period is standardized to zero mean and unit variance. Note that Turku-Pori, a Finnish province, includes both Satakunta and Varsinais-Suomi.

population size) $m = w\mu$, where $w \geq 1$. With zero or negative population change we have $m = \mu$, where $\mu$ is the base level of dispersal (here 10%). In the simulations, populations in the local habitat units were initiated with random numbers drawn from a uniform distribution (between 1 and 20). The system was left running for 2000 generations but we sampled only the final 100 generations for calculating our descriptive statistics: the coefficient of variation of the log(population size). Large values of this

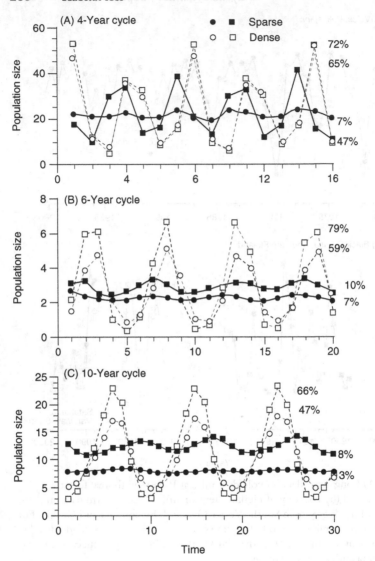

*Fig. 8.16.* Trajectories for populations obeying cyclic dynamics with different period lengths. The interpatch distances for the $n = 10$ subunits are: 1, 1, 1, 1, 2, 4, 6, 8, 10, 12 units. The open symbols are for subunits $1 = \square$ and $2 = \circ$ (at the dense end) and the closed symbols for subunits $9 = \bullet$ and $10 = \blacksquare$ (at the sparse end). Throughout $c = 0.5$, $\mu = 0.05$ and $w = 3$. Inserted values are coefficients of variation of the displayed time series, CV%. The parameter values that yield the various periodicities are: 4-year period: $r = 2.7$, $a_1 = -0.07$, $a_2 = -0.05$; 6-year period: $r = 1.45$, $a_1 = -0.1$, $a_2 = -0.45$; 10-year period: $r = 0.53$, $a_1 = -0.05$, $a_2 = -0.05$.

coefficient indicate that the periodicity of the cyclic dynamics is clearly pronounced, while small values suggest weak or absent cycles. We have experimented with a suite of different combinations of the landscape and dispersal. Therefore, we are confident that the results reported here are general, with the reservation that the system requires interconnectedness among the habitat subunits.

The modeling clearly shows that habitat fragmentation and differential dispersal in both the increase and decrease phase of the cycle reduce the amplitude and periodicity of the population dynamics (fig. 8.16). It is worth noting that with uniformly structured habitat there are no signs that the cycle of the population dynamics will disappear. In addition, our analyses show that dying-out of cyclic dynamics is not particularly sensitive to the period of the cycle, nor to the parameters controlling dispersal. The cycle die-out in the 1980s is not limited to Finland. There are observations of fluctuations becoming tamer in population numbers of field and bank voles in forest areas in northern Sweden (Hansson 1999). These are, at least partly, attributed to changes in forestry practice. Usage of large clear-cut areas became more common practice in Sweden too after the 1970s. The altered landscape structure, with associated changes in food availability and risk of predation, is the proposed reason why the northern Swedish vole cycle is losing its high peaks (Hansson 1999). We thus suggest that the documented loss of the vole and grouse cycle in forest areas in Finland may also be accounted for by habitat fragmentation.

In the previous section, we proposed an alternative explanation for presence and loss of cyclic dynamics. It is built on the assumption that the skeleton of the underlying dynamics is damped. Temporal loss of suitable areas forces residents of those areas to disperse. This, in turn, affects the density-dependence feedback so that the dynamics become cyclic. Common to both hypotheses is that changes in habitat availability can explain the observed changes in dynamics. We have also previously shown (Chapter 5, Ranta *et al.* 1997a) that a more explicit spatial structure together with dispersal coupling may cause loss of cyclic dynamics. We therefore conclude that landscape configuration, and possibly changes in it, is a strong candidate for regionwide and temporal changes in population dynamics.

## Summary

Landscape heterogeneity is a topic that draws ecologists' attention. Spatial heterogeneity can change population and community dynamics considerably. It is also important from a more applied point of view. In this chapter we reflect on the meaning of habitat loss and fragmentation for population persistence and dynamics. We show that the interaction between different patterns of fragmentation, and local and global stochasticity, have far-reaching consequences for population persistence and the co-existence between competing species. We also introduce the hitherto largely unnoticed "refugee" effect, i.e., that individuals in a landscape patch abandon it as it deteriorates and are relocated to other patches. This also has effects on overall persistence and co-existence, as well as on local and global dynamics. The loss of patches has further indirect effects, such that losses in one end of the distributional range of a species may translate into changes far away from where the loss occurs. This chapter also shows that changes in landscape structure may explain changes in large-scale dynamics, e.g., the disappearance of cyclic behavior over large geographic areas. We conclude that care must be taken when attempting to predict the population consequences of habitat loss. First, habitat loss and fragmentation require careful definition. Second, depending on where and how that loss is manifested, the resulting dynamics, persistence, and co-existence may be largely affected, or not at all. This is indeed a rich field for further studies.

# 9 · Population harvesting and management

Apart from giving us a general and fundamental understanding of dynamic processes at the population and community level, population ecology also has obvious practical connections and associations. It helps us to analyze and manage natural populations. In this chapter, we address selected aspects of population management. We first consider problems related to conservation issues, i.e., dealing with small and/or declining populations. The other side of the coin is pest management, not dealt with here however. We then focus on exploited populations, where there is generally a trade-off between the number of individuals (or amount of biomass) removed and the viability of the target population, i.e., the problem of sustainability. We do this by emphasizing the need for rigorous risk analysis procedures.

From a conservation point of view, we become concerned when the number of individuals of a species declines to low numbers, or when a particular population of a species becomes small. There is, of course, an ongoing debate about what the appropriate conservation units really are – should we, e.g., only focus on species preservation when there is so much biological diversity (genetic and phenotypic) also within a species? We take no stand in this debate here, and we boldly neglect the concern about genetic diversity. Here, it will be assumed that the population is a useful and reasonably well-defined entity, and that the population is an appropriate management unit. Depending on circumstances, a single population may represent the entire distribution range of the species, or represent a local part of a species that might be abundant, unthreatened or unexploited elsewhere. Conservation biology and harvesting theory typically have focused on single-species problems. Clearly, no species lives in isolation and both direct and indirect effects of human activities on larger parts of the community in which the focal species is embedded are now getting more attention. One such both direct and indirect community effect is the nontarget problem in fisheries, and we are going to have a closer look at the problem in this chapter.

## The problem with small numbers

As we noted in the introductory chapters, most population theory rests on the mean-field or a single-population approach. This means that we can view the population as consisting of a very large number of individuals, whose behavior can be averaged such that, e.g., population density and population growth rate are accurate descriptions about the population as a whole. However, when a population becomes sufficiently small, the mean-field approximation may break down. Chance events in births and deaths, easily averaged out when a large number of individuals are involved, may now become important. The inherent stochasticity in births and deaths (see Chapter 2, p. 26) that manifests itself when the number of individuals involved becomes small is called demographic stochasticity. It is a well-studied problem (Lande 1987, 1993, 1998; Burgman et al. 1993; Kokko and Eberhardt 1996; Engen et al. 1998; Sæther et al. 1999) and we will not dwell on it much here. Just recall the population renewal process outlined in Chapter 2. Equation 2.14 gives the probabilities of births and deaths, respectively, when they are truly stochastic variables. When the population is large, an excess of deaths over births due to chance may reduce population size considerably. Should the number of individuals be small, however, such an event may lead to extinction. It is also obvious that when the population is small a sufficiently distorted sex ratio might have irreversible consequences. A population relying on the possible reproduction of one female is far more vulnerable to chance than a population with a larger number of mature females, albeit in a population possibly dominated by males. The above birth and death problem is, of course, only relevant in a situation when the population is indeed the only extant one, or in fact isolated from other ones, and there is neither immigration nor emigration, possibly both reinforcing and mitigating the demographic chance events.

The issue of demographic stochasticity also highlights another important problem in population ecology, namely whether we should represent populations as the number of individuals or population density. The mean-field approach has no problem with the density measure. It is in fact the only operational one, since it is explicitly assumed that there should be density-dependent feedback mechanisms in the population. However, as mentioned in Chapters 2 and 8, density is in fact ambiguous and useless for populations with very few individuals. When caring about a rare species or population, it is not necessarily primarily its low density that matters. A population with high density and few individuals might be far worse off than a population with low density but with very many individuals. The Allee effect (Stephens *et al.* 1999) is also an interesting and potentially important factor at low population density, but not always when the population is small (Stephens *et al.* 1999; Petersen and Levitan 2001). A low-density population might withstand demographic stochasticity, but not Allee effects, whereas a high-density (but small) population might experience the reverse.

## Declining populations

As Caughley (1994) pointed out, conservation biology has to consider two principal problems: populations that are already small (and therefore threatened) and populations that are declining (but not yet small). In a sense, declining populations are far more problematic than populations that are already small. There are, of course, a number of both practical and theoretical problems associated with really small populations, except one: we usually are already well aware that they are small. Declining populations are more problematic because it can often be difficult to decide that a population actually is declining. Although an appropriate monitoring scheme may report that the number of individuals sampled (or some index thereof) has decreased over the years, it does not necessarily mean that the population in question is in fact under more threat than previously.

When dealing with declining populations we need to take a few steps. First, the trend has to be detected. That is, a time series of the population size has to have a significant negative slope over some relevant time interval. Second, given that a trend can be detected, it then has to be interpreted. Is this trend due to some inherent problem in the population such that the demography has changed significantly over the year? Is the trend a reflection of some larger changes in the relevant environment of the population, such as habitat alterations or climatic changes? Alternatively, is the trend actually a reflection of any problematic change in the first place? How do we ascertain that the putative trend due to some external and unwanted forces is not just part of the stochastic dynamics the population exhibits? Imagine, for example, that the population lives in a positively autocorrelated (red) environment, and that the population responds to the environment in a more or less direct manner (cf., the visibility problem in Chapter 2, p. 34), where trends are integral parts of the dynamics. Then a negative trend may not be informative at all about any immediate threats to the population. Figure 9.1 shows such a time

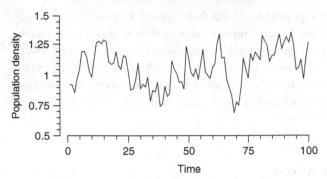

*Fig. 9.1.* A hypothetical (but very realistic) population time series. Although the series is stationary over a long time [i.e., the series is generated with a stationary AR(1) process, eq. 2.12], there are periods with both negative and positive trends, illustrating that short time series may reveal trends, but not that they are part of stationary time series with strong long-term fluctuations.

series. The series is long, but we may only have information about it in short time windows, as we usually do. This piece-wise knowledge is obviously not very informative about any long-term trends. The detection of trends also hinges on the assumption that whatever change in the environment that affects the demography of the population or population size directly (e.g., habitat alteration) is unambiguously manifested in the time series of the population. In other words, this is a practical visibility problem. Naturally, sufficiently big changes must inevitably be seen given enough time. For shorter time periods, however, the response by the population may be lagged to the extent that the detection of the underlying trend is not possible in time.

### Detecting trends and sampling – the spatial dimension

In addition to temporal aspects, there is also a spatial problem in sampling populations for the detection of changes. Sampling a resident population is usually straightforward. Naturally, one cannot avoid direct measurement errors and other sampling biases, but at least one knows what is sampled. Migratory species with large distribution ranges may be trickier. We will illustrate this with a simple example. Imagine we are interested in monitoring a species that has a rather wide distribution such that its range comprises areas experiencing different environmental variability. Assume further that we monitor the abundance of this species by sampling the population as it migrates away from its summer range to wintering areas elsewhere. This is

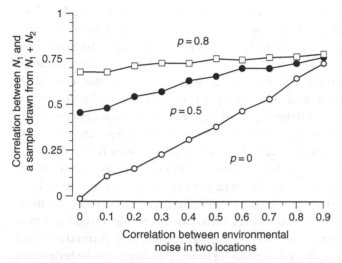

*Fig. 9.2.* An illustration of the problem of the information in sampled time series. On the *x* axis is the correlation between the environmental noise in two locations, and on the *y* axis the correlation between the population size in one of the populations, $N_1$, and the sample drawn from $N_1 + N_2$. This correlation is illustrated for three cases: when the fraction from the target population $N_1$ is 0, 0.5, and 0.8. Hence, when a sample originates from several populations (as in, e.g., migratory birds), the measured fluctuations in the sample may not reflect the changes in the environment one is interested in monitoring.

the situation for many monitoring schemes of birds that are based on samples (banded birds or counts of the visible migration). To make things a little simpler, we now assume that the total summer range of the focal species consists of two areas experiencing environmental variability that may or may not be correlated. Denote this correlation $r_E$. Suppose now that at the sampling site, a fraction *p* of all individuals originates from one of the areas, and a fraction $1 - p$ from the other. To what extent does the sample reflect the population dynamics, possibly including trends, in one of the areas? This, of course, depends on the environmental correlation $r_E$ and the fraction *p* that originates from the target population. One of the problems is that *p* is rarely known. A high correlation between the sample and the focal population can be due to two things. If the environmental correlation between areas is very high, then whatever happens to one population also happens to the other. The detected changes in the sample are true of the populations in both areas. Also, if *p* is large, then of course the sample gives a rather accurate picture of what is happening in area one (fig. 9.2). If *p* is large, then the correlation is overall high (but not perfect due to stochasticity, both measurement error and environmental variability along the migratory route). As the proportion of the sample actually originates from the population we want to keep track of, naturally the correlation vanishes if there is no correlated environment between the two areas. Increasing environmental correlation is matched by the correlation between the sample and the target population. Clearly, the correlation between the sample and the target is dependent on the balance between the proportion actually originating from the target population, and the environmental correlation between the two (or more) areas from where the sample stems. If none of those factors is known, then severe bias in our estimates of the dynamics in the target population may be introduced.

Finally, given that a trend can be ascertained, then the problem is to do something about it. We stress here again that observing a trend in a time series is quite another thing to revealing the causes of it. A trend detected by monitoring is nothing but a trailing index and there may be little room for action once the trend has started. Should we also have overcome the problem of deciding about the factual nature of the trend, one still has to make sure where it comes from.

### Decline – slow or sudden?

It is normally assumed, and data often speak in favor of the assumption, that populations on their way to becoming rare (or even becoming extinct) decline in number relatively slowly. Dynamically, the route to extinction can, however, be both complicated and unexpected.

In many population models, extinctions tend to occur from very high population densities, often from far above the deterministic equilibrium population density (or expected population density in stochastic models, e.g., Ripa and Lundberg 2000). This observation is counter to the commonly observed extinction process when in fact extinctions tend to occur from very small population sizes. This has also cast doubt on whether current models actually are appropriate tools for understanding the route to extinction in natural populations. The alternative – that our models indeed are doing a good job, but that our data are too poor for generalizations – would make us re-think much of conservation ecology theory. Ripa and Heino (1999) showed nicely where the problem lies. Models producing extinctions predominantly from very high population densities are generally set so that the endogenous dynamics are over-compensatory. Overcompensatory dynamics tends to be "blue" (see Chapter 2) and alternating from (very) high to (very) low numbers even

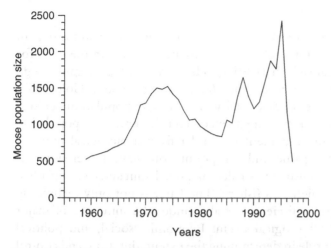

*Fig. 9.3.* The times series of moose population size on Isle Royale showing the dramatic decline over the last few years. This is a true crash! Redrawn from Peterson (1997).

in the absence of any stochasticity. Hitting sufficiently small numbers for extinction to occur is thus preceded by very high population densities. Extinction in models with undercompensatory dynamics, in contrast, would appear more gradually. Undercompensatory dynamics are inherently positively serially autocorrelated, and once at low numbers the populations tend to stay there for several time steps in a row. Any stochasticity now gets several chances to knock the population over the extinction edge. Ripa and Heino (1999) argued that strong overcompensatory dynamics are unlikely in most organisms; hence, such catastrophic extinctions from high densities are also unlikely. Hence, data, intuition, and theory are no longer in conflict. It is worth noting though that there are some remarkable examples of very swift population crashes from high numbers that resemble the previous model outcomes. The moose population on Isle Royale in Lake Superior in North America (Peterson 1997) declined ≈80% in 1995–1997 (fig. 9.3). This crash happened for a population near or rather above the expected carrying capacity. The reason for the sudden decrease is not clear, but Peterson (1997) argued that the cause might have been over-browsing on the island. Another example of possible overcompensatory mechanism working in the population dynamics is the Soay sheep (Clutton-Brock *et al.* 1997). Regardless of what the mechanism turns out to be, it is clear that dramatic catastrophes actually do occur in natural populations.

# Harvesting

The exploitation of natural populations is as long as human history. In fact, much of modern population ecology has its roots in the need for controlled or sustainable harvesting. The early studies of game (e.g., Leopold 1933) and marine fishes (e.g., Beverton and Holt 1957) initiated much of the fundamental work on basic population ecology concepts such as density dependence, recruitment, and population sampling. Although fundamentally similar from a biological point of view, the exploited game and fish populations have rather different dynamics. Hunting is usually a small-scale, small-unit and less controlled activity than, e.g., high seas fishery. The latter is not only technically advanced and highly efficient, but also under the influence of major factors outside the biological realm. Economic, social, and political forces play a major role in determining the extent, duration, and control of the harvesting activities. There is also generally a rather weak and less direct response to short-term changes in the exploited stocks, whereas hunters tend to have a more immediate and flexible response to game availability. However, both activities do share, at least in principle, the interest of sustainability, i.e., the idea that exploitation is a long-term activity that should ensure future availability of resources. Alas, the severe over-exploitation of many marine fish stocks and most of the larger whale species does not give that thought much credit (Hutchings 2000).

In pace with the increasing concern about over-exploitation particularly of marine resources, conservation issues have emerged (Reynolds et al. 2001). An appropriate tool for the management of rare, vulnerable, and exploited populations is risk analysis (e.g., Burgman et al. 1993). Risk analysis is the quantitative assessment of the probabilities of certain outcomes given a set of actions and given a set of hypotheses about the state of the system (e.g., Kokko et al. 1999). For example, we may be interested in the probability that a certain fish stock drops below some critical level. In order to evaluate the risk, we need to have some estimates of the stock size, the processes that determine stock size (i.e., both the biological population process and the harvesting activity), and the alternative actions, e.g., harvest rate (e.g., Restrepo et al. 1992). Because all this is usually associated with a fair degree of uncertainty, the risk analysis requires some statistical tools. Very useful introductions to the field are found in Hilborn and Mangel (1997), Burnham and Andersson (1998) and Wade (2001).

## Harvesting in temporally varying environments

In a stochastic world, populations fluctuate. The nature of these fluctuations has been dealt with in previous chapters. Albeit population fluctuations are interesting from a theoretical point of view, they are a nuisance from a practical one. Population size, density or age and size composition may be difficult to measure with enough precision, and the sampling inaccuracy introduces uncertainty about the very process behind the dynamics. In fact, it can be argued that stochasticity and uncertainty about the state and dynamics of populations have, together with economic and various political factors, contributed to over-exploitation of many target populations (Lande *et al.* 2001).

Given all this uncertainty, is there a harvesting strategy that is better than others in coping with the trade-off between maximum yield and the avoidance of extinction? This problem has been addressed by Lande, Engen and Sæther in a series of important articles (Lande *et al.* 1995, 1997; Sæther *et al.* 1996; see also Kaitala *et al.* 2003). Unlike most previous approaches, their work is free from the critical assumption that the populations in question have stationary population size distributions (see also Ludwig 1998). That is, the population is allowed to drift from say high numbers prior to harvesting, down to very low numbers as the exploitation rate increases, without compromising the accuracy of the statistical analyses of extinction risk. A way to handle that problem is to use diffusion approximations of the population process instead of the traditional discrete time renewal outlined in Chapter 2. Lande and colleagues also implemented four different harvesting strategies, all commonly applied in real systems and in theoretical investigations: a constant harvest independent of population size, a proportional harvest linearly dependent on population size, a threshold harvest with no harvest below a certain threshold population size and with some maximum harvest rate above that, and a proportional threshold harvest which is a combination of the latter two (fig. 9.4). In their models, Lande and his colleagues were able to show that the proportional threshold harvest resulted in the highest cumulative (over a certain time horizon) and average yield. Interestingly, this was also associated with the lowest risk of population extinction. In population management, fixed exploitation rate strategies have often been preferred over threshold harvesting (Hilborn and Walters 1992; Walters and Parma 1996; Ludwig 1998), a policy that has recently been questioned by Lande *et al.* (1997, 2001). An important argument for fixed exploitation rate strategies is that it reduces the variance of the yield

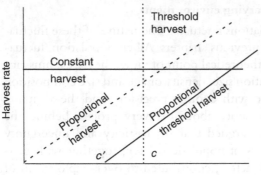

Fig. 9.4. Differing harvesting strategies as a function of the estimate of target population size. The three classical strategies are constant quota, proportional harvesting, and harvesting all after a given threshold, $c$. Harvesting in proportional threshold harvesting commences after a threshold $c'$ population size is reached.

compared to a threshold strategy (often called constant escapement in the fisheries literature).

An important aspect of work by Lande and colleagues is also how harvesting can affect the average size of the exploited population as well as influence the population dynamics. They argue that the proportional threshold harvesting strategy is a stabilizing strategy. The extent to which harvesting activity is indeed stabilizing, or not, also depends on the overall harvesting rate and the underlying deterministic dynamics of the population. One approach to this problem is illustrated in the following example from Jonzén *et al.* (2003). They used a nonlinear autoregressive model structurally modified from Royama (1992) and with an added harvesting term

$$N(t + 1) = N(t) \exp\left[1 - \frac{1}{N(t)^{a_1} N(t - 1)^{a_2}}\right][1 - H(t)]. \qquad (9.1)$$

The parameters $a_1$ and $a_2$ give the strength of density dependence at lags 1 and 2, respectively. The model is scaled such that the equilibrium without harvest is 1. Two different harvesting strategies were investigated. First, the fraction of the population harvested, $H(t)$, was assumed to be beta distributed, hence restricted to [0, 1], which must be the case regardless of the distribution of annual kills (Lauck *et al.* 1998). The mean and variance of a beta distribution are determined by two parameters, $\alpha$ and $\beta$. If $\alpha$ and $\beta$ are equal, the probability density function is symmetric around 0.5 (for further details about the beta distribution, see, e.g., Gelman *et al.* 1995).

Harvesting was taking place after reproduction and all individuals in the population were assumed equally vulnerable. In this version, all stochasticity stems from the harvest term.

The other version of the model incorporated environmental stochasticity as follows

$$N(t+1) = N(t) \exp\left[1 - \frac{1}{N(t)^{a_1}N(t-1)^{a_2}} + \sigma\varepsilon(t)\right][1 - H(t)],$$

$$(9.2)$$

where $\varepsilon(t)$ is a random normal deviate with mean zero and unit variance. The value of $\sigma$ sets the magnitude of the environmental stochasticity. Without harvesting, this model generates fluctuations with a period of 5–7 years for certain values of $a_1$ and $a_2$. Now, instead of using a beta-distributed harvest rate, the harvest rate is dependent on whether the population size before harvesting has increased or decreased compared to the previous year. This seems to be the pattern, e.g., in grouse hunting in Finland (Lindén 1991). This can be called an "adaptive" harvesting strategy, and we set $H(t) = 0.10$, 0.15 or 0.25 after an increase and set $H(t) = 0.05$ after a decline.

The effect of $a_1$ and $a_2$ on the population renewal process in the delayed Ricker model was recently treated in detail by Kaitala et al. (1996a, 1996b). The parameter values used here were taken from Lindström's (1996) estimation based on black grouse data from a Finnish province (Turku-Pori). Hence, $a_1$ was set to $-0.12$ and $a_2$ to $-0.71$. The two models were run for 800 generations and each parameter combination was repeated 1000 times. Periodicity in the time series produced was judged from the autocorrelation function.

Harvesting, now acting as an extra external disturbance to the system, clearly makes the dynamics more fluctuating (fig. 9.5). There is nothing new in suggesting that damped internal dynamics topped with noise can produce quasi-cyclic oscillations (Leslie 1959; Nisbet and Gurney 1982; Potts et al. 1984; Kaitala et al. 1996a, 1996b; Stenseth 1999), but harvesting has rarely been treated as a stochastic variable (but see Lauck et al. 1998; Patterson 1999; Mangel 2000b) and the potential role of harvesting as an external noise factor keeping periodic fluctuations has rarely been analyzed in detail (e.g., Kendall et al. 1998). It has, however, been noted that if there is a lagged response of harvest rate to population size, the effect may be periodic fluctuations of the exploited resource (Botsford et al. 1983). In fact, Botsford et al. (1983) touched upon some of the general results presented

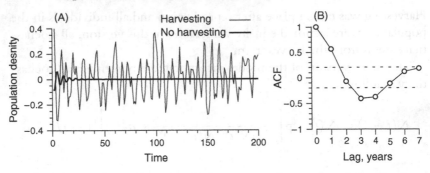

*Fig. 9.5.* The population dynamics and the corresponding autocorrelation function (ACF) of a simulated population with parameters matching Finnish grouse dynamics (approximately 6- to 7-year cyclic fluctuations) with (thin line) and without (thick line) harvesting. The figure illustrates how harvesting (as an external force) can alter the dynamics of the exploited population.

here in more detail. Also, Beddington and May (1977) and May *et al.* (1978) demonstrated that harvesting may affect population responses to environmental noise. Harvesting may be a stabilizing or destabilizing factor depending on the demography of the population.

Hence, apart from the obvious reduction in population size, the effects of harvesting on the dynamics of exploited populations may be intricate. This was further investigated in some detail by Jonzén *et al.* (2002b). They generalized the problem of harvesting effects on population dynamics by studying linearized versions of arbitrary stochastic nonlinear population models. By doing so, they were able to show that the harvesting effect on population variability hinges on three important properties of the population and the environment: the strength of the density-dependent feedback in the population, the variability in harvest rate (due to measurement errors, or bad control), and the degree of autocorrelation in the environmental variability. The results (fig. 9.6) were derived by Jonzén *et al.* (2002b) by analyzing the following modified Ricker model

$$N(t+1) = N(t)\exp[r - \beta N(t) + u(t+1)]\{1 - H[1 + w(t+1)]\},$$
(9.3)

where $\beta$ is the strength of density dependence, and $H$ is the harvest rate (proportion of the population removed each time step). Both population renewal and harvesting were assumed to be stochastic variables, $u$ and $w$, respectively. Environmental variability ($u$) was also allowed to be temporally autocorrelated, whereas harvesting variability ($w$) was white noise.

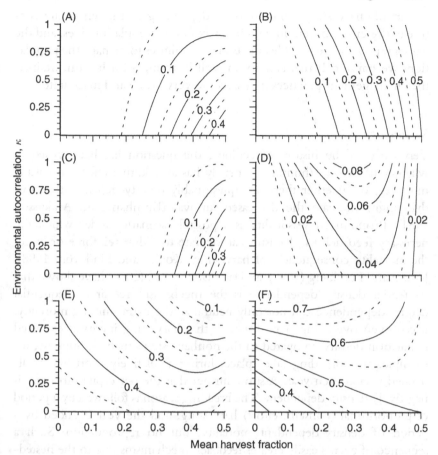

*Fig. 9.6.* The proportion of the variance in population density after harvesting ((A), (B)), before harvesting ((C), (D)) and the yield ((E), (F)) due to harvesting as a function of environmental autocorrelation ($\kappa$) and the mean harvest fraction of the population ($h$). The standard deviation was set to 0.1 for both environmental and harvest variability. Modified after Jonzén *et al.* (2002b).

   The response to harvesting depends on whether the underlying demography results in over- or undercompensatory dynamics. Thus, the variance of a population with undercompensating dynamics will increase as the autocorrelation of the environment increases. At the same time, the relative importance of variable harvest will decrease. A population with overcompensating dynamics will, however, be relatively more affected by a variable harvest in a positively autocorrelated environment than in an uncorrelated environment (fig. 9.6).

Variable harvesting can therefore – depending on the temporal structure of the environment in which the exploited population lives, and the demography – have notable effects on our abilities to manage the populations we harvest. If that activity in itself introduces further uncertainty, then management practices may become very crude and inaccurate.

## Seasonality

Very early in the history of ecology, the question has been posed of whether harvesting (hunting) generally has an additive effect on natural mortality, or whether there are (presumably density-dependent) factors that might mitigate the decreased survival (Burnham and Andersson 1984). Pronounced seasonality is a possible scenario under which the necessary feedback mechanisms can operate to either reinforce or buffer the mortality consequences of harvesting (Kokko and Lindström 1998). Jonzén and Lundberg (1999), and later Boyce *et al.* (2000) have shown that sequential density dependence is the mechanism required. Sequential density dependence can not only neutralize the extra hunting mortality, it can even overcompensate for it so that harvesting increases expected population density compared to the nonharvested situation. Suppose, for example, that breeding takes place during a relatively short and well-defined period each year. Assume further that the per capita birth rate is negatively density dependent. The breeding season is followed by a period of proportional (constant effort) harvesting, and the year is ended by a period of density-dependent mortality, but no reproduction. Such a sequence of events easily creates feedback mechanisms due to the nestedness of the density-dependent (and density-independent) processes that are capable of absorbing the harvesting mortality, or even overcompensating for it (fig. 9.7). The capacity to overcompensate for the harvesting mortality is of course dependent on the strength of the density-dependent natural processes, the actual sequencing of events, and the harvest rate that must not be too large. Although these results are not necessarily generally valid, they show that the effects of harvesting on average or expected population size are not always straightforward and intuitive (Kokko and Lindström 1998).

## Spatial heterogeneity

Despite the strong emphasis on spatial aspects in theoretical ecology, and the evidence for the importance of spatial structure in natural populations, most harvesting theory is built on the assumption of continuously

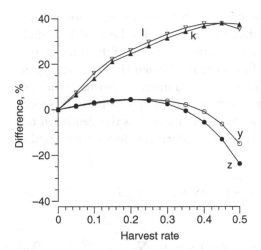

*Fig. 9.7.* The difference in mean population density (%) for different values of the harvest rate, maximum per capita birth rate ($\beta$), and for two different sequences of events, HBM and BHM (H = harvest, B = birth, and M = mortality). The key for the different lines is, z: BHM, $\beta = 2$; y: HBM, $\beta = 2$, k: HBM, $\beta = 3$, and l: BHM, $\beta = 3$. Modified after Jonzén and Lundberg (1999).

distributed populations in uniform environments (exceptions are Bisonette 1997; and short sections in Clark 1990; Hilborn and Walters 1992; Mangel 1994, 1998, 2000a, 2000b; Lauck *et al.* 1998; Quinn and Deriso 1999). There have been, however, a few recent attempts to inject harvesting theory with spatial ecology (see references in Quinn and Deriso 1999), e.g., metapopulation dynamics (Tuck and Possingham 1994, 2000; McCullough 1996; Supriatna and Possingham 1998; Cooper and Mangel 1999), source-sink dynamics (Lundberg and Jonzén 1999a; Tuck and Possingham 2000), and habitat selection theory (MacCall 1990; Lundberg and Jonzén 1999b). One may argue that the spatial aspects of harvesting theory are still premature. In practice, how-ever, spatial regulation has a long history in conservation and manage-ment of terrestrial systems (e.g., Leopold 1933) and is receiving immense interest also among contemporary scientists and managers (Joshi and Gadgil 1991; McCullough 1996). This trend towards spatial control of harvested populations as an alternative or complement to quotas and temporal restrictions is most obvious in fisheries management (e.g., Botsford *et al.* 1997; Acosta 2002; Lockwood *et al.* 2002).

Although notoriously difficult to document (Watkinson and Sutherland 1995; Diffendorfer 1998), sources and sinks are integral parts of the landscape of any species (Chapter 3). The whole idea behind the source-sink theory is that although certain habitats are very poor in terms of survival and reproduction, they may nevertheless be used by individuals of a population and are in a sense the extension of the fundamental niche of a species (Holt 1997). Lundberg and Jonzén (1999a) investigated what implications such a source-sink environment might have for harvesting. They let this be illustrated by a very simple population model where $S$ is the density in the source habitat and $N$ is the density in the sink. The rate of change in the respective habitat can now be expressed as (Lundberg and Jonzén 1999a)

$$\frac{dS}{dt} = rS\left(1 - \frac{S}{K}\right) - eS + iN - E_S S \qquad (9.4)$$

$$\frac{dN}{dt} = eS - mN - iN - E_N N. \qquad (9.5)$$

In the source, the population grows logistically where, as before, $r$ is the maximum per capita growth rate and $K$ the carrying capacity. Individuals emigrate from the source at a rate $e$ and immigrate to the sink at a rate $i$. The only net input into the sink is the number of individuals emigrating from the source. Since mortality exceeds reproduction in the sink, the population decreases intrinsically at a rate $m$. In addition, individuals leave the sink at a rate $i$. Harvesting is also introduced as a fishing effort in the source, $E_S$, and in the sink, $E_N$. Lundberg and Jonzén (1999a) showed the outcome of attempting to optimize the fishing effort in the above situation. It turns out that, under the assumptions specified above, two principal situations emerge. To maximize sustainable yield, either: (i) the sink should be harvested at an optimal effort (a value of $E$ that maximizes the product of the fishing effort, $E$, and the corresponding equilibrium population density) and the source be left alone, or (ii) if the sink is a very poor habitat and there is little back migration into the source, the sink should be harvested at maximum effort and the source at its optimal effort (fig. 9.8).

Not all spatial heterogeneity is manifested as sources and sinks. The resource matching across habitats of different qualities is, however, a ubiquitous phenomenon. In Chapter 10 this problem will be dealt with in some detail. However, we have already come across the notion of habitat selection. This process of habitat selection may have interesting and important ramifications for harvesting. One example is the possible effects of reserves on both the harvesting decisions and the exploited populations. Not least in marine fisheries, the idea of no-take areas (e.g., marine reserves) has become much emphasized (Lauck et al. 1998).

YIELD ISOCLINES

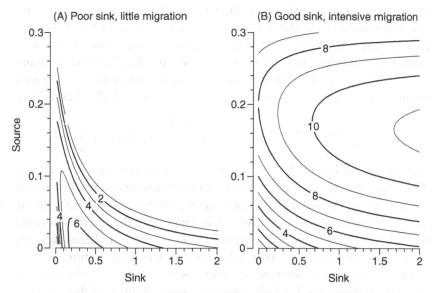

*Fig. 9.8.* The yield (the numbered isocurves) as a function of the fishing effort $E_S$ in the source ($y$ axis) and $E_N$ in the sink ($x$ axis). In (A), the sink is of low quality and the dispersal between the source and the sink habitat is limited. In (B), the sink is a relatively better habitat (still a sink) and the dispersal between the two habitats is high. Maximum yield is achieved if the sink is harvested at maximum rate and the source at an intermediate (here between 0.1 and 0.2) rate. Redrawn from Lundberg and Jonzén (1999a).

Suppose now that we distinguish between two habitats in the landscape or part of the ocean. The two habitats may or may not be inherently different. If we denote the total area $A$ and the fraction of that area set aside as a reserve $c$, then we have two habitats with area $(1 - c)A$ and $cA$, respectively. To make things simple, the habitats are characterized by only two parameters: the maximum per capita population growth rate, $\lambda_i$, and the strength of the density dependence, $a_i$. The change in population density in the two habitats can now be written as

$$X_1(t + 1) = X_1(t)\lambda_1 \exp[-a_1 X_1(t)] \tag{9.6}$$

$$X_2(t + 1) = X_2(t)\lambda_2 \exp[-a_2 X_2(t)] - EX_2(t), \tag{9.7}$$

where $X_i$ is population density in the respective habitats, and $E$ is the fishing in the area outside the reserve (Lundberg and Jonzén 1999b). Again, harvesting is assumed to be a fixed fraction (constant effort) of the population. According to the ideal free distribution (IFD) theory (Fretwell and Lucas 1970; Sutherland 1996), the per capita growth rates in the two habitats should be the same at equilibrium. That is, $X_1(t+1)/X_1(t) = X_2(t+1)/X_2(t)$. Under this condition, the equilibrium densities in the two habitats can be solved. Noting that $X_1 \equiv N_1/(cA)$ and $X_2 \equiv N_2/(1-c)A$, where $N_i$ is population *size*, the proportion of the total population that is occupying the reserve can now be calculated as follows

$$\frac{N_1^*}{N_1^* + N_2^*} = \frac{a_2 c \ln(\lambda_1)}{a_2 c \ln(\lambda_1) + a_1(1-c)\ln[\lambda_2/(1+E)]} \quad (9.8)$$

(Lundberg and Jonzén 1999b). Figure 9.9 summarizes the main results. Note that: (i) optimal fishing effort (with respect to maximum sustainable yield) does not change with the fraction set aside as reserve ($c$), or with the quality of the reserve; and (ii) the size and quality of the reserve affect the possibility of protecting a large proportion of the population. Hence, large fitness hot-spots may be needed for satisfactory protection.

The above results apply to situations when there is a cost-free and continuous flow of individuals across the reserve border. If there is a net migration in either direction, i.e., if the pure IFD does not apply, things will change. Imagine, for example, that the recruits within the reserve are partly exported to the outside and that only mature individuals are harvested. Lundberg and Jonzén (1999b) showed that under such circumstances, the fishing effort actually changes somewhat depending on the design of the reserve (fraction allocated to the reserve and its quality relative to the harvested areas, fig. 9.9). Although the habitat selection models used here may have little resemblance to real management situations they nevertheless further elucidate the problem of spatial heterogeneity in harvesting theory. This is true also for situations when habitat heterogeneity is created as a management tool (such as reserves).

Reproductive success is not, however, the only factor on which IFD may be based. An important alternative is distribution based on resource matching (Chapter 8). We shall now look, for comparison, at a modification of the model where the fish redistribute in space according to the availability of resources (Kaitala *et al.* 2004). In a homogeneous environment, this means that the population densities in the two areas will be equalized after harvesting.

After harvesting, the population densities will be given by eqs. 9.5 and 9.6, and the population sizes are

$$N_1'(t) = cAX_1(t)\lambda_1\exp[-a_1X_1(t)] \quad (9.9)$$

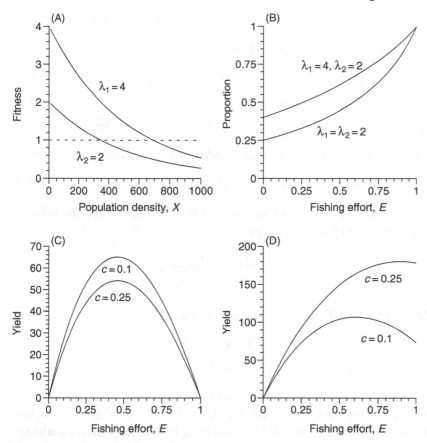

*Fig. 9.9.* A system where a fraction of the environment is a reserve ($N_1$) without harvesting, and the other ($N_2$) is open for exploitation. It is assumed that individuals are distributed across the two habitats according to either the fitness equalizing ideal free distribution (A)–(C) or the resource matching ideal free distribution (D). (A) The fitness in the absence of harvesting decreases with increasing population densities, $X = X_1 = X_2$. When the population values are such that the fitness values in each area are equal to 1, the fitness equalizing ideal free distribution is reached. (B) The proportion of the entire population found in the reserve habitat increases with increasing fishing effort outside it, but differently so depending on the relative intrinsic quality of the reserve compared to the exploited habitat. (C) The optimal fishing effort does not change with the size of the reserve (the fraction $c$, here either 10% or 25%, of the entire habitat set aside). (D) Under the assumption of the resource matching ideal free distribution, the optimal fishing effort will depend on the size, $c$, of the reserve. Panels (A)–(C) are modified after Lundberg and Jonzén (1999b).

$$N_2'(t) = (1 - c)A\{X_2(t)\lambda_2 \exp[-a_2 X_2(t)] - hX_2(t)\}. \qquad (9.10)$$

Now, after the densities are balanced, we get

$$\begin{aligned} X_1(t+1) = X_2(t+1) &= [N_1'(t) + N_2'(t)]/A \\ &= cX_1(t)\lambda_1 \exp[-a_1 X_1(t)] \qquad (9.11) \\ &\quad + (1 - c)\{X_2(t)\lambda_2 \exp[-a_2 X_2(t)] - EX_2(t)\}, \end{aligned}$$

from which we get the following condition for population equilibrium

$$1 = c\lambda_1 \exp(-a_1 X) + (1 - c) [\lambda_2 \exp(-a_2 X) - E]. \qquad (9.12)$$

For illustrative purposes we assume that $a_1 = a_2 = a$, yielding the following equilibrium population size

$$X = \frac{\ln\left[1 + \frac{(1-c)E}{c\lambda_1 + (1-c)\lambda_2}\right]}{-a}. \qquad (9.13)$$

At equilibrium, the fitnesses in area 1 and 2 are

$$\begin{aligned} &\lambda_1 \exp(-a_1 X_1), \\ &\lambda_2 \exp(-a_2 X_2) - E, \end{aligned} \qquad (9.14)$$

respectively. The equilibrium densities are denoted as $X_1$ and $X_2$. A comparison of the two IFD models shows that the yield is much higher in the resource matching IFD model than it is in the fitness equalizing IFD model (fig. 9.9(C),(D)). This appears to depend on the fact that in the resource matching IFD model the growth potential of the whole population can be better utilized than in the fitness equalizing IFD model. Moreover, in the resource matching IFD model, the maximum sustainable yield will be reached at considerably higher fishing efforts than in the fitness equalizing IFD model. Thus, for the fisheries economy, it is of paramount importance to understand the dispersal mechanisms in the fish populations. For example, for the 25% protection policy ($c = 0.25$), the resource matching IFD model would suggest that maximizing the yield would require a fishing effort of about 0.85, which in the fitness equalizing IFD fishery would eventually result in a substantial crash in the yield. Finally, we can recognize a crucial difference between the models. Increasing the area of the refuge will decrease the harvest in the fitness equalizing IFD fishery whereas increasing the area of the refuge may increase the harvest.

# Harvesting and evolution

So far, we have had a look at the two perhaps most obvious and immediate effects that harvesting can have on exploited populations. Harvesting changes population size, usually negatively, but sometimes positively, and it influences population stability. Harvesting may also change the genetic composition of a population. Selective harvesting of certain types of individuals, or just the reduction in population size, can both affect, e.g., genetic variability and selective forces. The exploitation itself may also be a strong direct selective force. Hunting is rarely random but often affects old and large males more than females (trophy hunting). Marine fisheries are generally strongly size selective both for management and market purposes (Myers and Hoenig 1997).

The evolutionary responses to harvesting have attracted surprisingly little attention (Getz and Kaitala 1993; Law 2000; 2001; Kokko *et al.* 2001; Ratner and Lande 2001; Olsen *et al.* 2004). However, it is now established that the high and selective mortality due to exploitation can cause evolutionary changes in size-dependent life history traits in fish (Law and Grey 1989; Stokes *et al.* 1993; Kaitala and Getz 1995; Heino 1998). It is reasonably easy to show that life history changes have occurred (e.g., changed age of maturity) under high exploitation rates, but it is less trivial to show that they are evolutionary responses (Rijnsdorp 1993; Heino 1998).

# Multi-species harvesting

No species lives in isolation from other species: all are embedded in a complex network of interactions at the same trophic level and across trophic levels. In a multi-species community of competitors, we have already seen the effect of targeting harvesting on the most abundant species (Chapter 7, p. 176). In such a system, harvesting can easily alter species abundance relationships (fig. 7.14, p. 177), and may also lead to extinction of some species, even those that were not targets for harvesting (fig. 7.15, p. 178). The aim of most fisheries is to capture species that are of financial value. Often these target species are associated with other species that may not be the intended catch of the fishery. However, many fishing gear are not selective enough to avoid nontargeted species.

We shall illustrate here the impact of the fishing of targeted and nontargeted species that are in a resource–consumer interaction. For our purposes, we use the discrete time resource–consumer model (Leslie and Gower 1960; Box 8.3, p. 193). For the model, we use the

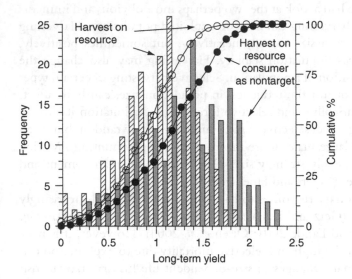

(A) Resource species as target

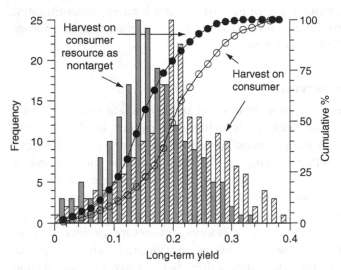

(B) Consumer species as target

*Fig. 9.10.* Long-term yield (the frequency distributions summarize results of 1000 replicated runs of the model, p. 193) of the target species is a resource–consumer interaction. Four different fishing scenarios are indicated. (A) Resource species is the target without nontarget catch (hatched bars), and with nontarget catch (darkened bars). (B) Consumer species is the target without nontarget catch (hatched bars), and with nontarget catch (darkened bars). The parameter values used for the resource consumer dynamics (Box. 8.3) are: $\lambda_R = 2$, $\lambda_C = 1.2$, $\beta = 0.1$, $\alpha_R = 0.1$, $\alpha_C = 0.5$.

parameter values given in fig. (9.10). The system is set running from the expectations of the resource population size and the consumer population size (Poole 1974, p. 157). The population dynamics of both the resource and the consumer are subject to common white noise. The system was set running for 500 time steps, and harvesting rate is 10% of population size. At the end, we scored the average yield of the target species over the previous 100 time steps. For each harvesting scenario, we repeated this process 1000 times. The following harvesting scenarios were used: (i) resource species ($R$) is the target, no nontarget, (ii) $R$ is the target, the consumer species ($C$) is the nontarget harvested at the same rate as the target, (iii) $C$ is the target, no nontarget, and (iv) $C$ is the target, $R$ is the nontarget harvested at the same rate as the target.

The results are simple: when $R$ is the target, the highest long-term yield is achievable by harvesting both the resource and the consumer species (fig. 9.10(A)). When $C$ is the target, the highest long-term yield is a result of harvesting only the consumer species (fig. 9.10(B)). The explanation is simply that when resource is harvested, harvesting acts as extra mortality, and therefore reducing consumer population via nontarget harvesting enhances the resource population. When the consumer is the target, harvesting consumers only gives them an enhanced resource level upon which to build the consumer population. Thus, when trawling, e.g., on herring (*Clupea harengus*), it does not harm the target population if cod (*Gadus morhua*) comes in as a nontarget catch. However, for maximizing cod yield, its major prey, herring, should not be trawled as nontarget.

## Summary

In this section we have shortly dealt with a few selected aspects of harvesting and resource management. In resource management, conservation issues are of high importance. Here we are usually dealing with small or declining populations, or populations that are exploited, probably on a commercial basis. Sometimes these two go together. Obviously, the means to handle the conservation issues usually vary. When working with small populations, we are worried about the declining trends, and whether these can be detected reliably. In addition, we need to take into account the effects of demographic stochasticity (Chapter 2). When working with harvested populations, we should consider our resource as a structured unit. We have shown that when addressing the harvesting problem using a spatial population ecology paradigm, many new aspects will arise. We gave examples of how harvesting may destabilize population dynamics and how stochasticity in harvesting will be diluted into population dynamics. We also gave examples of how spatial conservation reservoirs may yield different outcomes and policies depending on the ecological system with which we are dealing.

# 10 · *Resource matching*

This chapter addresses the problem of how individuals are distributed in space and time. Space is assumed to consist of areas differing in terms of profitability, some being more productive or otherwise of higher quality than others. The theory of ideal free distribution (IDF), or in more general terms resource matching, was developed to address the issue of how individuals are expected to be distributed across areas differing in availability of relevant resources. We shall first discuss resource matching in terms of distribution of foragers over their renewable resources under various circumstances. We end by extending our exploration at the level of population dynamics in areas with differing carrying capacities.

## Ideal free distribution

Ecology is the scientific exploration of the distribution of individuals and species in space and time (Krebs 1972). This is also the central theme of this chapter. We are specifically addressing the following question: how should individuals be distributed in an environment consisting of a number of habitat patches varying in resource availability? This is a question studied in the framework of the ideal free distribution (Fretwell and Lucas 1970; Fretwell 1972; or the theory on resource/habitat matching in general, Parker 1974; Morris 1994). According to the IFD theory (fig. 10.1), assuming virgin habitats, the first arriving individual should occupy the most rewarding area. From then on, its presence and activity there devalue that particular habitat patch. The next arrival should also go to a place where the highest reward can be extracted. It may be the same patch as for the first individual, or another one with lower initial rank. The decision of where to settle depends on how much the earlier individual has reduced the quality of the initially best patch. If resources per individual in the richest patch are higher than in the second best patch, the new individual should go there. Thus, the filling up of the different habitat patches will be dictated by initial resource availability and how sensitive the patch profitability and

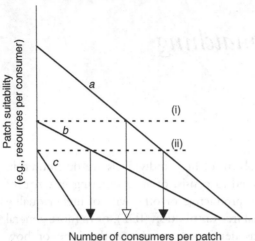

*Fig. 10.1.* The idea of the ideal free distribution. Patches *a*, *b*, and *c* differ in profitability ($a > b > c$) when no consumers are around. An increasing number of consumers devalues the profitability of the three patches. (Differing slopes indicate differences in the way increasing consumer numbers reduce the value of each patch. The functions of patch quality against consumer number do not necessarily need to be linear.) Individuals will first settle into patch *a* until level (i) is reached where the suitability of the patch *b* is reached. The number of consumers in patch *a* is indicated by the open arrow. From then on, the next individuals will gain matching benefits by going to patch *b*. With the competitor numbers still increasing point (ii) is reached (consumer numbers in patches *a* and *b* indicated with solid arrows). From this point onwards the consumers will start to use also patch *c* (modified from Fretwell and Lucas 1970).

individuals occupying it are to increased population density. Eventually, all the available patches will be filled up so that resources monopolized per individual should be the same regardless of the initial quality of the patch. The model is *ideal*, as every individual knows the value of each patch and every individual is *free* to go to the patch giving the best fitness advantage. This yields to an ideal free distribution (Fretwell and Lucas 1970).

The IFD model is an example of game theoretical approach in ecology. These models assume that the fitness benefit of an individual's active choice, e.g., whether to stay in the current patch or to move elsewhere, depends not only on the focal individual itself, but also on the actions taken by other individuals in the population (Maynard Smith 1982). The result of games of this kind often is, as with the IFD, that at one point in time a situation is reached when it does not pay anyone to move. If so, an evolutionarily stable habitat occupancy strategy, ESS (Maynard Smith 1982), is met.

The theory of resource matching predicts that individuals should be distributed across the landscape so that the percapita resource usage matches the resource distribution over all patches. The model has been a focus of intensive theoretical and empirical research (Fretwell and Lucas 1970; Milinski 1979; Sutherland 1983, 1996; Morris 1994; Tregenza 1995). In many empirical tests, animals have been given two or more choices of patches (often feeders) differing in food delivery rate. For example, Milinski (1979) used an aquarium system with two feeders and six sticklebacks of equal competitive ability. Prey input of the better feeder was twice the input rate in the poor one. The IFD expectation, two sticklebacks feeding at the poor feeder and four at the good feeder, was closely met. The main finding in this experiment, and in numerous other empirical tests, has been that the foragers distribute themselves roughly matching with the prediction of resource availability, but not quite so (Kennedy and Gray 1993).

## Distribution of unequal competitors

What happens to the IDF if we allow individuals to differ in their competitive ability? This question has been addressed by Sutherland and Parker (1985) and Parker and Sutherland (1986). The question can be specified by asking: what is the intake rate and optimal distribution of an individual with a given competitive ability across sites differing in profitability and with other individuals of varying competitive ability? Sutherland and Parker assumed a range of competitive abilities and a range of patch productivities and let individuals of varying competitive abilities settle where their intake rates were maximized. After letting the system run for long enough, the final distribution was roughly that the good competitors eventually settled down into the most productive patches, whilst the least competitive individuals were found in the leanest patches. Thus, individuals differing in competitive ability, while maximizing their food intake rates, tend to be found in habitat patches where profitability correlates with the competitive ability of the individuals. However, the answer is not that straightforward, as one might guess that even a small number of individuals in a few competitive phenotypes can be combined several different ways, and yet individuals will enjoy food intake rates matching their competitive ability (Milinski and Parker 1991). Let us illustrate emergence of such multiple IFD distributions with the following example.

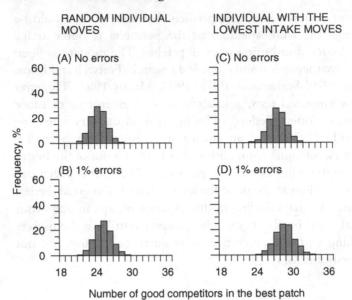

Number of good competitors in the best patch

*Fig. 10.2.* The number of good competitors in the more profitable patch. The system has two patches and the total number of good competitors is 36 (competitive weight = 2) and there are 36 poor competitors (competitive weight = 1) in the system. The frequency distributions are the outcome of 10 000 replicated runs. It is evident that in the four configurations of the model, the majority of good competitors are aggregated at the best patch. Note, however, that in every case there are a number of differing good/poor combinations yielding the IFD outcome (modified from Ruxton and Humphries 1999).

Ruxton and Humphries (1999) followed the classical IFD setting of two patches, one twice as good as the other. They assumed a population of 72 individuals in two classes, 36 good competitors (competitive weight = 2) and 36 poor competitors (competitive weight = 1). In the beginning, individuals of the two phenotypes were randomly allocated between the two patches. In their simulations, each lasting 10 000 turns, a genuine IFD movement took place, i.e., an individual that could improve its intake rate shifted from one patch to another one. There was also a constant probability for a non-IFD movement. This involved 1 out of the 72 individuals being selected and displaced from its current patch to the other one. The individual that moved was either selected randomly from among all its kinds, or it was the individual having the lowest intake rate. The insight into this is that the system has, in theoretical terms, in total 18 different ways in which an IFD could be achieved: the more productive patch could have all 36 good competitors, or 35 good ones and 2 poor competitors, or 34 good and 4 poor, and so on until finally 18 good and 36 poor competitors. The outcome of the study by Ruxton and Humphries (1999) is consistent with Sutherland and Parker's predictions. At the end of the simulations, regardless of the movement rule, most of the individuals of the better competitive rank were found in the high-quality patch (fig. 10.2). Interestingly enough, when the error rate increased from 0 to 1%, the good competitors tended to aggregate more into the better-quality patch. This was accounted for by the erroneous moves (i.e., non-IFD moves) triggering a cascade of good-quality IFD moves (Ruxton and Humphries 1999). The frequency distributions in fig. 10.2 clearly demonstrate that phenotype-matching food intake rates can be achieved with a multitude of individual combinations even in a case of two competitive abilities and two patch qualities.

## Natural ways of achieving undermatching

Most experimental results show slight biases towards the poorer areas (Kennedy and Gray 1993). This finding is known as undermatching. Several explanations have been suggested to account for this (Kennedy and Gray 1993; Milinski 1994; Tregenza 1995; Ranta et al. 1999c). We will here show why, in fact, the resource matching expectation should be undermatching rather than the perfect IFD match (Ranta et al. 1999c, 2000b).

### Limited knowledge

In the IFD theory, individuals are assumed ideal in the sense that they know the status of every patch in their environment and the number of individuals exploiting them. The matching of individuals to available resources is a result of this knowledge. A realistic modification is to assume that individuals have limited knowledge of their environment, on which they base their decisions to stay or move.

To model a system with spatially limited information of resource availability for the foraging individuals, we shall make use of a cellular automaton. Cellular automata are lattice-based models where the status of a focal cell in the next step in time is based on the status of the cell itself and the status of the neighboring cells. In the foraging context of resource matching the rule how the status changes is rather simple. A focal individual knows its food intake rate, and the food intake rate of other individuals in the neighborhood. If the focal individual's food intake rate in the current cell is less than the average intake rate in the neighborhood, it leaves the current patch; otherwise, it stays put. The individuals leaving will disperse in the neighborhood, but as they do not have knowledge of the quality of the surrounding patches, they will select the destination randomly from the neighboring patches.

In the system studied by Ranta et al. (1999c), the number of discrete food items delivered per unit time per patch is $r_i$, the environment consisting of a squared grid of $k$ cells totaling a ration of $R = \sum r_i$. A population of $N$ foragers is exploiting these resources, $n_i$ per patch at every time unit. By letting $N = R$, the expectation of the distribution of individuals among patches after resource matching follows a linear model

$$n = a + br. \tag{10.1}$$

The perfect resource matching expectation in this system is: $a = 0$ and $b = 1$. At every time unit foragers score their food intake rate relative to the intake rate of others in the neighboring cells. The neighborhood size is determined by $s$ cell rings around the focal cell. In fact, $s$ is the only parameter in this model. With $s$ matching the entire world we have the classical IFD with perfect knowledge, while with $s = 1$ only the nearest neighborhood is familiar to the foragers. If a forager's current intake rate is less than the local average, the forager moves to another cell in the neighborhood; otherwise, it stays put. The results shown below are for a $100 \times 100$ lattice initialized for $R$ drawn from uniform random numbers between 1 and 20, the distribution of foragers, $N$, was initialized by shuffling $R$ randomly in the lattice (ensuring that $N = R$). The automaton was allowed to run for 200 updating events for each value of $s$ examined.

As expected, convergence of the slope, $b$, and the intercept, $a$, towards the IFD expectation is very dependent on the neighborhood size, $s$ (fig. 10.3). With a large area of reference, IFD resource matching in forager distribution is almost – but not exactly – met, whilst with small $s$ there is no convergence whatsoever. More interesting, however, is that the model of limited knowledge yields undermatching ($a > 0$, $b < 1$): low-productive patches have more individuals than expected whereas high-productive patches have a lower number of exploiters than expected by resource availability (fig. 10.3). This is understandable: if foragers do not have any knowledge of the surroundings, each patch, regardless of its $r_i$, will harbor a random number of foragers, with the consequence that the slope $b = 0$. Undermatching, the major result observed in a large number of empirical tests of the IFD (Kennedy and Gray 1993), appears very easily to be the outcome with limited knowledge of the environment.

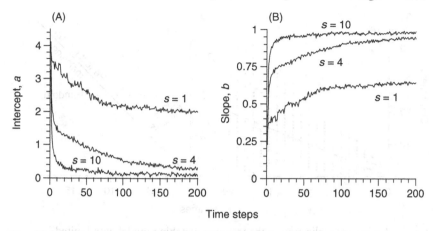

*Fig. 10.3.* Intercept *a* and slope *b* of eq. 10.1 graphed against time in cellular automata realizations with differing neighborhood sizes (1, 2, and 10 cells). The IFD expectation for the slope is 1.0 and 0 for the intercept with perfect resource matching (modified from Ranta *et al.* 1999c).

### Environmental grain

For most animals, their foraging environment consists of a network of patches. In random environments, there is no spatial autocorrelation at all, while in fine-grained systems positive autocorrelations flip to negative ones and back again with distance. With increasing grain size, the turnover rate of spatial autocorrelation slows down. Thus, the grain size refers to how rapidly the environment changes relative to the movements of the consumer (Levins 1968). Life in a coarse-grained world would make all patches look alike. Despite the fact that a consumer individual switches patches, little – if anything – will change, and the resource level remains much the same. On the other hand, movements in a fine-grained world would, move-after-move, transfer the consumer to a patch likely to be different from the previous one.

In order to study the effects of grain size on the resource matching process, we modify the cellular automaton system to form a closed loop of $k = 1000$ cells (Ranta *et al.* 2000b). As previously, an individual knows its own food intake rate and the average food intake rate of other consumers in the neighborhood of size *s*. The decision to stay or to move is based on this information. Those leaving a patch will disperse to the surrounding *s* patches. The parameter *s* measures how well the consumers know the resource environment. In our simulations, we let *s* range from 0.2 to 30% of the size of the automaton. For the grain size *G*, we used the

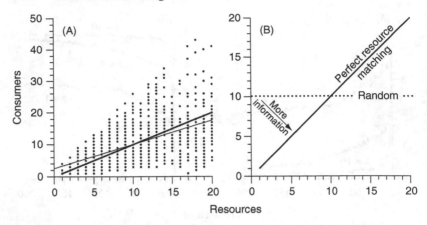

*Fig. 10.4.* (A) One realization of the resource matching model with limited knowledge ($s = 2$). The thick line gives the realized relationship between consumer numbers and resource availability. The thin line is the perfect match expectation. (B) With no knowledge of resource availability consumers will distribute randomly across the food patches irrespective of resource availability in them. With increasing information, the dotted line starts to turn counter clockwise until the perfect resource match is met (modified from Ranta *et al.* 1999c).

following: random, $G = 2, 5, 10, 20$, and $50$ with $r_i$ ranging from 1 to 20. The value of $G$ indicates how many subsequent cells it takes to travel from a top-productive patch to another top-productive patch (fig. 10.4). The automaton was left to run for 500 updating events for each value of $s$ and the resource grain $G$. At the end of the runs we estimated $a$ and $b$ in eq. 10.1. For every $s$ and the resource grain combination the procedure was repeated 100 times. Spatial autocorrelations of consumer distribution across the entire landscape were calculated for various grain sizes. In the random world, spatial autocorrelation was virtually nonexistent, while in the world with the alternate patch sizes correlations flipped in each pair of cells from −1 to 1. Increasing the grain size even further makes the spatial autocorrelation functions smooth (fig. 10.5). That is, by definition, an increase in grain size improves the predictability of the resource world: what a forager finds in the current patch is likely to be what it will find in the neighborhood too. Thus, even with limited information the consumers know a lot of the structure of their feeding environment when grain size is large.

However, large grain size produces a good match between resource and consumer distribution only if knowledge of the neighborhood size is also very large (fig. 10.5). Generally, the smaller the grain size, the more

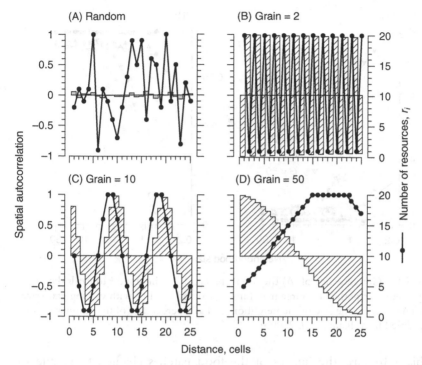

*Fig. 10.5.* Spatial autocorrelation functions (shaded areas, scale on the left $y$ axis) against increasing distance in four worlds differing in terms of resource-patch grain structure (from random patches to grain size of $G=50$ cells). The overlays (thick lines and dots) are examples of the numbers of resources in neighboring patches (scale on the right $y$ axis). The sample of 25 consecutive cells is picked from a random location of the cellular automaton (modified from Ranta *et al.* 2000b).

accurately the consumers can match their resource distribution, and the values of the intercept $a$ and slope $b$ will approach the IFD expectations (Ranta *et al.* 2000b). In environments with various structures the consumer distribution almost matched that of the resources only in the $G=2$ world (fig. 10.5). Note that the IFD expectation will be met with all grain sizes when $s$ matches the landscape size.

### Noisy world

In the preceding section we assumed that space may be structured but not varying in time (other than by the actions of the individuals in the population). It is also quite likely that the information the consumers acquire while harvesting resources in the patches is slightly erroneous.

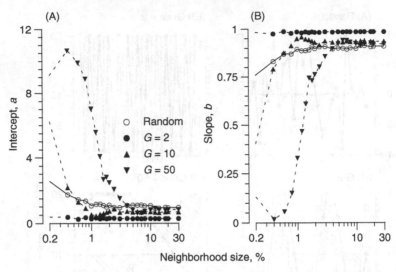

*Fig. 10.6.* Convergence of (A) the intercept $a$ and (B) the slope $b$ towards the expectation of perfect resource matching ($a = 0$ and $b = 1.0$) with varying grain size, $G$. The $x$ axis is the size of the neighborhood as a percentage of the automaton size (modified from Ranta *et al.* 2000b).

This is because the quality of the food patches changes (e.g., due to unpredictable resource renewal, or local hazards) without the foragers immediately knowing it. The question is hence to what extent the IFD model can be used as the ESS expectation when explaining the distribution and abundance of consumers relative to their resources when the information is wrong.

For those purposes, we shall use the closed loop version of the cellular automaton ($k = 2000$). Again, the number of discrete food items delivered per unit time per patch $r_i$ is drawn from uniform random numbers between 1 and 20. Also, as previously, the distribution of consumers, $c_i$, was initialized by shuffling $r_i$ randomly in the loop. By letting $C = R$, the expectation after resource matching under IFD follows eq. 10.1 with the expectations of $a = 0$ and $b = 1$. The parameter $s$ is a measure of how well the consumers know the extant resource environment. On top of this, we shall implement environmental noise as follows. For each cycle a given proportion of the cells, $w$, will be randomly selected from the $k$ cells to swap their contents. This makes the local resource level change unexpectedly, but keeps the overall level of resources $R$ constant in the system. Both $s$ and $w$ were selected so that the entire range from 1% to 30% of the entire landscape was covered. For increasing $s$, the foragers approach

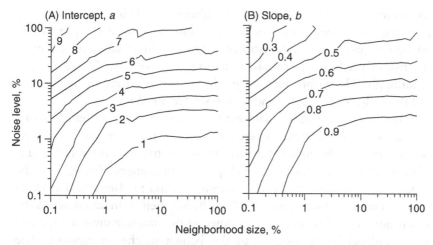

Fig. 10.7. (A) Intercept $a$ and (B) slope $b$ isoclines graphed against the neighborhood size ($s$ as a percentage of $k = 2000$) and noise level ($w$ as a percentage). The expectation of perfect resource matching is $a = 0$ and $b = 1.0$.

perfect knowledge; for increasing $w$, the disturbance level increases from a silent world towards complete disorder. The automaton updating rules are as previously and 500 updating events run for each value of $s$ and $w$ examined. At the end of the runs we estimated $a$ and $b$ and for every $s$ and $w$ combination the procedure was repeated 100 times.

The results are straightforward: the perfect match between resource availability and consumer numbers is hard to achieve in a noisy world (fig. 10.6). With increasing noise level the estimate for the slope $b$ rapidly degrades from values close to 1.0, and the intercept $a$ starts to deviate increasingly from zero. It is interesting to note that increasing neighborhood size $s$ interferes with the noise level (bending isoclines in fig. 10.7). However, even with a perfect knowledge of the resource environment, the increasing noise soon takes the $b$ and $a$ estimates far out of their IFD expectations.

## Resource matching and trophic interactions

Here, we will extend the IFD problem by studying consumer–resource matching in a community context (Jackson *et al.* 2004). Let the community in focus be a linear food chain with three trophic levels: sessile resources ($R$), which are preyed upon by consumers ($C$), who are themselves prey for top predators ($P$). The question of interest is whether the consumers will be able to match their resources with or without considering the risk of being preyed upon by their predators. The predators have to chase a prey which attempts to maximize both resource acquisition (reproduction) and avoid the risk of predation (survival) by moving and selecting habitat patches accordingly. The mobile prey and predators may then have varying degrees of knowledge of the global and local resource distribution.

The three-species food chain exists in an environment consisting of a closed loop of $k$ identical discrete patches. The resource level, $r_i$, in each patch is drawn independently from a uniform distribution between 0 and 1. The patch-specific resource level persists unchanged throughout the simulation. The total numbers of consumers and predators in the whole system remain constant (also $R = C = P$), but their distribution across the environment is liable to vary over time. At the start of a simulation, $c_i$ and $p_i$ are independently assigned to the habitat patches at random. The simulation then consists of a fixed number of movement events.

The suitability of the current patch of location for predators is their intake rate – the number of consumers divided by the number of predators, $c_i/p_i$. They assess the consequences of moving by calculating the mean of the intake rate of all individuals in their current patch and in all the patches up to some constant number of positions ($w_p$) either side of the current patch. If this mean $\mu$ is greater than $c_i/p_i$, then the individual will move with probability

$$P(\text{move}) = \frac{\mu - c_i/p_i}{\mu}. \tag{10.2}$$

The greater the disparity between the situation in the current patch and how well others are doing in the neighborhood, the more likely an individual is to move. If it moves, then its new patch is determined randomly from the subset of patches (including its current one) it used to estimate the performance of others.

Movement works in an analogous fashion for consumers, with the difference that the parameter $w_c$ now describes the size of consumers' sampling window. Consumers will also use different measures of habitat suitability. They are referred to as "resource maximizers," when considering just the availability of resource for them, $r_i/c_i$. The consumers are "risk averse," when they aim to minimize their risk of predation hazard, $p_i/c_i$, per unit of resource that they can gain access to, $r_i/c_i$. This simplifies to minimizing local predator number divided by local resource level, which is equivalent to maximizing local resource level divided by local predator number.

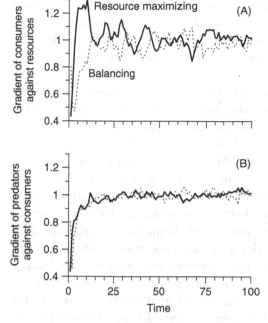

*Fig. 10.8.* Plots of the gradients of (A) consumers against resources and (B) predators against consumers for both the resource-maximizing and risk-averse models. Parameter values: $C = 600$, $P = 600$, $L = 23$, $w_c = 11$, $w_p = 11$ (modified from Jackson *et al.* 2004).

As earlier (p. 242), the measure of resource sharing match is, for consumers and predators, respectively

$$c = a_C + b_C r$$
$$p = a_P + b_P c,$$
(10.3)

where $a_C$, $b_C$, $a_P$ and $b_P$ are parameters of a regression model. Because we have $R = C = P$, it also follows that the IFD expectations are $a_C = a_P = 0$, and therefore we fit these regression lines using a reduced major axis (Fowler and Cohen 1990). A gradient ($b_C$ or $b_P$) of 1 indicates perfect IFD matching; <1 indicates undermatching; and >1, overmatching. The system was left running for 500 updating rounds and we calculated means and standard deviations over the last 100.

First, the information and movement windows for both consumers and predators will be assumed identical and equal for the whole of the system. The expectation is that the resource-maximizing model would lead to perfect matching of consumers to resources and predators to consumers, and this is the finding (fig. 10.8). We would expect perfect matching to be achieved in the risk-averse model too. Note that in this model consumers seek to maximize $r_i/p_i$ rather than $r_i/c_i$, so we would expect the predator population distribution to mimic that of the consumers; thus, the two currencies for the consumers will be equivalent. We again find perfect matching (fig. 10.8).

We now explore the consequences for the risk-averse behavior by reducing the size of the information and movement windows. The effect on the resulting distribution of keeping the consumers' window large and reducing the predators' window is shown in fig. 10.9(A). For medium-sized windows, predators are still able to track the consumer population effectively, and we see perfect resource matching. Reducing the window size leads ultimately to limited knowledge and mobility. It follows that the predators generally undermatch the consumers. Under these circumstances patches with high resource levels harbor fewer predators than expected due to perfect matching to their consumer numbers. In turn, this encourages more consumers to aggregate into these patches, leading to overmatching of consumers on resources.

The reverse situation (a large window for predators but a small one for consumers) leads to a rather different pattern. Limited knowledge and mobility tend to produce overmatching to resources, and high movement rates. These high movement rates make it hard for predators to perfectly match consumers. This undermatching of predators on consumers tends to induce overmatching of consumers on resources. The different and opposing pressures on consumers often interact, yielding very good matching even when the consumers' window is very small (fig. 10.9(B)). However, this is not always the case, and in extreme situations there generally is overmatching of the consumers to resources and undermatching of predators to consumers (fig. 10.9(C)).

The results above hold when the resource levels on neighboring patches are not correlated. We shall now introduce spatial autocorrelation into resource availability by setting the resource level in patch $i$ to be

$$r_i = \frac{1 + \sin(2\pi i/\tau)}{2}. \tag{10.4}$$

The larger some positive constant $\tau$, the more gradually resource levels change in space (see fig. 10.9). As one would expect, when $\tau$ is low, the situation with overmatching of consumers to resources and undermatching of predators to consumers is recovered. However, with increasing $\tau$, the limitations of the consumers' small window size become more and

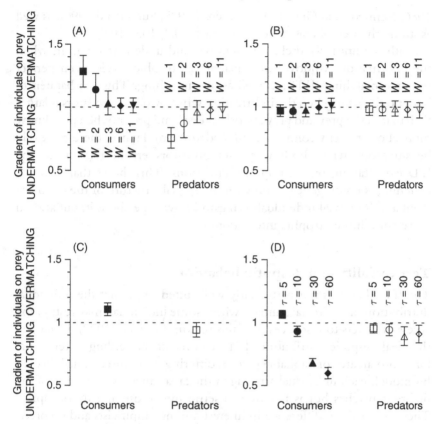

*Fig. 10.9.* Plots of the gradients of consumers against resources (solid symbols) and predators against consumers (open symbols) for (A) several values of $w_p$ (parameter values: $C = 600$, $P = 600$, $k = 23$, $w_c = 11$) and (B) of $w_c$ ($C = 600$, $P = 600$, $k = 23$, $w_p = 11$). Panel (C) gives the gradients of consumers against resources and predators against consumers for a very low value of $w_c$ ($C = 600$, $P = 600$, $k = 60$, $w_p = 29$, $w_c = 1$). In panel (D) the consumer and predator gradients are given for spatially structured resource distribution ($C = 600$, $P = 600$, $k = 60$, $w_p = 29$, $w_c = 1$; modified from Jackson *et al.* 2004).

more pronounced, leading to stronger and stronger undermatching of consumers to resources (fig. 10.9(D)).

These results confirm the already known effects of limited knowledge. If the consumers have poor estimates of the global resource distribution and/or limited possibilities to realize that knowledge through constrained movement among patches, then the general consequence is undermatching. There is ample empirical evidence for that conclusion (e.g., Abrahams

1986; Kennedy and Gray 1993; Tregenza 1995; Sutherland 1996), as well as strong theoretical reasons for it (see figs. 10.3, 10.6–10.9).

With the linear food chain system we find undermatching, or, when the information windows are equal across trophic levels, even perfect resource matching (IFD), as well as overmatching. This can happen to both predators $p$ (when the information window of the predators is larger than that of the prey and prey are risk-averse) and prey $c$ (when prey know more about the environment than predators do). Hence, there can easily be situations when both prey and predators err in relation to the IFD expectation, but in opposite directions. This shows that resource matching is contingent not only on the spatial structure of the environment and how well individuals can gain knowledge about it, but also on more complicated trophic interactions.

## Territoriality and despotic behavior

The IFD theory is not particularly well suited to predict the individual distribution of territorial animals, where some individuals may restrict the access of others to resources. For this reason, Fretwell (1972) developed the ideal despotic distribution, IDD, to predict the settling of territorial birds into an area of habitable patches differing in resource availability. In his model, each individual arriving in the area can assess the value of the different patches but is not free to settle into those already occupied. There is a striking difference in interest among empiricists and theoreticians between the two theories. IFD has been a focus of intense research, and numerous variations of the basic theme have emerged. Most of the IFD modifications are specifically relaxing the two central aspects (*ideal* and *free*) of the theory (Milinski and Parker 1991). This is particularly interesting as despotism, or interference in any other form, is common in many natural situations (e.g., Huntingford and Turner 1987).

Against this background, it is interesting to note (Tregenza 1995) that in the basic form of the IDD the only prediction not common to the IFD is that territory ownership will lead to the differential success of otherwise equal competitors. Another reason why IDD has remained largely undeveloped is that the outcome it predicts is intuitively obvious and hence regarded as naïve. Consider, e.g., a number of habitats ranked in resource availability and a number of individuals (making the population) also being ranked in terms of competitive ability. The IDD theory predicts linear matching between the two rankings. This simple model is shown to fit distribution of salmonids occupying territories in rivulet pools with better access to drift food in upstream pools than in pools of the lower end (Hughes 1992; Nakano 1995). However, not all habitats housing territorial animals suit this scenario (see also Box 10.1 below on related individuals).

### IDD in variable environments

Ruxton *et al.* (1999) used an individual-based model to address IDD in a two-dimensional world with variation in territory quality. Their world is composed of a lattice, where cells refer to the territories. Each territory has its own resource production rate that is harvestable only by the top-ranked individual in that particular location. The reward rate of each territory is a function of its resource production and the relative rankings of the other individuals in that territory. After each round of environmental change, each of the $N$ individuals is considered in turn (in random order). If the intake rate of an individual at the current territory is higher than its expectation it stays put, otherwise it moves to a randomly chosen neighboring territory. If the focal individual cannot improve its food intake in any of the adjacent territories, it stays put. The model has an implicit assumption that individuals cannot estimate the resource production rate of the territories but that they can detect the competitive rank of individuals seated in any of the neighboring territories. The movement rules in the model (Ruxton *et al.* 1999) have similarities to the biased diffusion model used by Farnsworth and Beecham (1997).

In the simulations, after each individual has been allowed to make one environmentally triggered move, each receives a reward, $F$, which is either equal to the resource production rate in the current territory $E$ (provided it is the top-ranked individual there) or zero (otherwise). Each individual then updates its expected foraging return $E_{i(NEW)} = \alpha F + (1 - \alpha)E_{i(OLD)}$. Here $\alpha$ is a constant ($0 < \alpha < 1$); the higher it is, the more quickly the individual discounts previous experience. With each parameter combination (Ruxton *et al.* 1999, p. 115) the model was simulated for 10 000 time steps and both intake rates and movement rates were assessed for the last 5000 steps.

The model generated three main predictions. When the number of foragers exceeds the number of territories, there is a sigmoid relationship between dominance rank and resource gain (fig. 10.10(A)). Middle-ranking individuals in the competitive hierarchy are more mobile than those with higher or with lower rank (fig. 10.10(B)). Finally, the geometry of the landscape where the territories are located has a major effect on resource gain and movement patterns only when the array reduces into one dimension of a single territory wide (fig. 10.10(A),(B)). The

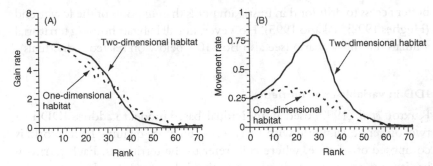

*Fig. 10.10.* Mean gain rate (A) and mean movement rate (B) in the IDD model by Ruxton *et al.* (1999). The outcome is not as sensitive to manipulation of competitor density as it is to habitat shape. Especially in terms of movements, the territorial foragers behaved differently when territories were aligned into single row (simplified from Ruxton *et al.* 1999).

territory occupancy system is driven by three fundamental elements. An individual moves, or attempts to move, when its intake rate falls below the average it can remember having experienced. An individual cannot move into a territory used by a more dominant individual, and when subordinate individuals are forced to share space with a dominant, they gain no resource. Ruxton and his co-authors suggested that the key predictions of the model (fig. 10.10) are rather robust and generally observable in nature. Hence, the predictions serve as a valid discriminator between alternative explanations of the IDD. In addition, these results should encourage empirical research into IDD, as they are so clear-cut.

## Population structure and resource matching

Almost all IFD theory rests on the assumption of individual choices. For many organisms, this is not necessarily an appropriate premise. Morris and his associates (Morris *et al.* 2001) suggested that among relatives there is no free choice. They based the argument on the theory of kin selection (Hamilton 1964a, 1964b), assuming purely selfish decisions to be less favored than those maximizing inclusive fitness. The core of the argument by Morris *et al.* (2001, p. 921) is that, "Individuals maximizing inclusive fitness may thus distribute themselves so as to over-exploit habitats where each individual has little negative effect on fitness (low fitness loss) while their relatives under-exploit habitat where each individual has a large effect (high fitness gain)." The inclusive fitness-based habitat selection model is given in Box 10.1.

**Box 10.1.** · *An inclusive fitness theory of habitat selection*

After Morris *et al.* (2001), let us assume that the per capita population growth rate in habitats 1 and 2 is a function of population size. Individuals in the population are identically related to each other. The inclusive fitness $I$ of an individual occupying either of the two habitats is

$$I_1 = f_1(N_1) + r(N_1 - 1)f_1(N_1) + rN_2 f_2(N_2)$$
$$I_2 = f_2(N_2) + r(N_2 - 1)f_2(N_2) + rN_1 f_1(N_1),$$

where $f_i(N_i)$ is the density-dependent fitness function in habitat $i$ and $r$ is the coefficient of relatedness. With a further assumption that the costs of movement are negligible, the decision functions of departure from one habitat to the other one are

$$B_1 = I_2(N_1 - 1, N_2 + 1) - I_1(N_1, N_2)$$
$$B_2 = I_1(N_1 - 1, N_2 + 1) - I_2(N_1, N_2),$$

respectively. The term $B_1$ is the inclusive fitness of an individual if it moves and $B_2$ is its inclusive fitness if it stays, and an individual should move whenever $B_i > 0$. The decision function for an individual in habitat 1 can be approximated by

$$B_1 = I_2 - I_1 - \frac{\partial I_2}{\partial N_1} + \frac{\partial I_1}{\partial N_2},$$

which, after substitution, becomes

$$B_1 = f_2 - f_1 + r(N_2 f_2') + f_2'(1 - r),$$

(note that the arguments of the functions $f_i$ are dropped, and the $N_i f_i'$ is the change in fitness in a habitat with changes in population size). The final term will be very small relative to the others, and, for any reasonable value of $N$, will be a small fraction of fitness. Let $g(N_1, N_2)$ represent the remaining terms

$$g(N_1, N_2) = f_2(N_2) - f_1(N_1)$$
$$+ r[N_2 f_2'(N_2) - N_1 f_1'(N_1)].$$

After performing the same procedure for $B_2$, Morris *et al.* (2001) arrived at the decision rule for an individual tempting to move as $g(N_1, N_2) > 0$ in habitat 1 and $g(N_1, N_2) < 0$ in habitat 2. If all individuals are unrelated, $r = 0$, then these conditions reduce to comparison between $f_1$ and $f_2$ such that if $f_1 = f_2$ no migration is expected and one has the classical IFD.

We can illustrate the result derived in Box 10.1 by letting the fitness of an individual decline with increasing density of individuals in habitat $i$ as

$$f_i = a_i - b_i N_i \qquad (10.5)$$

($a_i$ is the maximum fitness achievable and $b_i$ describes how it decreases with an increasing number of individuals in that particular habitat). When solving $f_1 = f_2$ for $N_2$ we have

$$N_2 = \frac{a_2 - a_1}{b_2} + \frac{b_1}{b_2} N_1. \qquad (10.6)$$

This defines the isodar (Morris 1988), the set of population densities in both habitats yielding IFD for unrelated individuals. Along the isodar, there is no net migration between the two habitats. When individuals in the population are related, the above equation holds for the per capita growth rate but not for the fitness (Morris *et al.* 2001). Now the condition for no migration is satisfied when $g = 0$ (Box 10.1) yields the isodar

$$N_2 = \frac{a_2 - a_1}{b_2(1 + r)} + \frac{b_1}{b_2} N_1. \qquad (10.7)$$

The conclusion is simply that the equal per capita growth isodar does not coincide with the zero-migration isodar if $r > 0$ (fig. 10.11(A)). The combinations of population densities in the two habitats giving no net population growth (when $r > 0$), i.e., $N_1 f_1 + N_2 f_2 = 0$, is (Morris *et al.* 2001)

$$N_2 = \frac{a_2}{2b_2} + \sqrt{\left(\frac{a_2}{2b_2}\right)^2 + N_1(a_1 - b_1 N_1)}. \qquad (10.8)$$

This function intersects the $N_2$ axis at the equilibrium density for that habitat, $N_2^* = a_2/b_2$, and the per capita growth-rate isodar at the joint population equilibrium in both habitats ($a_1/b_1$, $a_2/b_2$), and it intersects the zero-migration isodar (fig. 10.11(B)).

These results have interesting implications. When $r > 0$, each equilibrium occurs at a different combination of population densities. There is no simple ESS (fig. 10.11) because the system cannot settle on a single pair of $N_1$ and $N_2$. In systems with fast migration relative to population growth, the pattern of habitat occupancy will move towards the zero-migration isodar. In the reverse case, the system settles towards the IFD

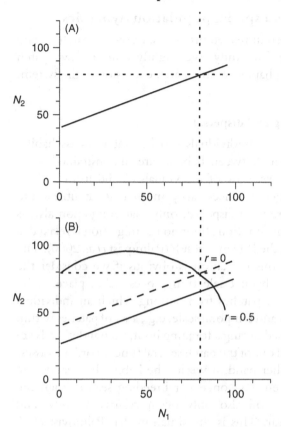

Fig. 10.11. (A) An isodar (solid line) corresponding to ideal free habitat selection when fitness declines linearly with an increase in density $N$ in habitats 1 and 2. Zero population growth rate occurs when the population density in habitats 1 and 2 is at their $K$ (dashed lines, $a_1/b_1$, $a_2/b_2$). (B) An example of the inclusive fitness "no migration" isodar for the two habitats when individuals are related ($r = 0.5$; the IFD isodar $r = 0$, long dashes, is shown for comparison). An unstable equilibrium occurs at the intersection with the arched zero-total-growth isodar ($N_1 = 87.65$; $N_2 = 70.40$). Migration attempts to maintain the equilibrium, which is opposed by positive population growth in habitat 2 and negative population growth in habitat 1 ($a_1 = 80$, $a_2 = 60$, $b_1 = 1$, $b_2 = 2$). Modified from Morris et al. (2001).

equilibrium. The result with $r > 0$ is that the resources across will be mismatched against the IFD expectation (fig. 10.11). Thus, relatedness among habitat selectors adds one more explanation to the list of factors yielding mismatch between the theoretical expectations and empirical observations (Kennedy and Gray 1993).

## Resource matching and spatial population dynamics

Here we combine two often unrelated processes, resource matching and population dynamics. We first study two slightly differing two-patch systems and will finish this chapter by an $n$ patch extension of the system.

### Linking resource matching and dispersal

The IFD theory assumes that individuals can accurately assess habitat quality and that movements between habitats are unconstrained. As a result, there should be no *net* flow of individuals to and from a given habitat. Note that this does not necessarily mean that (at fitness and dynamic equilibrium) there is *no* dispersal, only that emigration always equals immigration. If the assumption about no net migration is relaxed, a number of deviations from the IFD may result (Palmqvist *et al.* 2000). The biological justification for this relaxation is obvious if we consider the spatial scales at which the habitat selection process takes place. If by "habitats" we mean a set of patches from among which an individual can choose at a short spatial and temporal scale, e.g., a set of patches within a homerange, or patches used during a foraging bout, then individuals are (potentially) moving in and out of the patch several times in order to assess habitat quality. On the other hand, if we let the habitat be a landscape element that is only chosen, say, once per breeding season, then the migration among habitats will also only occur relatively rarely, and at distinct times of the year. This is the situation that Palmqvist *et al.* (2000) studied. They let the population renewal process in habitat 1 be described by

$$N_1(t+1) = N_1(t) - E_1(t) + E_2(t)$$
$$+ r_1[N_1(t) - E_1(t) + E_2(t)]\left[1 - \frac{N_1(t) - E_1(t) + E_2(t)}{K_1}\right].$$

$$(10.9)$$

Note that this is a sequential process within a year (from time $t$ to $t+1$). $E_i$ are the migration rates, where $E_1$ is total migration from habitat 1 to habitat 2 and $E_2$ is migration from habitat 2 to habitat 1. Hence, in this model migration takes place before population renewal. Equation 10.9 also tells us that dynamic equilibria in both habitats are functions of $E_1$ and $E_2$. If net migration, i.e., $|E^*_1 - E^*_2|$ (* indicating equilibrium rates), is greater than zero, both post- and pre-reproductive population equilibria will differ from the IFD solution. That is, the result is over- or

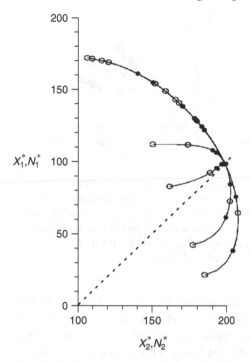

*Fig. 10.12.* Positions of stable pre- and post-reproductive equilibria, in the $N_1 - N_2$ and $N_1 - N_2$ plane, respectively, for a two-habitat system with unequal carrying capacities $K_1 = 200$, $K_2 = 100$. Population equilibria are shown for systems with different $r$ but with equal differences in $K_i$ ($K_1 = 200$, $K_2 = 100$ and $m$ ranges from 0 to 0.6 for all curves). Broken straight line is the isodar. The uppermost curve shows the positions of pre-reproductive population equilibria ($n_1^*, n_2^*$) for the two habitats, for all $r$ ($r = 0.6$, 0.9, 1.3, and 1.5). The lower curves show the position of post-reproductive equilibria ($N_1^*, N_2^*$) for different $r$ ($r$ increasing downwards). Dots indicate equilibria from systems with density-dependent fractions of emigrants and circles represent equilibria from systems with constant fractions of emigrants. For each emigration type, the symbols from ($K_1$, $K_2$) and outwards, represent equilibria resulting from $m = 0$, 0.3 and 0.6. (modified from Palmqvist *et al.* 2000).

undermatching. Those general results are robust to any assumptions about the dispersal rules (Palmqvist *et al.* 2000), i.e., what determines the number of individuals leaving a habitat each year.

Since the model is explicit about the timing of events within a year, both pre- and post-reproductive population equilibria are shown in relation to the IFD isodar for the system (fig. 10.12). The figure also shows that the deviations from the IFD solution are affected by the maximum per capita growth rates in the two habitats. The problem of establishing the expected

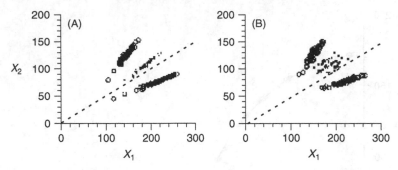

*Fig. 10.13.* Distribution of population sizes in a two-habitat system with environmental stochasticity (global stochasticity in (A), local stochasticity in (B)) affecting habitat-specific carrying capacities. The broken lines are the isodars. Each panel displays results for three systems with a constant fraction of emigrants (from above: $m_1 = m_2 = 0.4$, $m_1 = m_2 = 0.1$, and $m_1 = 0.2$, $m_2 = 0.6$), always $K_1 = 200$, $K_2 = 100$ and $r = 1$ (modified from Palmqvist *et al.* 2000).

resource matching when there is net migration between habitats is further illustrated in fig. 10.13. In a stochastic environment, whether the stochasticity is local or global, the point equilibria are now stretched out along straight lines (or at least close to straight lines). Those lines are the *realized* isodars, all with slopes and intercepts different from the null expectation with no net migration. Hence, we could easily arrive at "erroneous" conclusions about the habitat selection process should we rely on population estimates in the two habitats only. This should be taken as a caveat for overinterpreting isodars thus derived. As is often the case, patterns only rarely reveal unambiguous information about underlying processes.

## Population dynamics and IFD dispersal

We shall continue with a two-patch system. For simplicity of our argument, the two patches are taken to be equal ($K_1 = K_2 = 1$), and they are in close enough proximity to cause no mortality or fitness costs for dispersing individuals. Individuals in the population are assumed ideal by knowing the status both of patches in the environment and the size of the populations ($N_1$ and $N_2$) exploiting resources in them. A perfect match with resource availability and population size is met with (Pulliam and Caraco 1984; Fagen 1987; Kennedy and Gray 1993; Morris 1994)

$$\frac{K_1}{N_1} = \frac{K_2}{N_2}.$$

(10.10)

In order to build up a patch-departure rule obeying the IFD theory we shall take the departure rule for leaving to be

$$\Delta_i = N_i - \bar{N}$$

$$\begin{cases} \text{if } \Delta_i > 0 \text{ then move} \\ \text{otherwise stay put.} \end{cases} \qquad (10.11)$$

The individuals dispersing, $\Delta_i$, are a random sample of the $N_i$ individuals resident in the patch $i$ and where $\bar{N}$ is the average. For population renewal, we used the Ricker model

$$N_i'(t+1) = N_i(t) \exp\left\{ r\left[ 1 - \frac{N_i(t)}{K_i} \right] \right\}. \qquad (10.12)$$

Here $N'$ refers to population size before any IFD-based dispersal adjustments have taken place and $r$ is the maximum per capita growth rate. This is the system explored by Ranta and Kaitala (2000).

Experimentation shows that both subpopulations – even though initiated out of phase – will eventually fluctuate in step (fig. 10.14(B),(C)). In fact, even chaotic fluctuations will become synchronized with the IFD dispersal rule (fig. 10.14(E)). Experimentation also allows us to conclude that dispersal of individuals between the two patches after the perfect resource-matching movement rule neither dampens nor destabilizes fluctuating population dynamics. Rather, the subpopulations will start to fluctuate in perfect synchrony.

A question now emerges: will the conclusion remain valid if the patch departure does not obey the perfect IFD rule? Fuzziness into the patch-departure process can be implemented, e.g., as follows: for each step a random number $v$ was drawn from a uniform distribution ranging from $v_{min}$ to $v_{max}$ (with $\bar{v} = 1.0$) by which $\bar{N}$ was multiplied. Another random number, $w$, was drawn from the same distribution and $\Delta_i$ was multiplied by it. The navigation success of the departing individuals was decided by comparing a random number, $h$, drawn from a uniform distribution [0,1]. For $h > \frac{1}{2}$ the dispersing individuals returned back to the patch of their origin; otherwise, their navigation succeeded. As there were many stochastic elements in the system, they were all taken to be in action at the same time. The Ricker renewal in the two patches, linked via the fuzzy IFD patch-departure rule, yields synchronous dynamics (fig. 10.14(C),(F)). A word of caution is appropriate here, however. When the growth rate is in the chaotic range, an IFD match in population sizes in the two patches becomes harder to achieve for very long periods (more details in Ranta and Kaitala 2000).

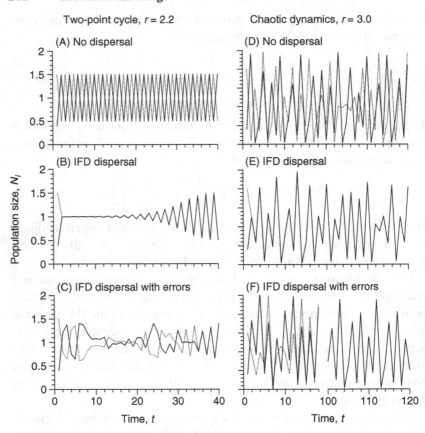

*Fig. 10.14.* Each of the panels displays dynamics of two populations obeying the Ricker dynamics in their renewal ((A)–(C) with two-point cycle, (D)–(F) with chaotic dynamics). At $t_1$ the populations are initiated in opposite phase. In panels (A) and (D) the populations behave as two independent units. In panels (B) and (F) they are linked together by dispersing individuals obeying the IFD dispersal rule with perfect knowledge. The populations – initially out of phase – rather soon come to fluctuate in step (the length of the transient phase in (B) depends on the initial population sizes). In panels (C) and (F) the IFD departure rule is prone to errors ($v_{min} = 0.5$, $v_{max} = 1.5$; see text for more details). Nevertheless, when a long enough period has elapsed, the two populations will eventually fluctuate in synchrony. Modified from Ranta and Kaitala (2000).

We have now extended research on the relationship between resource supply and population size under the conditions of fluctuating population dynamics. For the dispersal the IFD rule was applied: leave if local conditions will be worse than elsewhere in the environment, on average. This dispersal rule synchronizes local population dynamics. Unlike many other dispersal rules (Ruxton 1996b; Ruxton *et al.* 1997; Kendal and Fox 1998; see also fig. 4.8, p. 83), the perfect IFD patch-departure rule does not stabilize or destabilize the dynamics. It simply leaves the underlying skeleton of the population dynamics unaltered! This holds even in the chaotic range of population fluctuations. An explanation is that since the IFD patch-departure rule utilizes information derived from short-term predictions, then long-term uncertainties in chaotic dynamics do not affect the functioning of this departure rule. These results are rather robust for effects of biased information and failures during the dispersal phase.

### Resource matching and synchronous population fluctuations

The results of the previous section prompt us to explore the tempting implication that synchronous population dynamics and resource matching have something in common. In fact, it has been shown in various simulations, with a number of population units coupled by dispersing individuals (Chapter 4), that dispersal easily synchronizes the dynamics of local populations. In these systems, synchronous population fluctuations are achievable via redistribution of individuals, or if the system of $n$ subpopulations is disturbed by a common external modulator (Moran 1953b), or if both of them are in action simultaneously (Ranta *et al.* 1995a, 1997c, 1999a). The achieved synchrony, when gained via redistribution of individuals, does not call for dispersal rules after the IFD. Often the dispersal is taken to be negatively distance dependent with a constant fraction of dispersers leaving the natal population unit, but other rules (positively density dependent, negatively density dependent, spatially implicit dispersal, etc.) have been shown to give largely matching results of synchrony as long as the number of population subunits $n$ is large enough (Ylikarjula *et al.* 2000).

Characteristic of the explorations discussed above is that the $n$ local subunits explored have equal profitability. When synchrony is achieved in fluctuations of local population numbers, it implies that individual numbers at every patch are on average the same. Thus, in the frame of the IFD theory it would not pay individuals to disperse elsewhere as all is matching everywhere. Yet, under local disturbance, the system in synchrony would eventually drift away from synchronous fluctuations without any dispersal.

A simple extension of the spatially structured population systems explored in Chapter 4 is to assume that the local subunits differ in their carrying capacity $K_i$. Assuming no dispersal, the density-dependent

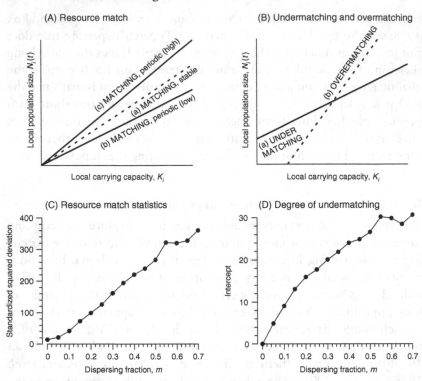

*Fig. 10.15.* Measures of resource matching for population sizes, $N_i(t)$, in a spatially structured heterogeneous (in terms of local carrying capacity, $K_i$) world. (A) The dashed line (a) indicates resource matching expectation (regression line between $K_i$ and $N_i(t)$) under stable population dynamics, while the two solid lines (b and c) indicate resource matching expectation under periodic dynamics. (B) Illustrations of (a) undermatching, i.e., larger populations sizes relative to local resource availability in areas with smaller carrying capacities, and (b) overmatching. Examples of deviation from resource matching in a 25-population system obeying eq. 10.13 in population renewal under constant fraction dispersal, $m$ ($x$ axis). Panel (D) gives the sum of squared residuals (standardized by dividing by $n$), while (C) is the intercept of the regression line between $K_i$ and $N_i(t)$, cf. panel (B).

feedback system keeps the subunit-specific population fluctuations around the $K_i$. Thus, the resource matching is roughly met. In this system, there are two different measures of resource matching: deviation from the resource matching line (fig. 10.15(A)), i.e., squared standardized residuals, and the deviation of the regression slope (between $K_i$ and $N_i(t)$) from zero (fig. 10.15(B)). Redistribution of individuals among the subunits corrupts the match between resources and local population size. This is because units with higher $K_i$ (with fixed proportion dispersal) will send out more dispersers than they will receive and units with lower $K_i$ will receive more individuals than they send out. Thus, contrary to a system where each subunit is alike, in a heterogeneous and spatially structured world under-matching (i.e., deviation of the intercept (fig. 10.15(B)) from zero) is the expectation. A demonstration of this will be given next.

We let the $n = 25$ population subunits be randomly distributed into a $10 \times 10$ co-ordinate space. Each subunit has its renewal process (match-ing in all units) influenced by local disturbance, $u_i(t)$, i.e., local densities multiplied by uniform random numbers between 0.95 and 1.05. All the units are also disturbed by the global Moran effect, $\mu(t)$, i.e., all units multiplied by uniform random numbers between 0.9 and 1.1. For sim-plicity, we assume that a fixed proportion of individuals ($0 < m < 1$) from the local subunits redistribute annually. The number of immigrants arriving to patch $i$ from patch $j$ follow dispersal kernel I (p. 53) with $c = 1$. After Moran (1953b), we have selected an AR(2) process for the popula-tion renewal function

$$N_i(t+1) = (1-m)[K_i + a_1 N_i(t) + a_2 N_i(t-1)]u_i(t)\mu(t)$$
$$+ \sum_{s,s\neq i} M_{s,i}(t). \tag{10.13}$$

The values of $K_i$ are selected so that the expectations (carrying capacities) of local population sizes range between 5 and 50, while $a_1 = 1.25$ and $a_2 = -0.75$ yield, when the system is disturbed, roughly cyclic dynamics with 7- to 8-year period length. The populations are initiated out of phase with random numbers between 5 and 50. The system is left running for 5000 generations and, for our purposes, we shall sample the next generation to assess the match between resource availability and population size. As an index of resource matching, we scored two measures: (i) the mean squared deviance between subunit-specific popu-lation size and resource match at the end of the simulation, and (ii) the intercept of the regression between $K_i$ and $N_i(5000)$. The results of this exercise are as anticipated above: with *increasing* fraction of dispersing individuals the deviation from perfect resource matching increases (fig. 10.15(C)) and the deviance is towards a higher degree of undermatch-ing (fig. 10.15(D)). Let us remind ourselves that experimentation with different values of $a_1$ and $a_2$ in eq. 10.13 yield matching outcome in terms of undermatching and deviation from resource matching.

We conclude that in a spatially structured heterogeneous world (in terms of local carrying capacity) perfect resource matching is achievable via perfect IFD redistribution rules. However, if we take the position that redistribution rates are substantial and the departure/arrival rules do not resemble the IFD rules (such as with constant fraction dispersal), there is no perfect resource matching. Rather, undermatching will be the outcome. However, it is also obvious that much more research is needed into combining population dynamics and resource matching theories.

## Summary

The distribution and abundance of individuals is contingent on the distribution and abundance of the essential resources for the population. The IFD theory provides us with the template for our understanding of how and to what extent the distribution of individuals matches the distribution of resources. In this chapter, we review the processes underlying the IFD, and we scrutinize why we generally observe consistent deviations from the IFD predictions. We show that, in fact, the IFD gives biased predictions and that "undermatching" (the overuse of poorer patches in relation to the richer ones) should be the null-model, not a deviation from the IFD. This chapter takes a closer look at some of the factors that are responsible for this pattern, e.g., environmental stochasticity, poor knowledge, and environmental grain size. We also show that the trade-off between resource acquisition and predation risk may be a delicate one sometimes resulting in deviations from the expected null-models. In this chapter, we also analyze the role of dispersal for the resulting resource matching. Populations with closely genetically related individuals distort the expected dispersal pattern. Also, should there be a net flow of individuals (immigration exceeding emigration, or the reverse) between patches, significant deviations from expected patterns are expected. Finally, we closed with a brief excursion between resource matching and population dynamics.

# 11 · *Spatial games*

Individuals in natural populations encounter each other in numerous different ways. Such encounters include mating, conflicts over food or other resources, or the joint and co-operative acquiring of resources. The behavioral adaptations to such situations are often studied by evolutionary game theory. In this chapter, we will review some classic behavioral games: the Hawk–Dove, the Prisoner's Dilemma (including the evolution of co-operation), and the somewhat more obscure Rock–Scissors–Paper game. We also extend those problems to spatially heterogeneous environments. Towards the end of this chapter, we will combine the game theoretical analyses with dispersal-coupled population models.

Many, but far from all, encounters between individuals are pairwise. If the encounter involves a conflict, there is generally a winner and a loser. Take, e.g., two male deer fighting for the chance of mating with a female. The fight may be furious and last for a long time, possibly resulting in injuries to one or both contestants. Eventually one of the males will retreat and the winner will gain the mating. Such behavioral and ecological problems have inspired the development of evolutionary game theory (Maynard Smith and Price 1973; Maynard Smith 1982).

Most evolutionary theory assumes selfishness-driven adaptations (Dawkins 1976). It does not pay an individual to be nice or altruistic and helpful towards others unless there is a guarantee for not being cheated. Hence, altruistic behaviors are susceptible to selfish cheaters and will disappear from the population. In a striking contrast to this, social behavioral patterns, such as co-operation and helping, are frequently observed throughout the animal kingdom (Clutton-Brock and Parker 1995; Connor 1995; Dugatkin 1997; Komdeur and Hatchwell 1999; Clutton-Brock 2002). The evolution of co-operative behaviors, such as shared vigilance for predators in birds and fish, nonparental care for young in birds, and co-operative hunting in primates, has therefore received much attention in modern evolutionary ecology (Hamilton 1964a, 1964b).

Evolutionary game theory has often been used to address both animal conflict and co-operation (Maynard Smith 1982; Axelrod 1984; Dugatkin and Reeve 1998). The evolution of co-operation has also been a particularly interesting challenge for evolutionary game theory (Trivers 1971; May 1981, 1987; Maynard Smith 1982; Sigmund 1993; Dugatkin and Reeve 1998; Hofbauer and Sigmund 1998; see also recent summaries by Dugatkin 1997, 1998). Before having a closer look at that research, here we briefly review some of the classic behavioral games.

## Classic games

### The Prisoner's Dilemma

The Prisoner's Dilemma game gives an elegant setting in which to introduce the problems that evolutionary game theory is addressing. Imagine two players in a game with two options: to either co-operate or be selfish (defect). Initially the Prisoner's Dilemma game was framed to address the behavior of two prisoners having committed a crime (Axelrod and Hamilton 1981), therefore *Prisoner's* Dilemma. Defecting would mean that you accuse the partner of being solely guilty of the crime. Co-operating, on the other hand, would mean not revealing anything. Denoting defecting as $D$ and co-operating as $C$, we have four possible outcomes: $DD$, $DC$, $CD$, and $CC$. In the $DD$ case, both prisoners will accuse each other of having committed the crime. Should both do so, they will both get a sentence of, say, 5 years in prison. In the $DC$ case, the cheater will be released and the one admitting nothing will be sentenced for 7 years in prison. If both prisoners collaborate ($CC$) then they will both get a sentence of 2 years. Thus, the payoffs (the rewards for the respective person) would be $DD$: $(-5, -5)$; $DC$: $(0, -7)$; $CD$: $(-7, 0)$; and $CC$: $(-2, -2)$. We can now normalize the payoffs (by adding 7 to all), and we have: $DD$: $(2, 2)$; $DC$: $(7, 0)$; $CD$: $(0, 7)$; and $CC$: $(5, 5)$. The question now is whether there is a behavioral strategy ($C$ or $D$) that is the "best" solution to this dilemma. The decision matrix for *Player* 1 with the above payoffs is:

|  |  | *Player* 2 | |
|---|---|---|---|
|  |  | $D$ | $C$ |
| *Player* 1 | $D$ | 2 | 7 |
|  | $C$ | 0 | 5 |

Naturally, the payoff matrix for *Player* 2 is the same as for *Player* 1 (for a more general version of the payoff matrix in the Prisoner's Dilemma game,

**Box 11.1** · *The Prisoner's Dilemma*

In the Prisoner's Dilemma, two players (prisoners) have been involved in an illegal act which they in fact have committed jointly. Consequently, they are accused of having carried out a crime. While being kept in custody, they are isolated from each other. In this troublesome situation, they have two options: defect or co-operate. If both players co-operate they can avoid most of the charges, at least. In this case, the payoff for *Player* 1 is R. If both players choose to defect then the payoff for *Player* 1 is P. If *Player* 1 decides to co-operate when *Player* 2 defects, she gets S. And finally, if she defects when *Player* 2 co-operates her payoff will be T. The payoff matrix for *Player* 1 can be presented as follows

|  |  | *Player* 2 | |
|---|---|---|---|
|  |  | *Defect* | *Co-operate* |
| *Player* 1 | *Defect* | P | T |
|  | *Co-operate* | S | R |

If $P > S$ and $T > R$ the only stable outcome is defection by both players, even if the sum of the payoffs for co-operation ($2R$) would be greater than the sum of the payoffs for defection ($2P$).

see Box 11.1). Both players need to carry out the following reasoning in isolation. If *Player* 2 decides to defect then the best response for *Player* 1 is to defect as well. If *Player* 2 decides to co-operate then the best response for *Player* 1 is to defect. Thus, whatever *Player* 2 decides to do the best *Player* 1 can do is to defect. Because the game is symmetrical the same applies for *Player* 2, and the outcome of the game (the "best" solution) is defection by both players. The Prisoner's Dilemma game illustrates the following profound problem in evolutionary ecology. Nature seems to waste resources: individuals would be better off by co-operating. However, individual fitness maximization drives the evolution towards selfish behavior, since co-operation represents an unstable behavior – it is vulnerable to cheating (hence Prisoner's *Dilemma*). For this reason, the evolution drives the populations towards evolutionarily stable behavior, ESS (Maynard Smith 1982), where no individual can improve its benefit or fitness by deviating from the ESS behavior.

### The Hawk–Dove game

In the Hawk–Dove game (Maynard Smith 1982), a population is composed of two behavioral phenotypes: Hawks and Doves. In pairwise encounters,

> **Box 11.2** · *The Hawk–Dove game*
>
> In the Hawk–Dove game, the players fight for a resource using two different tactics: Hawk and Dove. In an encounter, Hawks display and escalate, and Doves display and retreat. The payoff matrix for *Player 1* can be presented as
>
> |  |  | *Player 2* | |
> |---|---|---|---|
> |  |  | Hawk | Dove |
> | *Player 1* | Hawk | $(1/2)(V-C)$ | $V$ |
> |  | Dove | 0 | $(1/2)V$ |
>
> where $V$ is the reward and $C$ is the cost for two Hawks fighting.

two individuals compete for a resource, whose value is $V$. A Hawk first displays and, if the opponent does not retreat, the encounter is escalated into a fight in which the loser will suffer a cost of $C$. If a Dove is the first displayer and if the opponent is a Hawk, the Dove will retreat. Two Doves have an equal opportunity to get the resource without anyone suffering costs. In a Hawk–Hawk encounter, both competitors have an equal probability of winning the resource and to suffer costs from fighting. Thus, the payoff will be $0.5(V-C)$. In a Hawk–Dove encounter, the payoff for the Hawk is $V$ and 0 for the Dove. In a Dove–Dove encounter, the payoff for each will be $0.5V$. Which strategy is the unbeatable (ESS) one in the long run?

Obviously, it is always better to play Hawk independent of whether the opponent is a Hawk or Dove. Thus, the Hawk strategy is an ESS. This means that, in a population of Hawks, the Dove strategy cannot increase. In this case, Hawks can also invade a population of Doves. The problem becomes more challenging if the cost from possible injury is higher than the reward obtained from the resource, i.e., $V < C$. It is not evident that the individuals should all play the Hawk strategy. In particular, in a population of Hawks, it pays to play Dove against a Hawk, and we may predict that, in a population of Hawks, individuals playing the Dove strategy would gain an advantage and therefore the Dove strategy will start to increase in frequency. However, in a population of Doves, Hawks will gain advantage against Doves, and the Hawk strategy would start increasing. Under these circumstances, both strategies are "best," depending on the frequency of Hawks and Doves in the population. That is, neither Hawk nor Dove is an ESS, and the population will be composed of a mixture of Hawks and Doves.

Assume now that a fraction $p$ of the population plays the Hawk strategy and the remaining ones, $(1 - p)$, play the Dove strategy. If a Hawk is engaged in an encounter with an opponent the probability of gaining $0.5(V-C)$ will be $p$, and the probability of gaining $V$ will be $(1-p)$. The total expected payoff will be $0.5p(V-C)+(1-p)V$. Following the same reasoning, the payoff for an individual obeying the Dove strategy is $0.5(1-p)V$. Is there a value of $p$ such that both strategies have equal payoffs, i.e., no individual would like to change its strategy? Thus, we get the following condition for an ESS: $0.5p(V-C)-(1-p)V=0.5(1-p)V$. Solving for the frequency we get $p=V/C$. The frequency of Hawks is exactly equal to the ratio between the value of the resource and the cost from the injury.

The Hawk–Dove game illustrates another fundamental phenomenon in evolutionary ecology: frequency-dependent selection. It states that if a rare phenotype gains advantage from being rare it will increase in frequency until the gain has been lost.

### The Rock–Scissors–Paper game

The Rock–Scissors–Paper game represents a situation where multiple phenotypes play against each other and the competitive ability is ordered in a circular manner: Rock blunts Scissors, Scissors cut Paper, and Paper wraps Rock. Assume that the payoff for a winner is always equal to 1; for the loser, $-1$. However, each type can also play against itself. Let us denote the score against the same type by $-\varepsilon$ (Box 11.3). This game has no pure strategy ESS solution. Thus, we need to look at mixed strategies. Maynard Smith (1982) showed that a frequency distribution of one-third of each type in the population is an ESS when the diagonal of the payoff matrix is negative. When the diagonal of the payoff matrix is positive, there is no ESS.

---

**Box 11.3** · *The Rock–Scissors–Paper game*

In this game (Maynard Smith 1982), the players may apply three different tactics: Rock, Scissors, and Paper. Rock beats Scissors, Scissors beats Paper, and Paper beats Rock. The payoff matrix for Player 1 is

|          |          | Player 2 |          |          |
|----------|----------|----------|----------|----------|
|          |          | *Rock*   | *Scissors* | *Paper*  |
|          | *Rock*     | $-\varepsilon$ | 1        | $-1$     |
| Player 1 | *Scissors* | $-1$     | $-\varepsilon$ | 1        |
|          | *Paper*    | 1        | $-1$     | $-\varepsilon$ |

where $-\varepsilon$ is the payoff playing against self.

## Co-operation: early history

Hamilton's (1964a, 1964b) idea revolutionized the research on evolution of co-operation in two ways. First, he formulated the problem by using very intuitive mathematical notation (Box 11.4). Second, with the help of the model, he suggested that for co-operation to evolve calls for not only the cost and benefit of the nonselfish behavior for the donor but also for the level of relatedness between the donor and the recipient (the one receiving the benefits of the donor's act). In its simple form Hamilton says that a gene linked with the co-operative behavior will increase in frequency when the relatedness between the players and the benefit-to-cost ratio are high enough (Box 11.4). Thus, according to Hamilton's rule, genetic relatedness should help co-operation evolve among players. The problem remains for genetically unrelated individuals. Two ways to solve the evolution of co-operation have been suggested.

Axelrod and Hamilton (1981) and Axelrod (1984) suggested that repeated interactions could solve the situation for the Prisoner's Dilemma (originally only played once by each player). This led to the tit-for-tat strategy (TFT; i.e., co-operate first, thereafter follow whatever your opponent did last turn; Axelrod and Hamilton 1981) and its various descendants (Dugatkin 1997). Now, players can adjust their strategies according to previous plays. Second, in the 1980s, trait-group selection models (Wilson 1980) replaced the traditional group selection models (Wynne-Edwards 1962; Williams 1966) to explain co-operation. A thorough description of the ins and outs of trait-group models is provided by Dugatkin (1997) and Dugatkin and Reeve (1998). The underlying idea is that a population is divided into small groups, and those with co-operative individuals are better off than

---

**Box 11.4** · *Hamilton's Rule*

Hamilton's rule (1964a, 1964b) states that a gene linked with the co-operative behavior will increase in frequency always when the inequality

$$b/c > 1/r, \text{ or } rb - c > 0$$

is true. Here $b$ is the benefit the recipient gains and $c$ is the cost the donor pays. Hamilton's rule clearly says that there is no option for co-operation unless the relatedness $r$ between the players is higher than zero. Thus, with $r = 0$, no co-operation is expected to occur.

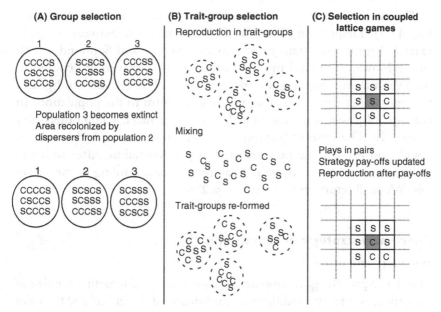

**(A) Group selection**

**(B) Trait-group selection**

**(C) Selection in coupled lattice games**

*Fig. 11.1.* Reference to space in models explaining the evolution of co-operative behavior. In panels (A) and (B) space is implicit, while in (C) it is explicit. (A) Group selection models refer to spatially separated populations that are reproductively isolated. The probability that a group becomes extinct (here group 3) is in proportion to co-operative $C$ individuals ($S$ = selfish individuals) in the local population. Extinct populations are colonized by individuals in extant populations (here from group 2). (B) In trait-group models groups are embedded in the population and individuals reproduce in these trait-groups to redistribute into the population and to form new trait groups ((A) and (B) modified after Dugatkin and Reeve 1998)). (C) In lattice games individual(s) in a focal cell (shaded) interact with individuals in neighboring cells (here the neighborhood size is one cell layer around the focal cell). Rewards of these interactions are given after a specified payoff table. The cell occupancy is updated after the payoffs of individuals gained during one round of games played. The updating rules may vary.

groups including selfishly acting individuals. Thus, groups favoring co-operation will produce more offspring to next generations than groups with selfish behavior dominating. These models are spatial, but the reference is implicit (fig. 11.1). The different trait-groups, within which interactions among individuals take place, are separately located. All individuals in the population, or their descendants, come into contact in the mixing phase, and new trait-groups are formed. Trait-group models have been rather powerful in explaining the evolution of co-operation (Wilson 1983; Dugatkin 1997, 1998).

The two approaches are related when the games are played in space (fig. 11.1). Despite this, there is a clear distinction between spatially extended games and trait-group models (Nowak and Sigmund 1998). In the Nowak–Sigmund model, random pairs of players are chosen in succession from a population, one as a potential donor and another as a recipient. After the interaction, individuals return to the population. In this type of game, some individuals may never meet, while others play repeatedly. The ratio of playing to nonplaying individuals depends on population size and the number of plays per generation. After all interactions have been played, individuals reproduce according to their payoffs. We shall return to this issue on p. 294.

## Space and strategy co-existence

### Boundary and order

In order to play the game in space and time, we need to define the size of the lattice, boundary conditions on the borders of the lattice, and the rules for scoring the payoffs in each cell. Two different boundary conditions have been used. With fixed boundaries, the cells on the border of the lattice have fewer neighbors with which to interact. With periodic boundaries, cells on the lattice margins assume that cells on the opposite edge are their neighbors as well (i.e., the lattice is a toroid, a doughnut-like topological object). There are also several options to be used for scoring the payoffs. Firstly, we need to define the neighborhood for each cell. One option is to assume that the game is played against a minimum number of neighbors. One choice could be that a focal cell plays against the four orthogonal neighbors. For our purposes, we assume that the game is played against all eight neighboring cells in the lattice.

On a lattice, the interior cells have eight neighbors, the edge cells have five neighbors, while the corner cells have three neighbors. Consider first a cell sitting in the interior of the lattice. The scores in pairs are taken from the payoff matrix on p. 268. Assume that such a focal cell (cell (2, 3) in fig. 11.2(A)) is occupied by an individual playing $C$, and has eight neighbors, three playing $C$ and five neighbors playing $D$. Thus, the payoff (middle row of panels in fig. 11.2) for the focal individual will be 4. Instead, if the focal cell with the same number of different neighbors (cell (3, 3) in fig. 11.2(A)) is playing $D$ its score would be $4b$. Updating depends now on the parameter value $b$. Assume that $b > 5/4$. Updating the lattice results in global takeover by the defecting strategy (fig. 11.2(C)). We shall next illustrate the

(A)

```
C D D C
C D C D
D C D C
C D C D
```

(D)

```
D D D D
D D D D
D C C C
D C C C
```

(G)

```
D D D C
D D D D
D C C C
D C C C
```

(B)

```
1  2b 2b 2
3  4b 4  3b
3b 5  4b 3
2  3b 3  3b
```

(E)

```
0   0   0  0
b   2b  3b 2b
2b  4   6  4
2b  4   6  4
```

(H)

```
0   0   1  0
b   2b  4b 3b
2b  5   6  4
2b  5   6  4
```

(C)

```
D D D D
D D D D
D D D D
D D D D
```

(F)

```
D D D D
C C C C
C C C C
C C C C
```

(I)

```
D D D D
C D D D
C D D D
C C C C
```

*Fig. 11.2.* Three examples (A, D, G) of a 4 × 4 lattice configuration with the corresponding scores (B, E, H). The last row (C, F, I) gives the updated configuration for the first row using fixed boundary conditions.

importance of order in the space. Assume that $5/4 < b < 2$, and consider a configuration where an individual of the $C$ strategy occupies the lower right corner (fig. 11.2(D)). The ones playing $C$ are able to utilize their synergy and hence they are able to take over a neighboring cell from individuals playing the $D$ strategy (fig. 11.2(F)). Ultimately, the co-operative strategy will occupy the whole lattice in the third step. The importance of the order can be illustrated by the following example. Assume, in the previous example, that $4/6 < b$, and further that the upper left cell (1, 4) $D$ is replaced by $C$ (fig. 11.2(G)). Thus, one might predict that the more individuals there are playing $C$, the better the co-operative strategy is doing. However, in this example it helps the defection strategy more than the co-operative strategy. We see that the lonely $C$ will in fact destroy the effective order of the co-operative strategy (fig. 11.2(I)). In this example, the co-operative strategy is driven into extinction in the third step!

### Prisoner's Dilemma

We observed in the Hawk–Dove game that under certain conditions two different strategies might co-exist. We will now show that the conditions for strategy co-existence may be much relaxed in a spatial context. The first explicit spatial extension of evolutionary games was done by Nowak and May (1992). Using coupled lattices (fig. 11.1(C)) as spatial reference, they played the Prisoner's Dilemma game with two kinds of players: $C$, who always co-operates, and $D$, who always defects. Each lattice cell is occupied by a single individual playing either $D$ or $C$. The neighborhood was selected to be one cell layer around the focal cell. In their plays, mutual co-operators each scored 1, mutual defectors 0 and $D$ scored $b$ benefit (exceeding unity) against $C$, who scored 0 in such encounters (see the payoff matrix below). The fitness function for each player is the sum of payoffs in the encounters. Each generation is updated by re-occupying the lattice cells by the player type with the highest score between the previous owner and the immediate neighbors.

Varieties of this game were played by Nowak and May (1992; lattice size varying, initial proportion of $C$ and $D$ varying, and $b$ varying). The main conclusion of their study was that co-operation, once evolved, can be maintained in a spatially structured population. The system stabilizes for an asymptotic frequency of co-operators with a great range of initial frequencies of $C$ and parameter values of $b$. Another alternative is that the frequency of co-operators oscillates in a complicated manner around some level. In both cases, however, the system yields complex (often aesthetic) spatial clusters of $C$ and $D$ traveling through space over time.

Nowak and May (1992) considered the patterns arising when the Prisoner's Dilemma game is put into a spatial context. They chose the following payoff matrix for the game

|          |   | Player 2 |            |
|----------|---|----------|------------|
|          |   | $D$      | $C$        |
| Player 1 | $D$ | $P=0$  | $T=b, b>1$ |
|          | $C$ | $S=0$  | $R=1$      |

We observe that the parameter values do not satisfy exactly the assumptions $T>R>P>S$ and $R>(S+T)\times 0.5$ posed to the Prisoner's Dilemma (e.g., Dukatkin 1997). However, one can see immediately that to defect is an ESS. Whatever strategy *Player* 2 plays, $D$ is always the best choice by *Player* 1. Thus, to defect is an ESS in a single shot game, and co-operation should not be maintained in the population.

A spatial version of the game is composed of a lattice where each cell contains one player. The players either defect ($D$) or co-operate ($C$). An individual in each cell plays the game with its neighbors. The payoff each player will get is the sum of the payoffs of the individual games the player will play with each of its neighbors. In the next generation (the game is extended also in time!), the cell will be occupied by an identical offspring of the individual playing the strategy with the highest payoff obtained in the neighborhood and in the focal cell.

## Nowak and May in space

The crucial importance of the spatial structure of the games has now been recognized for about 10 years (Nowak and May 1992). To illustrate the nature of the effect of the spatial dimension in games, we carried out a simulation using a $61 \times 61$ lattice with $b = 1.82$. Each focal cell plays against itself in addition to playing against each of its eight neighbors. We first assumed a random initial configuration. Such a game will lead to sequential spatial patterns where the strategies in space vary in a very unpredictable manner (Nowak and May 1992). The strategies often make single-strategy clusters of varying size (see fig. 11.9, p. 285). These clusters grow and diminish in a fractal manner, and they frequently disappear and reappear. Here, instead of attempting to understand the spatial-temporal changes in the strategy composition, we pay attention to the co-existence between the different strategies. Our focus is on the temporal dynamics of the frequencies of the two strategies. We observe that the frequency of co-operation first drops close to zero. This is because the aggregations of the co-operative strategy have had no time to build up yet such that the co-operative strategy could take advantage of the organization (fig. 11.3(A)). After a relative short transient phase, aggregations of co-operators will begin to emerge. Such aggregations are born and lost in place through time. In a spatial setting, such a dynamic process makes the co-existence of both strategies possible. After the transient phase, the frequency of co-operators will fluctuate around 0.34. It is tempting to use the same time series analysis tools as used in the population dynamics context (Chapters 2 and 5). In the population renewal process, population size fluctuates due to density-dependent feedback. In the evolutionary games, fluctuations in the numbers of players of different strategies in time are due to the payoff matrix and the temporary spatial constellation of the strategies. In the spatial plays of the Prisoner's Dilemma, as described here, the frequencies of $C$ and $D$ continue to oscillate forever. The oscillations are more violent at a regional level (a sample of 100 cells in the center of the lattice) than at the global level (frequencies of $C$ and $D$ summed over all cells). At the cell level, the strategies flip between $C$ and $D$ (fig. 11.3(C)). The power spectra of strategy-specific frequencies do not indicate any apparent periodicity. Instead, it suggests that the temporal dynamics of the frequencies of $C$ and $D$ are close to a power law

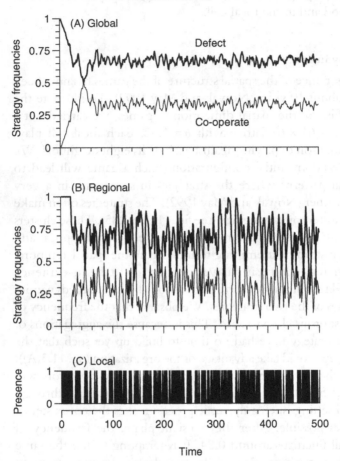

PRISONER'S DILEMMA

*Fig. 11.3.* The frequencies of the defecting and the co-operative strategies in a spatial Prisoner's Dilemma game (61 × 61 lattice) with random initial configuration ($b = 1.82$). (A) The global frequency of co-operative strategy stabilizes to fluctuate around 0.34. (B) The frequencies of the two strategies fluctuate much more widely in a 10 × 10 subset in the center of the lattice. (C) The local presence of defecting strategists is given for a randomly selected cell.

(fig. 11.4), with a slope indicating red dynamics. Also, when we scored the time for how long sequences a randomly selected cell was occupied by a given strategy, e.g., by $C$, the data appear to obey the power law (fig. 11.8(A)).

There is no doubt that the Nowak and May (1992) spatial Prisoner's Dilemma game yields spatial self-organization. Earlier (Chapter 5; Kaitala

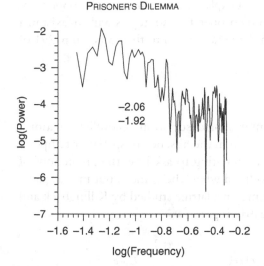

PRISONER'S DILEMMA

*Fig. 11.4.* Power spectrum for the frequency of the co-operative strategy in a spatial Prisoner's Dilemma game (fig. 11.3(A)). The power spectrum was calculated omitting the first 100 generations of the simulations lasting 500 generations. The slopes of the global (upper) and regional (lower) power spectra are inserted.

*et al.* 2001a, 2001b), we have proposed that when the temporal dynamics of a population follow the power law (Kauffman 1993; Halley 1995) it is a signature of the underlying spatial self-organization process. Here we have yet another hint at self-organization being involved.

We also simulated the spatial dynamics using a symmetrical initial configuration by initiating the game by one defector in the middle of the lattice that was otherwise occupied by co-operators. When the frequency of co-operators is high enough then the frequency of defectors increases rapidly until self-organization occurs in the configuration. Co-operative individuals remain on the lattice as aggregations and are able to persist in a population of mostly defectors. As in the random initial configuration with $C$ and $D$ in the space, we witness sustained oscillations of the frequencies of $C$ and $D$. The average frequency level of co-operators stabilized again around 0.34, indicating that strategy co-existence may occur on the lattice in a dynamical manner. Again, the power spectra suggest that temporal fluctuations of the two strategies match the power law. Thus, there appear to be no major differences between the games with random and symmetrical initial conditions. However, regardless of the initial configuration, the Nowak and May (1992) spatial

Prisoner's Dilemma game does not explain the evolution of co-operation; it only tells us that the selfish and co-operative strategies will co-exist in a spatially structured population. We shall return to this topic on p. 291 of this chapter.

### Hawk–Dove game

The Hawk–Dove game is an interesting description of conflict situations. It allows strategy co-existence even in the absence of spatial or temporal modification of the game. It is interesting to ask how the predictions of the Hawk–Dove game are modified when the game is put into a spatial context. Consider, e.g., the game on a lattice studied by Killingback and Doebeli (1998). The payoff matrix is given as

|  |  | *Player 2* | |
|---|---|---|---|
|  |  | *Hawk* | *Dove* |
| *Player 1* | *Hawk* | $0.5(V-C)=1-\beta$ | $V=2$ |
|  | *Dove* | 0 | $0.5V=1,$ |

where $0 < \beta$. Clearly, if a Hawk plays against a Hawk the payoff from the encounter will be $1 - \beta$. If a Hawk plays against a Dove, the payoff will be 2. For a Dove playing against a Hawk and a Dove the payoffs will be 0 and 1, respectively. When $\beta > 1$ we have in a single shot game a mixed solution: $p = V/C = 2/2\beta = 1/\beta$. With the above values we have

|  |  | *Player 2* | |
|---|---|---|---|
|  |  | *Hawk* | *Dove* |
| *Player 1* | *Hawk* | $-0.1$ | 2 |
|  | *Dove* | 0 | 1 |

We analyzed the Hawk–Dove spatial game assuming $\beta = 1.1$. From the single shot game we get the prediction that the frequency of Hawks will be $p = 1/\beta = 0.91$. A $61 \times 61$ lattice was initiated with equal abundance of Hawks and Doves. Starting from random initial conditions, the simulations result in strategy co-existence where the frequency of Hawks will be much less than predicted from a single shot game (0.67 versus 0.91). As in the spatial Prisoner's Dilemma game, in the spatial Hawk–Dove game the frequencies of the two strategies soon settle at a certain level (fig. 11.5), where they keep fluctuating. Again, the fluctuations are more violent at the regional than at the global level, and at a local single-cell level a strategy stays a short time only. Most interesting, however, is the finding that, as in

HAWK–DOVE

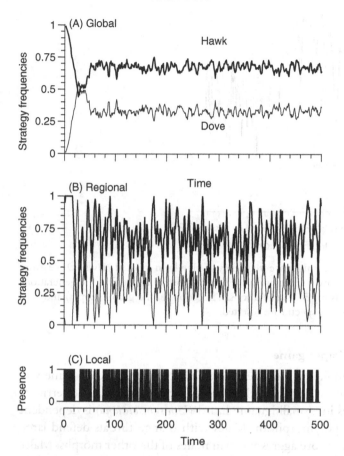

*Fig. 11.5.* The frequencies of Hawk and Dove strategies in a spatial Hawk–Dove game (61 × 61 lattice) with random initial configuration ($\beta = 1.1$). (A) The global frequencies of the two strategies soon stabilize, while (B) they keep on fluctuating more widely in a 10 × 10 subset in the center of the lattice. (C) The local presence of Hawk strategists is given for a randomly selected cell.

the Prisoner's Dilemma game, the strategy-specific frequencies show red dynamics, the power spectrum indicating a behavior close to a power law (fig. 11.6). Thus, the spatial organization in this game is also visible as the power law of the strategy frequencies. Also in this game we scored for how many sequences a randomly selected cell was occupied by a given strategy, e.g., by Hawk. The data appear to obey the power law (fig. 11.8(B)).

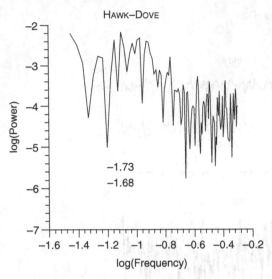

*Fig. 11.6.* Power spectrum for the frequency of Hawks in a spatial Hawk–Dove game (fig. 11.5(A)). The power spectrum was calculated omitting the first 100 generations of the simulations lasting 500 generations. The slopes of the global (upper) and regional (lower) power spectra are inserted.

## Rock–Scissors–Paper game

An interesting application of the Rock–Scissors–Paper (RSP) game was provided by Sinervo and Lively (1996). They showed that territory defense by males in a small iguanid lizard (*Uta stansburiana*) is dependent on throat-color polymorphism. Males with orange throats defend large territories and are more aggressive than males of the other morphs. Males with dark-blue throats defend smaller territories and are less aggressive. Males with yellow stripes on their throats do not defend territories – they are sneakers. Putting the problem of frequency-dependent selection in a game theory context, Sinervo and Lively (1996) concluded that the territory defense is an example of the RSP game. The aggressive strategy of yellow-throated males is defeated by the sneaker strategy; sneakers are defeated by the blue-throated males with small territories that again are defeated by the aggressive strategy. Thus, no morph is an ESS.

The term "territory" above refers to space. Therefore, we shall commence the exploration of the RSP game in a 61 × 61 lattice, initiated with equal abundance of *R*, *S*, and *P*, one strategy per cell. After an erratic initial phase, the dynamical changes of the strategy frequencies start to oscillate in a periodic manner. A closer look at the period length shows that it is about 9 years (fig. 11.7). As compared to the Prisoner's Dilemma

ROCK–SCISSORS–PAPER

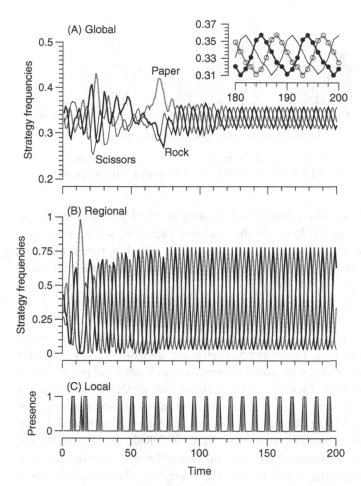

*Fig. 11.7.* The frequencies of Rock, Scissors, and Paper strategies in a spatial RSP game (61×61 lattice) with random initial configuration ($\varepsilon = 0.1$). (A) At the global level, the strategy-specific frequencies will start to oscillate in a regular cyclic manner (the inset) after an initial phase. (B) At the regional level (a 10 × 10 subset in the center of the lattice) the frequencies settle much sooner to a stable cyclic fluctuation, but with a considerably wider amplitude than in (A). (C) At the local level (in a randomly selected cell) a strategy (here displayed for Rock) stays a while, and then turns to Paper, only to be replaced by Scissors, to be replaced by Rock, etc.

and Hawk–Dove games, the Rock–Scissors–Paper game shows crucially different properties when put into a spatial context. With a proper selection of diagonal values of the payoff matrix (Maynard Smith 1982; Hofbauer and Sigmund 1998), cyclic dynamics of the three strategies are

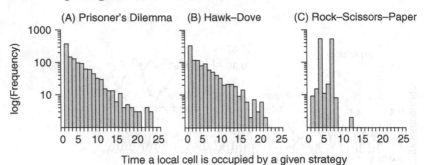

Fig. 11.8. Frequency distribution of the times for which a local cell is occupied by a given strategy in spatial versions of the three evolutionary games, Prisoner's Dilemma, Hawk–Dove and Rock–Scissors–Paper.

easy to maintain. The periodic fluctuations are well pronounced at the regional level (fig. 11.7(B)). At the cell level a strategy, say $R$, stays for a few time steps, is replaced by another strategy $P$, which in turn is replaced by another strategy $S$, and the original strategy $R$ is back again after a few more time steps (fig. 11.7(C)).

The frequency distribution of temporal cell occupancy by a given strategy (say $R$) is clearly bimodal (fig. 11.8(C)). This is in striking contrast to the spatial versions of the Prisoner's Dilemma and Hawk–Dove games. For these two games, strategy-specific cell occupancy time clearly obeys the power law (fig. 11.8). In a spatial context, a match with the power law is often taken as a sign of self-organization (Kauffman 1993; Halley 1995). To illustrate that all three games will display temporal spatial organization, we took samples of the cell occupancies by the strategies in the three different games. Not unexpectedly (Nowak and May 1992; Killingback and Doebeli 1998), both the Prisoner's Dilemma and Hawk–Dove games display clear spatial patterning (fig. 11.9(A),(B)). However, we also see clear spatial structuring in the RSP game (fig. 11.9(C)). This was also suggested by Sinervo and Lively (1996) with the color morph data of the iguanid lizard. Here the periodicity for the global lattice is 9 years while most lattice cells display either 3-year or 6-year periodicity in occupancy of a given strategy (fig. 11.9(C)). Thus, spatial organization is pronouncedly visible in the RSP game, though via cyclic fluctuations of the three morphs.

### Noise in the RSP payoff matrix

Based on an empirically derived payoff matrix, Sinervo and Lively (1996) suggested in their RSP model that the lizard morph frequencies oscillate

(A) Prisoner's Dilemma

(B) Hawk–Dove

(C) Rock–Scissors–Paper

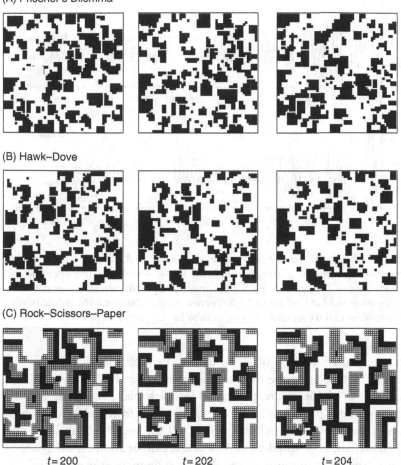

t = 200          t = 202          t = 204

*Fig. 11.9.* Temporal snapshots of cell occupancies by the playing strategists in the spatial versions of the (A) Prisoner's Dilemma game ($D$ = white, $S$ = black), (B) Hawk–Dove game (Hawk = black, Dove = white), and (C) Rock–Scissors–Paper game (Rock = Black, Paper = white, Scissors = gray). In each case, the game is played in a 61×61 lattice (cf. figs. 11.3, 11.5, 11.7). The snapshots are taken at $t$ = 200, $t$ = 202, and $t$ = 204 to show the changing spatial reconfiguration of cell occupancies.

in cycles such that the oscillations are damped but maintained by random perturbations. Here we would like to explore the influence of stochastic elements on the outcome of the RSP game. This is because the standard spatial RSP game tends to yield clearly pronounced sequential cyclic replacement of the strategies. A 20 × 20 lattice is used and is initiated

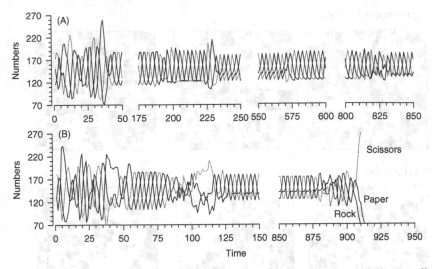

*Fig. 11.10.* The spatial Rock–Scissors–Paper game is played with noise in the payoff matrix. Note the emerging irregularities in the R→P→S→R ... sequence and in the appearance of the strategy-specific dynamics. Despite the irregularities the three strategies co-existed at least up to 10 000 generations in the game (A), while in (B) Rock becomes extinct around $t = 910$, soon to be followed by Paper.

with an equal abundance of *R*, *S*, and *P*. We shall use the same payoff matrix as used to generate the data displayed in fig.11.7. However, the payoff matrix elements are perturbed for each play by adding a small random element $\mu$ to them, so that $\mu$ takes values from uniform random numbers between $-w$ and $w$ (here $w = 0.05$; i.e., $\pm 5\%$ noise). The rationale is that, now and then and for one reason or another, the contestants are either in better or in worse shape at influencing the payoff values. Note, however, that the long-term expectations of the strategy-specific payoffs are as in the previous exploration.

The stochasticity introduces interesting features of the RSP game. First, omitting the initial transient phase, the interactions and the spatial configuration will yield lengthy periods under which the three strategies fluctuate in a stable cyclical manner, one strategy replacing another in a predictable sequence. However, now and then the system breaks down, and a new transient phase occurs (fig. 11.10(A)). These phases do not last long, and the system returns back to stable cyclic fluctuations. This will last until the system breaks down again, followed by a recovery of the stable cycles some time later on. During such periods the sequence order of the strategies may also change (fig. 11.10(B)). Another interesting feature is that the intervening transition phase may escalate to a take-over and extinctions (fig. 11.10). The fact that our simulated games did not reproduce the outcome by Sinervo and Lively (1996) is because we did not use their payoff matrix. Clearly, the RSP game is far from thoroughly analyzed.

## RSP game generalized to $\Pi$ strategies

The critical element for strategy co-existence and dynamics in the RSP game are the diagonal elements $\varepsilon$: with a proper selection of them, periodic strategy fluctuations follow (Maynard Smith 1982; Hofbauer and Sigmund 1998). We took the following as a general form of the payoff matrix:

$$\begin{bmatrix} \varepsilon & q & 0 & q & 0 & q \\ 0 & \varepsilon & q & 0 & q & 0 \\ q & 0 & \varepsilon & q & 0 & q \\ 0 & q & 0 & \varepsilon & q & 0 \\ q & 0 & q & 0 & \varepsilon & q \\ 0 & q & 0 & q & 0 & \varepsilon \end{bmatrix}.$$

To reveal the pattern, it is here shown in a complete form for six strategies, $\Pi = 6$, but there is no limit to extend it to any number of playing strategies $\Pi$. The matrix entries are, for simplicity, of three kinds: the diagonal elements $\varepsilon$ are payoffs when two individuals of the same strategy play against each other, the alternating elements, $q$ and $0$, ensure that we have the sequence: $R > P > S > T > U > \ldots \Pi > R$, the RSP game sequence. The diagonal of the payoff matrix $\varepsilon$ can take any value. We limited our analysis into three cases, $\varepsilon = -1$, $\varepsilon = 0$, and $\varepsilon = 1$; always $q = 1$. In our simulations a $30 \times 30$ lattice was used. We initially had $\Pi = 10$, and we initiated the lattice in random proportion to the $\Pi$ strategies. The system was left running for $10\,000$ replicated plays for each cell (neighborhood being the eight surrounding cells). Fixed boundaries were used. After each play session, we scored the number of strategies present. The system was replicated 100 times for the three differing values of $\varepsilon$. The results are clear-cut (fig. 11.11): with $\varepsilon < 0$ and $\varepsilon = 0$ the remaining number of strategies was always an odd number. Only when $\varepsilon = 1$ did the remaining number of strategies include even numbers as well (fig. 11.11(D)). An explanation for this is provided in Box 11.5

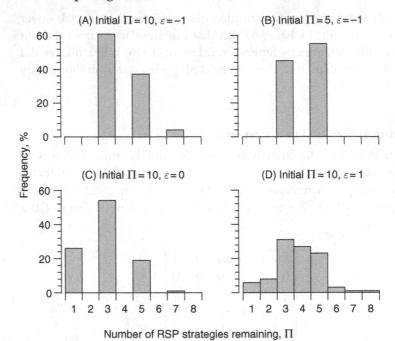

*Fig. 11.11.* Frequency distributions of the number of RSP strategies remaining with different initial strategy numbers $\Pi$, and the diagonal of the payoff matrix $\varepsilon$. Each panel is a result of 10 000 independently replicated runs (initially $\Pi = 10$).

---

**Box 11.5** · *Strategy dynamics in the Rock–Scissors–Paper game*

Assume there are $\Pi$ strategies in a generalized Rock–Scissors–Paper game (in the original game, $\Pi = 3$). The original game with three strategies is symmetrical, i.e., $A$ beats $B$, $B$ beats $C$, $C$ beats $A$. In a $\Pi$ strategy game, there are $\Pi(\Pi - 1)/2$ interactions. Let $x$ be the number of interactions in which each player is beating the other. Then, $\Pi x$ is the total number of interactions. Hence

$$\Pi x = \frac{\Pi(\Pi - 1)}{2},$$

i.e.,

$$x = \frac{\Pi - 1}{2}.$$

Integer solutions (required for an integer number of strategies) only hold for $\Pi$ being odd. That is, symmetrical games are only possible for an odd number of strategies.

We can also calculate the equilibrium proportions of the different strategies. The frequency of the $i$th strategy, $p_i$, varies over time as

$$p_i(t+1) = p_i(t)W_i(t)/\bar{W}(t),$$

where $W_i(t)$ is fitness of strategy $i$ at time $t$ and $\bar{W}(t)$ is mean fitness. Mean fitness is

$$\bar{W} = \sum p_i W_i.$$

At equilibrium, we know that $W_i(t)/\bar{W}(t)$ must equal 1, i.e., $W_i = \bar{W}$. $W_i$ is easily calculated from the payoff matrix. For example, from the payoff matrix

$$\begin{bmatrix} \varepsilon & a & b \\ c & \varepsilon & d \\ f & g & \varepsilon \end{bmatrix}.$$

$W_1 = \varepsilon p_1 + a p_2 + b p_3$. Using $W_i = \bar{W}$ and some algebra, we then have

$$p_1 = \frac{a(c-f) - cg + \varepsilon(f+g-\varepsilon)}{a(c\varepsilon - df) + b(f\varepsilon - cg) + \varepsilon(dg - \varepsilon^2)}$$

$$p_2 = \frac{a(d+\varepsilon) - b(g-\varepsilon) + dg - \varepsilon^2}{a(c\varepsilon - df) + b(f\varepsilon - cg) + \varepsilon(dg - \varepsilon^2)}$$

$$p_3 = \frac{b(c+f) + c\varepsilon - d(f-\varepsilon) - \varepsilon^2}{a(c\varepsilon - df) + b(f\varepsilon - cg) + \varepsilon(dg - \varepsilon^2)}.$$

Assume that a system with a stable co-existence of three (or any odd number) RSP strategies exists. An interesting question is: can the system be invaded by a new strategy? To address this question, we did some pilot invasion analyses (Box. 12.2). When $\varepsilon = -1$ or $\varepsilon = 0$, the invasion is not possible in small numbers. In large enough numbers of the invading new RSP strategist exist, two things can happen. The invader is capable of entering the system, but drives one of the extant strategies into extinction, or only one strategy survives. Invasion to such an odd-numbered RSP strategy system can occasionally succeed, but only if two alien strategies enter the system within a short enough time interval from the entrance of the first invading strategy. The outcomes of successful invasions we have seen have had periodic dynamics. When $\varepsilon = 1$, invasion is more often successful than in the previous case. However, with even-numbered co-existing RSP strategies, we quite often see stable dynamics instead of periodic fluctuations. It is obvious that the RSP game system, with its many extensions, clearly calls for more research.

## Evolution of co-operation

### Indirect reciprocity

We shall now return to the Nowak and Sigmund (1998) game (p. 277), in which random pairs of players are chosen in succession from a single population to interact, one as a potential donor and another as a recipient. If the donor acts as a co-operator, it will suffer a cost $c$ from helping the recipient, whereas the recipient – regardless of its type – will enjoy the benefit $b$ from being helped. Thus, in each interaction in pairs, the co-operating individual will create additional value in the group without being able to benefit itself at all. Instead, this individual suffers costs from being helpful as compared with the rest of the population.

Nowak and Sigmund (1998) proposed that indirect reciprocity through image scoring can provide a basis for the evolution of conditional co-operation. Unlike in most formulations of the evolution of co-operation (see Dugatkin 1998 for a review) direct reciprocity with repeated interactions cannot occur in the Nowak–Sigmund game. The crucial feature is that each individual has a reputation, or image, which increases every time the individual helps another one. The reputation is known to every other member in the population. It follows that the strategies, whether to co-operate or not, can be made to depend conditionally on the image scores of the recipients. Nowak and Sigmund (1998) showed that, starting with random distribution of strategies, co-operative action may overcome the selfish action in the population. The details of the outcome, of course, depend on the payoff structure of the game. In particular, individuals helping those who have helped others in the past may be favored in selection. The problem of reciprocity and image scoring has also been analyzed in some detail by Riolo et al. (2001) and Milinski et al. (2001), and also with spatial elements by Leimar and Hammerstein (2001).

## Direct reciprocity

We shall now show – with a slight modification of the Nowak–Sigmund game – that image scores are not necessary for co-operative behavior to become common in the population. In these games, the strategies are pure: an individual either co-operates, $C$, or acts selfishly, $S$. Consider first a single population of $N$, $N = C + S$, composed of two types of individuals playing two different unconditional strategies. Individuals of type $C$ always co-operate or help in interactions in pairs as donors regardless of the type of the recipient, whereas individuals of type $S$ always play selfishly as donors. In each generation, pairs of players are drawn at random in successions of $n$ times to interact with each other. Each individual interaction results in a change in the fitness scores of the individuals in the pair, while the fitness scores of the rest of the population remain unchanged (i.e., they are living in implicit solitude). The fitness scores of the interacting individuals increase according to the payoff table in fig.11.12. Thus, an individual may be chosen for interactions from zero up to $n$ times. At the end of each generation individuals produce progeny of their kind in proportion to the fitness scores. In each generation the population size is held constant ($N = 100$, $n = 25$ or $n = 75$, initially $S:C = 1:1$, $c = 0.1$).

Simulation outcomes with a range of benefit-to-cost ratios and with a fixed number of encounters in pairs within each generation are shown in fig.11.13(A). When the benefit-to-cost ratio increases, the proportion of the games with the co-operative strategy winning will increase. It appears that with small numbers of plays ($n = 25$, i.e., limited mixing) the proportion of games taken over by the co-operative players is larger than with $n = 75$. The re-formulated game shows that image scoring is not necessary for the maintenance of co-operation in this game. It appears that when the benefit-to-cost ratio is large enough one may observe frequent take-overs by co-operative strategies in the population. Nonetheless, despite the benefit-to-cost ratio being large enough co-operation never becomes invasion-proof against selfish individuals. The same is true for the invasion ability of selfish-only populations by co-operative individuals. In fact, within a rather wide benefit-to-cost ratio we find that with matching parameter values a proportion of the games will evolve to all-selfish populations, or to all-co-operative populations (fig. 11.13(B)).

We shall now extend this game into a spatial context by using a coupled lattice of size $100 \times 100$. The neighborhood size is taken to be one cell layer around the focal cell. The lattice is initialized with 1:1 ratio of $S$ and

Pay-off matrix

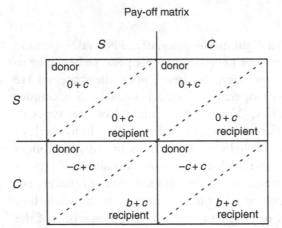

*Fig. 11.12.* Payoff matrix for the direct reciprocity game. There are two strategies: *S* is a selfish player, *C* is a co-operative player (donor payoffs above diagonal, recipient payoffs below diagonal in each box), *b* is benefit gained by the recipient of the interaction; *c* is the cost to the donor of the altruistic act.

*C* players totaling $N$, so that each cell is occupied and the average number of individuals per cell is 10 in one set of runs, and 20 in another one. For each round, every cell in the lattice will be scanned through. The number of games played in the focal cell is the number of individuals in that cell, $F_S + F_C$. A pair of individuals, a donor and a recipient, will be a random draw (with replacement) from these individuals. Payoffs of *S* and *C* players will be scored for the *S* and *C* strategists in the focal cell. As in the single-population games above, the *S* and *C* player composition in the focal cell will be replaced after their accumulated payoffs after each simulation round, when the entire lattice is updated. With the reproduction, we standardized population size to $N$ individuals. A total of 1000 generations were played before calculating the proportion of *S* and *C* individuals in the population. One more refinement: a given proportion $p$ of the $F_S + F_C$ plays in the focal cell can be played by drawing the donors and recipients from the focal cell itself. For a proportion of the plays, $q = (1 - p)$, donors and recipients will be drawn from among individuals in the neighboring layer of cells. The payoffs for *S* and *C* players in the $q$ plays will be scored for *S* and *C* in the focal cell, respectively. Thus, $q$ is a measure of redistribution of individuals.

We first set $q = 0.1$ and played an entire range of benefit-to-cost ratios from 2 to 100. Stable co-existence of *S* and *C* strategies emerges over almost the entire $b/c$ range (fig. 11.13(C)), though at low $b/c$ ratios only a

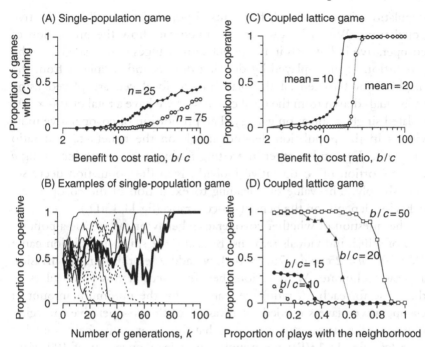

*Fig. 11.13.* (A) Proportion of single-population games ending in populations with all co-operating individuals against a range of differing benefit-to-cost ratios, $b/c$ (the number of replicated runs is 100 with initially 1:1 ratio of $S{:}C$, $N=100$, random draws of donors and recipients for each round numbers, either 25 or 75, 100 generations played). (B) Examples of outcomes of the single-population donor-recipient game with $b/c=30$, $n=25$. A proportion of the games end with all co-operating individuals, while other realizations with exactly same parameter values end up as all selfish populations. In some runs the fixation takes a long time (thick line) while in other runs the fixation to either end may last just a few generations. (C) The donor–recipient game is extended to space. Here we used a coupled lattice of $100 \times 100$ cells, 1000 plays (with 10% of plays by individuals drawn from the neighborhood before scoring the proportion of co-operative individuals, the average number of individuals per cell is either 10 or 20). A wide range of $b/c$ ratios is used in the plays. (D) Proportion of co-operative players in the population ($b/c$ ratio varies) in plays where donors and recipients are drawn in varying proportions from the neighborhood. The trajectories give the maximum proportion of neighborhood games tolerated when co-operation still persists (mean population size per cell is 10).

small proportion of individuals in the population are co-operative. It also appears that with larger population size the $b/c$ ratio has to be substantial for the co-operatives to succeed. Note also that there is a very sharp transition phase in the $b/c$ ratio (approximately 20 and 30 for the two population sizes examined) from mixed populations to all-co-operative populations. We shall now turn to investigate how the proportion of co-operative individuals in the population changes with changing $q$, the proportion of plays played by drawing donors and recipients from the neighborhood (when all the plays in the focal cell are played using individuals drawn from the focal cell ($q = 0$) we have a total of $100 \times 100$ isolated single-population games). The proportion of co-operative individuals in the population depends much on the benefit-to-cost ratio (fig. 11.13(D)), and another interesting finding is that with increasing $q$ the proportion of co-operative individuals in the population decreases very sharply. The value of $q$ leading to extinction of the co-operative behavior depends on the benefit-to-cost ratio (fig.11.13(D)).

The question of whether co-operative behavior can invade a population of selfish individuals remains. By using the single-population game ($N = 100$, $n = 25$ or $n = 75$, $c = 0.1$), we addressed this using an invasion analysis technique. With various benefit-to-cost ratios (1000 plays of the game for each $b/c$ ratio) we searched for the minimum amount of co-operative players needed for invasion of the co-operative strategies. Our score for a successful invasion is that the co-operative behavior takes over (cf., fig. 11.13(B)) the population at least once out of 100 trials. In this game, it is possible for individuals using co-operative behavior to invade a population of selfish individuals (fig. 11.14). However, when the benefit-to-cost ratio is small, the number of invaders has to be pretty large, while with increasing benefit-to-cost ratios the number of invaders needed for successful long-term co-existence of the two strategies becomes smaller and smaller—even to the point that when the benefit-to-cost ratio is very large only one co-operative individual is needed to invade the population of selfish individuals. Moreover, the invasion by co-operative players appears to be dependent on the proportion of the plays played with the neighborhood: a smaller number of invaders is needed for success when $n = 25$ than when $n = 75$ (fig. 11.14).

## Games and population dynamics

We shall finish this chapter by moving the evolutionary games towards population dynamics. While doing this, we shall follow the path cleared by Hofbauer and Sigmund (1998), albeit in much less technical terms. In population dynamics density-dependent feedback takes care of the population renewal process (Chapter 2). Survival and reproduction as a function of current and previous population sizes are the agents influencing

Single-population game

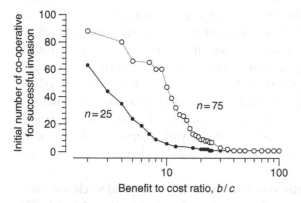

*Fig. 11.14.* The minimum initial number of co-operative individuals needed for successful invasion of an all-selfish population with varying benefit-to-cost ratios. For details of the invasion criterion, see p. 293–294. ($N = 100$, $n = 25$ or $n = 75$, $c = 0.1$).

changes in local population sizes. In dispersal-coupled populations, redistribution of individuals (Chapters 3 and 4) adds an extra flavor to the emerging dynamics of populations. In evolutionary games, the frequencies of the various strategies are due to the payoff matrix. In dispersal-coupled games the frequencies also depend on what strategies the neighbors are playing.

Consider a population in which the individuals use two or more interaction strategies. Each individual occupies one patch (or territory). The space is made up of a fixed number $N$ of such territories, which may be distributed evenly (lattice structure) or unevenly (heterogeneous environment). A territory may be occupied by only one individual with a certain strategy or it may be empty. The individuals produce a strategy-specific number of offspring, which then redistribute among the territories according to a given dispersal rule (dispersal kernel). After redistribution, each territory may attract a varying number of individuals of different strategies. However, before reproduction, the individuals queuing for a territory play a game among themselves. The strategy getting the highest payoff will win and occupy the territory for reproduction. The losers will not reproduce. The sequence is: payoff-dependent survival → reproduction → redistribution → payoff-dependent survival → and so on.

In the simplest case, reproduction is just strategy-specific offspring numbers produced by the survivors. However, it may as well be dictated by any of the renewal functions discussed in Chapter 2 (e.g., in Box 2.1) (with the restriction that population size should be in integers, in individuals). For simplicity, we here use the first option, and will demonstrate the idea with the Rock–Scissors–Paper game using the payoff matrix used for fig. (11.7). The $N = 400$ territories are distributed into a $5 \times 5$ co-ordinate space with uniform random numbers, $n_R:n_S:n_P$ being initially 1:1:1 and each territory occupied. The three strategies all produce equal numbers of offspring per reproduction, $\lambda = 10$. The dispersal follows kernel II (p. 54), with $d_{MAX} = 2.5$ (median distance among the territories in these conditions). The territory occupant will be decided from among all the potential invaders based on the payoff matrix, Rock > Scissors, Paper > Rock, and Scissors > Paper. Only one individual can possess the territory at a time.

Recall that in the lattice context the RSP game yielded clearly cyclic dynamics with 9-year period length (see the inset in fig. 11.7(A)). The new setting makes the frequencies of the three strategies fluctuate more irregularly (fig. 11.15), as is also the case with the total population size. Note that in the spatial RSP game (discussed on pp. 279–284) the population remains stable, while frequencies of the strategies $R$, $S$, and $P$ fluctuate in a cyclic manner. In fact, the deterministic cycle of $n_R$, $n_S$, and $n_P$ is replaced by long-term fluctuations (fig. 11.15). The game was repeated in a regular $20 \times 20$ lattice (with 0.55 units being the distance between neighboring cells; the median distance between the cells is about the same as above). The results echo what was found with the irregular placing of the territories: cyclic dynamics turn into long-term and less regular fluctuations obeying the power law. In these plays, all strategies being equal in terms of reproduction, $\lambda$, and redistribution, $d_{MAX}$, the long-term average numbers match (fig. 11.15).

In the second exercise, we let both strategy-specific $\lambda$ and $d_{MAX}$ vary. Now the population sizes for the different strategies very much depend on what the other strategies are doing. For example, letting $\lambda_R = 30$, $\lambda_P = 10$ and $\lambda_S = 10$, increased only slightly the numbers of $R$, reduced much the numbers of $P$, and increased the numbers of $S$ (fig. 11.16(A)). Changes in the maximum distances dispersed by the different strategies brought up similar surprises. When the $d_{MAX}$ for $R$ was reduced, their numbers went down to a half of what they were with matching dispersal, while the numbers of $P$ almost doubled and $S$ suffered the most (fig. 11.16(B)). When we had strategy-specific differences both in $\lambda$ and $d_{MAX}$ we observed similar changes in strategy numbers (fig. 11.16(C)). That is, when one strategy is doing well in terms of either reproduction or dispersal, it is not necessarily the strategy that is gaining the most. Rather,

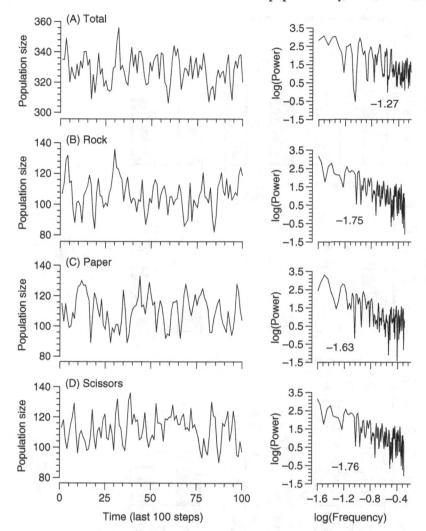

*Fig. 11.15.* The RSP game as a population renewal process. Population trajectories for the Rock (B), Paper (C) and Scissors (D) strategies (and for the pooled data (A)), with the corresponding power spectra (slopes inserted). The population data are displayed for the final 100 steps of a 500-steps-long simulation. The power spectra are calculated for data excluding the first 100 steps of the simulation.

the winning strategy, in terms of strategy-specific numbers, is the next one in the chain $R > P > S > R > \ldots$. This finding contradicts, to some extent, that of Frean and Abraham (2001). They used the RSP game to study competition among three species with the competitive hierarchy being $A > B > C > A$. In their models, the weakest competitor was the

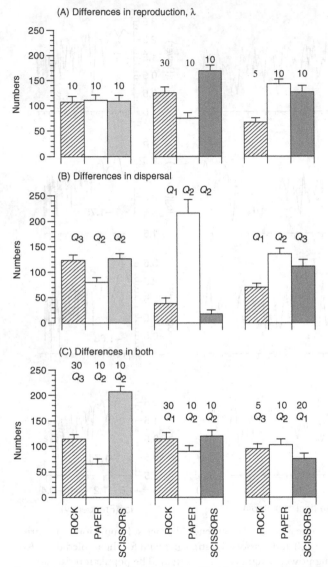

*Fig. 11.16.* Numbers of individuals of Rock, Paper, and Scissors strategists when the three strategies differ (A) in terms of reproduction, (B) in terms of dispersal, and (C) in both. In (A) maximum dispersal distance $d_{MAX}$ is always the median, distance ($Q_2$) among the population subunits. In (B) $\lambda = 10$ throughout while $d_{MAX}$ is either the lower quartile ($Q_1$), median or the upper quartile ($Q_3$) distance among the population subunits. In (C) the top row in the inserts is for reproduction, while the lower row is for dispersal. The simulations are run for 500 steps, and averages of 25 replicated runs (with 95% confidence limits) are given for each parameter combination.

one winning. In our case, the success is measured in terms of offspring production, and the strategy producing the most is not necessarily the one gaining the most in terms of strategy-specific numbers. A trade-off between reproduction and dispersal can balance the situation (fig. 11.16(C)), and the outcome is the expectation of the theory of allocation (Roff 1992; Stearns 1992).

## Summary

After introducing the classical evolutionary games, Prisoner's Dilemma, Hawk–Dove and Rock–Scissors–Paper, we extend them into space. The games show clear features of spatial self-organization: strategies aggregate and the aggregates move in space and time. Self-organization is also visible in strategy-specific frequencies aggregated over the population. These frequencies have periodic features matching the $1/f$ power law. The evolution of co-operation is discussed and it is shown that with large enough benefit-to-cost ratio co-operative behavior can invade a population of selfish individuals. The Rock–Scissors–Paper game is explored in more detail. It turns out that with negative entries in the diagonal of the payoff matrix (strategy playing against itself), only co-existing strategies are always odd-numbered. We also introduce a novel way to incorporate evolutionary games into population dynamics. We assume a territorial species, where individuals produce a strategy-specific number of off-spring, which then disperse. After redistribution, each territory may attract a varying number of individuals of different strategies. However, before reproduction, the individuals queuing for a territory play a game among themselves. The strategy getting the highest payoff will win and reproduce. With the Rock–Scissors–Paper game we are able to show, using these rules, that the deterministic cyclic fluctuation of the strategy frequencies turns out to be fluctuations matching the $1/f$ power law.

# 12 · *Evolutionary population dynamics*

The interface between the evolution of life history traits and population dynamics in temporally and spatially variable environments is the topic of this chapter. Thus, the frame for the life history processes is set by spatial and temporal fluctuations in population density. Here, we will focus primarily on modes of reproduction and we are especially interested in whether alternative reproductive strategies can co-exist in a population. We show that spatially structured populations may allow co-existence of various life history strategies that do not easily co-exist in a nonstructured environment. Also, intrinsic and external temporal fluctuations in the environment tend to enhance polymorphism in certain traits, e.g., iteroparity versus semelparity and whether monogamy or polygamy are favored reproductive strategies. In this chapter, we largely omit genetics, but a short comment on that aspect is found towards the end.

All life history problems are related to optimizing reproduction. One central question is how individuals allocate resources to survival and reproduction, and, for example, how offspring number, size, and sex are decided. When we put the evolution of life histories and optimizing behavioral decisions into the context of population dynamics, we will change our focus from optimizing to evolutionary stability. We may introduce, say, different reproductive behavioral patterns in our population models and ask which one of them will be an ESS. Technically, as is the tradition in the ESS literature, we will assume two or more distinct phenotypes competing (Maynard Smith 1982; Bulmer 1994). Naturally, we may also continue posing population ecology questions, and ask, for example, whether one or another behavioral pattern will stabilize or destabilize population dynamics (Fryxell and Lundberg 1997). Throughout this chapter, we are focusing on the role of spatial and temporal heterogeneity for the evolution of life history strategies (Levins 1968), and conversely how the dominance of one strategy over the other may change the population dynamics.

We will begin by considering sexual reproduction, mating systems, and reproductive allocation during the life span.

## Sexual reproduction

Various views exist about how males affect population dynamics. A most explicit position is that any population has enough potent males to fertilize all sexually mature females. This idea is implicitly reflected in the fact that, in standard population renewal models, the presence of sex is reduced to bookkeeping of female numbers (Chapter 2). The few studies that explicitly focus on the consequences of sexual reproduction on population dynamics are referring to dynamics of populations in isolation (Das Gupta 1972; Schoen 1983; Caswell and Weeks 1986; Dash and Cressman 1988; Doebeli and Koella 1994; Johnson 1994; Castillo-Chavez and Huang 1995; Ruxton 1995b; Doebeli 1996, 1997; Lindström and Kokko 1998). These authors have come to differing conclusions on the significance of sex in population dynamics: sex has no effect on population dynamics (Castillo-Chavez and Huang 1995), sex stabilizes the dynamics (Doebeli and Koella 1994; Ruxton 1995b; Doebeli 1996), or sex does not necessarily stabilize the dynamics (Lindström and Kokko 1998). One obvious reason for these conflicting views is the use of various approaches in different studies. Reduced propensity to complex dynamics has been observed in two-sex models with mixing of different genotypes (Doebeli and Koella 1994; Ruxton 1995b; Doebeli 1996). Polygyny and demographic sex differences become important if the role of males in reproduction and density dependence is explicitly considered (Lindström and Kokko 1998).

There is, however, one unifying aspect in the research on the consequences of sexual reproduction on population dynamics: the focus has been on dynamics of isolated populations. This contradicts other aspects of current research on population dynamics where the focus is on temporal behavior of populations coupled by redistributing individuals. In many species females and males differ in their dispersal ability and distances traveled (Cockburn et al. 1985; Matter 1996; Alonso et al. 1998). In addition, data exist to indicate that females and males may differ in their demography. Especially in polygynous, sexually sized dimorphic species males often suffer higher mortality risk than females (Clinton and LeBoeuf 1993; Anholt 1997; Jorgenson et al. 1997) while in monogamous species sex differences are often less dramatic and more variable (Berger and Cunningham 1995; Cooch et al. 1996; Gaillard et al. 1997).

There is also an often forgotten ecological aspect to the very existence of sexual reproduction. Sexual reproduction prevails despite the famous twofold cost of it compared to asexual reproduction. Doncaster *et al.* (2000) have shown that there are critical population dynamical circumstances that may or may not make sexual reproduction more or less likely, regardless of the genetic mechanisms promoting it. Thus, sexual reproduction not only affects the dynamics of populations – the dynamics of populations may affect the mode of reproduction.

**Sex in space**

The question we address here is whether spatial dynamics affect the dynamics of two-sex populations. In order to do this we will use a modification of the delayed Ricker model (Lindström and Kokko 1998) but we shall assume a large number of subpopulations that are coupled by dispersal (Ranta *et al.* 1999d). The mating system is either monogamy or polygamy. In our analyses, there are sex-correlated differences in dispersal rate and vulnerability to density dependence. The structure of the model is given in Box 12.1

---

**Box 12.1** · *A spatial model of sexual reproduction*

To track temporal changes in two-sex population dynamics in a spatial setting we write

$$\tilde{N}_{F,i}(t) = (1 - m_F)N_{F,i}(t) + m_F\bar{N}_F(t)$$
$$\tilde{N}_{M,i}(t) = (1 - m_M)N_{M,i}(t) + m_M\bar{N}_M(t)$$

and

$$N_{\bullet,i}(t + 1) = f_\bullet\big(\tilde{N}_{F,i}(t), \tilde{N}_{M,i}(t)\big).$$

Here $N_{F,i}(t)$ is the number of female and $N_{M,i}(t)$ the number of male individuals in the $i$th subpopulation at time $t$. The symbol $\bullet$ refers to either females or males. A given sex-specific proportion $m_\bullet$ of individuals leaves the natal population, which also receives immigrants from the other populations in the system, $\bar{N}_\bullet(t)$ is the mean density of females or males taken over all populations at time $t$, and $\tilde{N}_\bullet$ is the population size after dispersal which is spatially implicit. The dependence of the number of births $B$ on population size and the sex ratio is taken from the harmonic mean birth function (Caswell and Weeks 1986; Caswell 2001), which is modified to incorporate polygyny

$$B(N_{F,i}, N_{M,i}) = \frac{2\lambda N_{F,i} N_{M,i}}{N_{F,i} h^{-1} + N_{M,i}}.$$

The total number of offspring per female is $\lambda$ (here from 1 to 40). When $h = 1$ we have monogyny, while $h > 1$ refers to polygyny. The number 2 in the denominator is simply due to the harmonic mean. For the population renewal function $f_\bullet$ we shall use

$$N_{F,i}(t+1) = 0.5B\left[\tilde{N}_{F,i}(t), \tilde{N}_{M,i}(t)\right] \exp\left\{-\mu_F\left[\tilde{N}_{F,i}(t) + \tilde{N}_{M,i}(t)\right]\right\},$$
$$N_{M,i}(t+1) = 0.5B\left[\tilde{N}_{F,i}(t), \tilde{N}_{M,i}(t)\right] \exp\left\{-\mu_M\left[\tilde{N}_{F,i}(t) + \tilde{N}_{M,i}(t)\right]\right\}.$$

The parameter $\mu_\bullet$ is for sex-specific density-dependent responses. The simulations were done with 100 populations. They were initiated for female numbers drawn from uniformly distributed random numbers from the interval (0.5, 2). Initial male numbers per population were set to $N_{M,i} = N_{F,i} + \varepsilon_i$, where $\varepsilon =$ uniform random numbers from the interval (−0.05, 0.05). Here we define sex ratio as females/(females + males). The data to be displayed were sampled by randomly selecting one population from the 100 as the focal one. For the focal population, initial values were held constant for all values of $\lambda$. We let the system run for 1000 generations and displayed the next 100 generations in the bifurcation graphs.

Before we proceed, we set out a reference and explore the consequences of sex on population fluctuations in an isolated population. For that purpose, we used the model of Lindström and Kokko (1998; which is identical to the one given in Box 12.1, with single population). In terms of population variability (illustrated as bifurcation diagrams: fig. 12.1), polygamy yields far more complex population fluctuations than monogamy. Increasing the harem size in polygynous populations enhances the difference. The difference between the two breeding strategies also greatly affects the emergent sex ratio dynamics. However, to achieve this, one has to assume differential density dependence for the two sexes (details in Ranta et al. 1999d). In our particular example, we assume that males suffer more from density than females (fig. 12.1). Extending the system into space (more than one population; Box. 12.1) reveals an interesting feature. In monogamous populations, population fluctuations and sex ratio changes are much tamer (fig. 12.2(A),(B)) when compared to the single-population system. With a polygamous mating system, one can also

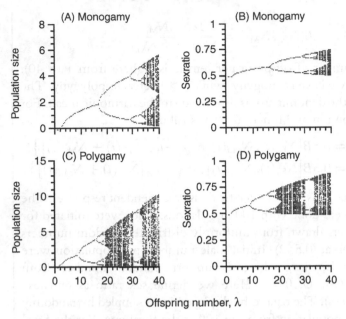

*Fig. 12.1.* Bifurcation diagrams of the monogamy–polygamy model (Box 12.1) of explicit sexual reproduction in an isolated population. Left-hand panels display population size variation while right-hand panels give sex ratio dynamics. Monogamous and polygamous breeding systems are treated separately (throughout $\mu_M = 1.2\mu_F$, in polygamy $h = 3$).

conclude (fig. 12.2(C),(D)) that details in population and sex ratio fluctuations will change according to the spatial extension. However, one can hardly conclude that spatial linkage will stabilize population dynamics.

These findings (Ranta *et al.* 1999d) suggest that sexual reproduction may be a more important determinant of the local dynamics than dispersal. An interesting result is the difference between monogamous and polygamous mating systems. A likely explanation is the variation in the population sex ratio induced by dispersal and the corresponding consequences of sex ratio variation on the local population growth rate. If the number of births occurring in the population at a given time is given by the harmonic mean birth function, the optimal sex ratio for population growth is given by $X_F = X_M \sqrt{h}$, where $X_F$ and $X_M$ refer to the number of females and males, and $h$ is the average harem size (Caswell and Weeks 1986). In monogamy, the population growth peaks at 50:50 sex ratio (see also Lindström and Kokko 1998 for a discussion of the interpretation of parameter values that relate to monogamy and polygamy). However,

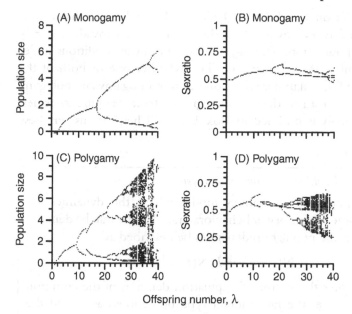

*Fig. 12.2.* Bifurcation diagrams for dispersal-coupled populations with explicit sexual reproduction (eq. 12.1) with monogamy (A), (B) and polygamy (C), (D) as the breeding system. Population size variation (A), (C) and sex ratio variation (B), (D) are displayed against net offspring number $\lambda$. Sex-specific dispersal rates are: $m_F = 0.05$, $m_M = 0.2$ and density-dependency values: $\mu_M = 1.2\mu_F$, in polygamy $h = 3$ (for a more detailed analysis see Ranta *et al.* 1999d).

in a polygamous system the number of females present becomes relatively more important than the number of males for population growth. Therefore, increased male mortality or deviations from a 50:50 sex ratio influence the population growth rate less than in a monogamous mating system.

### Monogamy or polygyny?

The rich variety of mating systems in nature is striking. It is also a great challenge for evolutionary explanations (Clutton-Brock 1991; Davies 1991; Kokko *et al.* 2002). For example, in birds, up to 90% of all species have (social) monogamy as the predominant mating system (Lack 1968). This observation alone is enough to raise our interest in the evolution of polygyny as a mating system (Ranta and Kaitala 1999). Polygyny is used here as a term for a mating system where one male can monopolize several females. We are not seeking the ultimate fitness benefits of polygyny or monogamy (Clutton-Brock 1991; Davies 1991). Rather, we shall examine whether a rare polygynous breeding strategy can invade a purely monogamous population. The objective is to analyze conditions under which a mutant strategy can invade, i.e., whether one or both of the strategies is an ESS, and under what conditions monogamy and polygyny can co-exist. In our study, the two strategies are treated as pure strategies. The invasion analysis outlined in Box 12.2 will help us answer these questions.

---

**Box 12.2** · *Evolutionary invasion analysis*

Consider a population dynamics model where the dynamics are affected by some life history or behavioral trait. Let the trait be denoted as $\psi$. Now, let the population dynamics be described as

$$\mathbf{N}(t+1) = F[\mathbf{N}(t), \psi],$$

where $\mathbf{N}$ denotes the vector of population densities of the common (resident) type. Let the rare mutant type be denoted as $\tilde{N}$ and the corresponding single-type dynamics as

$$\tilde{N}(t+1) = F[\tilde{N}(t), \tilde{\psi}].$$

The joint competition dynamics can be described as

$$N(t+1) = F[N(t), \tilde{N}(t), \psi]$$
$$\tilde{N}(t+1) = F[N(t), \tilde{N}(t), \tilde{\psi}].$$

Usually, the competition is assumed to be additive in population size, within age classes or within a population component, such as sex. Now, when studying the invasion dynamics, we assume that the common type has reached its typical attractor. Thus, in the simulations, we let the common type renew alone for a few hundred to several thousands of generations to remove the initial transient. When the dynamics of the common type are settled to its attractor we introduce a rare type into the system, initiating it with a small number. We let these two types compete for a certain period, and then we score the outcome. We have three options. First, the invasion does not succeed, and the rare type becomes extinct. Then, the common type is resistant to the invasion of this particular rare type. Second, the invasion succeeds, and the two types will co-exist. And, third, the invasion succeeds, and the rare type replaces the common type. These outcomes can be scored by the use of bifurcation diagrams.

An alternative for evaluating the invasion success is to calculate the invasion exponent (Metz *et al.* 1992). It gives an estimate of the growth rate of the rare type when the densities of the common type fluctuate within its attractor. Thus, this method is based on linearizing the population dynamics of the rare type at $(\bar{N}, 0)$ where $\bar{N}$ denotes the distribution of the population sizes of the attractor of the common type. If the (average) growth rate of the rare type exceeds 1 then invasion is considered to be successful. We prefer to use the bifurcation approach in this book, as it also gives information about the possible co-existence of different types.

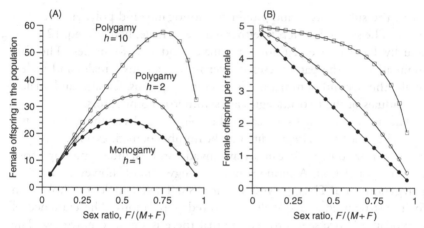

*Fig. 12.3.* The number of offspring per female after the harmonic birth function (Box 12.1) graphed against the sex ratio in a population of 20 individuals for monogamous and polygynous breeding strategies ($h$ is the number of fertile females a male can monopolize, $\lambda = 5$). For the numerical exercise we have selected 20 as the population size, $\lambda = 5$, and varied $h$ from monogamy ($h = 1$) to modest polygyny ($h = 2$) and to a relatively high harem size ($h = 10$). Modified from Ranta and Kaitala (1999).

The harmonic birth function (Box 12.1) provides a way to incorporate female $F$ and male $M$ roles into population dynamics. In that function, the parameter $h$ characterizes the mating system by indicating the number of fertile females a single male is capable of monopolizing. If $h = 1$ the population is monogamous and if $h > 1$ it is polygamous. The behavior of the birth function is explored under very simple settings (fig. 12.3). It is apparent that over a wide range of sex ratios, the polygynous mating strategy will beat monogamy, raising the question as to why there are monogamous species in nature at all. In an attempt to answer that question, we have to modify the last equation in Box 12.1 such that the invasion analysis can be undertaken. In practice, we shall set either monogamy or polygyny as the resident (common) strategy and examine under what conditions the alternative rare strategy can invade the population. The dynamics of the system, including dispersal, then become (Ranta and Kaitala 1999)

$$N(t+1)_{\bullet,m} = 0.5 \times B(N_{m,F}, N_{m,M})\exp\left\{-\mu_{\bullet,m}\left[\Sigma N(t)_{\bullet,m} + \Sigma N(t)_{\bullet,p}\right]\right\}$$

$$N(t+1)_{\bullet,p} = 0.5 \times B(N_{p,F}, N_{p,M})\exp\left\{-\mu_{\bullet,p}\left[\Sigma N(t)_{\bullet,m} + \Sigma N(t)_{\bullet,p}\right]\right\}.$$

$$(12.1)$$

Here, the subscripts $m$ and $p$ refer to monogamy and polygyny, respectively. The symbol $\bullet$ refers to either female or male, hence in eq. 12.1, we actually have two equations for females and two for males. The first summation in the parentheses is over monogamous females and males, while the second summation is over polygynous females and males. Individuals of the two strategies are symmetric competitors.

The invasion analysis revealed the following. Everything else being equal, i.e., the two strategies differ in the number of matings only, polygyny is a superior strategy. It can always invade and exclude a monogamous resident population. A monogamous strategy cannot, however, invade a polygynous one. Thus, polygyny is an ESS. These conclusions hold both for nonstructured and spatially structured populations. The existence of monogamy hence seems to require that there is a cost to polygyny. The costs of polygyny can be due to reduced fecundity, $\lambda_p < \lambda_m$, stronger density dependence, $\mu_{\bullet,p} > \mu_{\bullet,m}$, or higher dispersal costs, $d_{\bullet,p} < d_{\bullet,m}$.

If there is no spatial structure, invasion by polygynous (/monogamous) breeders into a population of monogamous (/polygynous) breeders is successful when $\lambda_p$ is sufficiently smaller than $\lambda_m$ (Ranta and Kaitala 1999). For a few polygyny levels ($h = 2$, 5, and 10) we sought for $\lambda_p$ values to make both invasion and co-existence possible for the two strategies. Co-existence is possible by reducing $\lambda_p$ relative to $\lambda_m$; the reduction has to be greater the larger the number of females a male can monopolize (fig. 12.4). Successful co-existence is possible also when the density dependence is stronger in the polygynous strategy than in monogamy. Long-term co-existence is possible only in the range of $1.17 < \mu_{M,p} \leq 1.2$ (fig. 12.4(A),(a)). Outside this range, monogamy and polygyny are exclusively ESSs (fig. 12.4(B),(b)). The border is strict and the value of $\lambda$ of demarcation depends on the value of $\mu_{M,p}$ (or of $\mu_{F,p}$ and

**Nonstructured population**

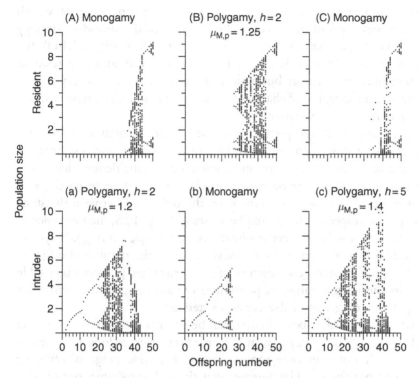

*Fig. 12.4.* Bifurcation graphs of resident (top row) and intruder (bottom row) reproductive strategies in a nonstructured population. The data are taken after 10 000 generations had elapsed since the introduction of the invading strategy. Relevant parameters are given in the different panels. The panels are paired so that (A) and (a), (B) and (b), (C) and (c) are from the same simulations. The dots in the panels indicate population size with each given offspring number, $\lambda$. For example, in (a) the dynamics are stable up to $\lambda = 12$, two-point periodic from then on up to $\lambda = 20$, from this point onward the period number increases until complex or chaotic dynamics are achieved from $\lambda > 25$. In the region of $\lambda$ where population sizes are graphed for only one strategy, this strategy is the ESS (e.g., in panel pairs (A) and (a) polygyny is an ESS with $\lambda < 35$). When population sizes are indicated for both reproductive strategies, long-term co-existence is possible (e.g., in panel pairs (A) and (a) the range of coexistence is $35 \leq \lambda \leq 43$). The resident strategy is left to establish itself for 10 000 generations. Then a rare $(1 \times 10^{-8})$ mutant of the other strategy attempts to invade the population. The success of invasion is checked after 10 000 time steps of the joint dynamics, eq. 12.1, have passed. A sample of the next 100 steps is taken for this analysis. For a more detailed account, see Ranta and Kaitala (1999).

$\mu_{M,p}$ if they both deviate from 1.0). The strategy-specific bifurcation graphs after the invasion are symmetric, i.e., independent of which strategy was resident or mutant. The invasion of the intruding strategy is not very sensitive to the polygyny parameter $h$ (fig. 12.4(c)) the parameter of relevance is $\mu_{\bullet,p}$. Also, polygyny can invade a resident monogamous population, but their stable co-existence is in a very limited range of $\lambda$ at the region of chaotic population dynamics, i.e., when there is strong temporal heterogeneity.

In spatially structured populations, invasion of polygyny into a monogamous population is possible. Also here the invasion system is symmetric. In the spatially structured population, differences in the density-dependent parameter are no longer necessary, provided the two strategies differ in their dispersal rates, $d_m > d_p$. However, the difference between the strategy-specific dispersal rates should be substantial (fig. 12.5; the difference in dispersal rates can be reduced if simultaneously also $\mu_{\bullet,p} > \mu_{\bullet,m}$). Polygyny is the ESS breeding strategy for low values of $\lambda$ (the number of offspring produced). Long-term co-existence of monogamy and polygyny is possible only in the chaotic range of population dynamics, again indicating that temporal heterogeneity enhances co-existence.

The results of the above analysis can be summarized as follows (Ranta and Kaitala 1999). Polygyny is an ESS breeding system as long as it imposes no costs. It can overcome monogamy over a wide range of offspring produced per female. This suggests that the polygyny superiority is due to the risk of females remaining unmated, this risk becoming smaller when the parameter $h$ increases. Likewise, the polygyny superiority assumes that monogamous systems suffer from difficulties of finding mates. Hence, Lindström and Kokko's (1998) suggestion that $h > 1$ might be appropriate for monogamy, but $h \gg 1$ for polygyny. Co-existence of monogamy and polygyny as pure strategies requires a cost of polygyny. One such cost is reduced offspring number $\lambda_p$. Alternatively, density dependence is stronger in the polygynous strategy ($\mu_{\bullet,p} > \mu_{\bullet,m}$). This yields, however, a very narrow range of $\lambda_\bullet$ and $\mu_{\bullet,p}$, where the long-term co-existence is possible. The third way to stable co-existence is to assume that both monogamous and polygynous breeders live in a spatially structured environment. Making the dispersal rate of polygynous breeders substantially lower than that of monogamous ones enables co-existence. There are few empirical studies of the potential costs of polygyny. A study of Spanish house sparrows (*Passer domesticus*) found, however, that polygynous males raised 30% fewer nestlings than monogamous males (Veiga 1990). In a Swedish tree sparrow (*P. montanus*) population, monogamous pairs raised on average 4.7

**Spatially structured population**

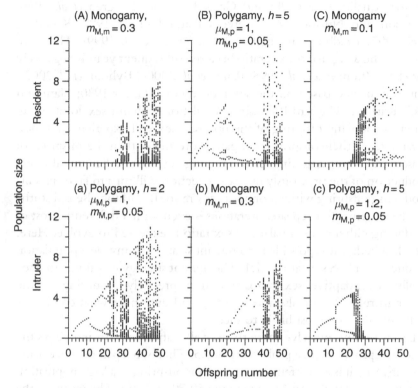

Fig. 12.5. Bifurcation graphs of resident (top row) and intruder (bottom row) reproductive strategies in a spatially structured population. See fig. 12.4 for more explanation.

fledglings, whereas polygynous pairs only raised 1.0 fledglings (Pekka Rintamäki, pers. com.). Kvarnemo *et al.* (2000) could also show that in an Australian seahorse (*Hippocampus subelongatus*), the polygynous male's inter-brood interval was about one-third longer than that of monogamous males. Thus, it appears that in these species polygynous males pay a fitness cost. It remains to be seen whether there is any evidence for the two other costs: stronger sensitivity to density dependence or poorer dispersal.

## Sex ratio dynamics

An increasing amount of evidence suggests that animals are capable of adjusting their progeny sex ratio in an adaptive fashion at the family level in relation to their own circumstances, such as attractiveness, condition or territory quality (Cassinello and Gomedio 1996; Ellegren *et al.* 1996; Svensson and Nilsson 1996; Bereczkei and Dunbar 1997; Komdeur *et al.* 1997; Bradbury and Blakey 1998; Kruuk *et al.* 1999). There are also data showing that the sex ratio between subsequent years is negatively correlated (Lummaa *et al.* 1998; Ranta *et al.* 2000c; Byholm *et al.* 2002). This is expected based on the sex ratio theory (Fisher 1930; Hamilton 1967; Karlin and Lessard 1986). In a population with one sex dominating, parents producing offspring of the opposite sex will experience a fitness advantage, as their offspring will gain more mates than the offspring of those parents not adjusting their offspring sex ratio. After initial over-production of the previously rare sex, selection will turn to favor parents producing offspring with an opposite sex ratio. If nothing else is disturbing the system, after repeated iterations – even with a differential cost of producing either sex – a balanced sex ratio is expected to evolve. Here, we ask whether it is possible for a mutant strategy to invade a population producing an ESS sex ratio of 1:1. The mutant strategy is assumed to obey locally the adaptive sex ratio allocation precisely according to the Fisherian rule. We are also interested to find out under what conditions such invasions are most likely to succeed.

Let us remind ourselves here that sex allocation in offspring obeys the Binomial process with a probability of 0.5. That is, if a pair produces only one offspring it is either female or male, but offspring in a large population of such parents will balance close to the 50:50 sex ratio. The results of the invasion analysis after the sex allocation model (Box 12.3) are rather straightforward: the parameter space (total number of breeding pairs, offspring number per pair, or, rather, the number of recruited offspring) enabling invasion of the mutant capable of locally adapting the sex ratio of their offspring is rather limited (fig. 12.6(A),(B)). The invasion is possible with relatively small population sizes and small offspring numbers per pair. Even within the narrow area of population sizes where invasions are possible, the success quickly goes down when the number of offspring produced per breeding pair increases (fig. 12.6(B)). The proportion of the mutants in the population after a successful establishment with one offspring per pair averages 43% for $n_{POP} = 10$; for $n_{POP} = 20$, it is 29%; for $n_{POP} = 50$, it is 15%. The corresponding values are 31%, 18%, and 5% when the offspring number per breeding pair is two. If the population consists entirely of the wild type obeying the global 1:1 sex ratio allocation rule, the frequency distribution of sex ratios within a single run is symmetrically distributed around 0.5, and there is no relationship

**Box 12.3** · *An inclusive-fitness model for sex ratio allocation*

Assume a diploid and bigametic population consisting of a number of breeding pairs or families, $n_{\text{WILD}}(t)$. Each season $t$, every pair $i$ will produce $b_{\text{WILD}}(t, i)$ offspring, and a proportion of these are daughters. Sex allocation for each breeding pair obeys binomial probability with the parameters $b_{\text{WILD}}(t, i)$ and 0.5. For the number of daughters $f_{\text{WILD}}(t, i)$ and sons $m_{\text{WILD}}(t, i)$ per breeding occasion per family we get

$$f_{\text{WILD}}(t, i) = \text{Binomial}[b_{\text{WILD}}(t, i), 0.5]$$
$$m_{\text{WILD}}(t, i) = b_{\text{WILD}}(t, i) - f_{\text{WILD}}(t, i).$$

Let $F_{\text{WILD}}(t) = \sum f_{\text{WILD}}(t, i)$ and $M_{\text{WILD}}(t) = \sum m_{\text{WILD}}(t, i)$, the summation being over all $n_{\text{WILD}}(t)$ families. For the population sex ratio we then get $s_{\text{WILD}}(t) = F_{\text{WILD}}(t) / [F_{\text{WILD}}(t) + M_{\text{WILD}}(t)]$.

Can such a population be invaded by a mutant strategy ($n_{\text{MUTANT}} = 1$), which adjusts offspring sex ratio using local information? For the intruding mutant family we shall write

$$f_{\text{MUTANT}}(t, i) = \text{Binomial}[b_{\text{MUTANT}}(t, i), p(t)]$$
$$m_{\text{MUTANT}}(t, i) = b_{\text{MUTANT}}(t, i) - f_{\text{MUTANT}}(t, i)$$

[$b_{\text{WILD}} \approx b_{\text{MUTANT}}$, and $p(1) = 0.5$]. The population-wide sex ratio, calculated over all wild type and mutant families, becomes

$$s_{\text{POP}}(t) = \frac{F_{\text{WILD}}(t) + F_{\text{MUTANT}}(t)}{F_{\text{WILD}}(t) + M_{\text{WILD}}(t) + F_{\text{MUTANT}}(t) + M_{\text{MUTANT}}(t)}.$$

The parameter $p(t)$ above is $p(t) = 1 - s_{\text{POP}}(t)$. Thus, if one sex is underrepresented in the population, the mutant is capable of producing more of the rare sex in the next generation.

At an invasion by the mutant strategy, the strategy-specific inclusive-fitness will be defined as the relative number of daughters plus the number of females inseminated by sons as follows (Bulmer 1994, p. 213)

$$w_{\text{WILD}}(t + 1) = \left[ F_{\text{WILD}}(t) + \frac{F_{\text{POP}}(t)M_{\text{WILD}}(t)}{M_{\text{POP}}(t)} \right] / n_{\text{WILD}}(t)$$

$$w_{\text{MUTANT}}(t + 1) = \left[ F_{\text{MUTANT}}(t) + \frac{F_{\text{POP}}(t)M_{\text{MUTANT}}(t)}{M_{\text{POP}}(t)} \right] / n_{\text{MUTANT}}(t).$$

These are the fitness measures for an average individual of the wild and mutant type, respectively. Let the number of breeding pairs in the population be $n_{POP}(t) = n_{WILD}(t) + n_{MUTANT}(t)$. In the next generation the number of breeding pairs formed is in proportion to the fitness of the wild and mutant type

$$n_{WILD}(t+1) = \frac{w_{WILD}(t)n_{WILD}}{w_{WILD}(t)n_{WILD} + w_{MUTANT}(t)n_{MUTANT}} n_{POP}(t)$$

$$n_{MUTANT}(t+1) = n_{POP}(t) - n_{WILD}(t).$$

It is now possible to explore whether a mutant, capable of adjusting its offspring sex ratio, manages to invade a population obeying the global 1:1 ESS sex ratio allocation rule. Our points of deviation from the majority of sex ratio theorizing (reviewed in Karlin and Lessard 1986) is that we are:

- dealing with populations of relatively small size (in our simulations $n_{POP}$ ranges from 10 to 200 breeding pairs),
- and with limited litter sizes ($b$ ranges from 1 to 10),
- and that the sex allocation process of the offspring born is governed by binomial stochasticity.

At the beginning of each simulation round we had $n_{WILD} = n_{POP} - 1$, and $n_{MUTANT} = 1$, and the offspring production was decided after the above equations for 1500 generations. At the end of each run, we scored the presence of the wild and the mutant type. To classify as a successful invasion, we accepted those cases where the proportion of the mutant increased from the initial of one breeding pair. The simulations were repeated 100 times to get an estimate of the probability of invasion by the mutant strategy.

between subsequent population sex ratios in time (fig. 12.6(C)). However, when the mutant is present (in our example ca. 30% the population) there is clear negative correlation between sex ratios separated by one time step (fig. 12.6(D)).

Why then should the success of invasion be highest in small populations and with small offspring numbers? The answer is simple. In large populations sex ratio is averaged over a large number of families and the sex ratio is close to 1:1 (Palmer 2000). With smaller populations, the stochasticity of the binomial probability process very easily deviates the sex ratio from the expected 1:1 and the deviation is bigger the smaller the offspring number is per pair.

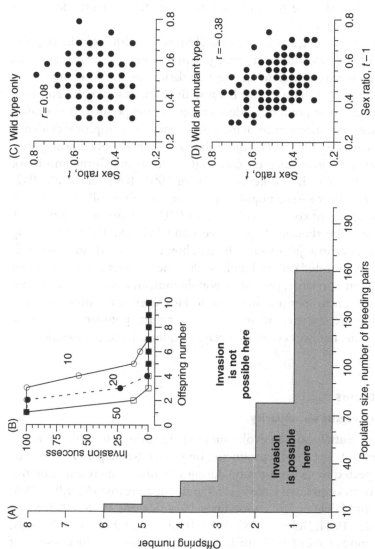

*Fig. 12.6.* (A) The parameter space (offspring number per pair against population size) enabling invasions by the mutant strategy, which is capable of locally adjusting the offspring sex ratio, into a population dominated by a wild strategy producing offspring in binomial probability of equal numbers of females and males. (B) The invasion success (results of 100 replicated runs each lasting 1500 generations) as a function of the number of offspring per pair for three different numbers of breeding pairs (viz., 10, 20, and 50) in the population. The panels (C) and (D) give examples of long-term sex ratio dynamics in a population (C) with only the wild type (obeying a 1:1 sex ratio) and in a population (D) where the wild and the mutant strategies co-exist in harmony (correlation coefficients inserted).

Note that we are not arguing that a mutant strategy could not invade a large population. For a successful invasion the large population has to be divided for smaller semi-independent subunits or neighborhoods from which the information about the prevailing sex ratio is gathered (Ranta *et al.* 2000c). Hence, our analysis suggests that if a mutation arises that can locally adjust the offspring sex ratio in a true Fisherian manner it can easily establish itself into a relatively small local population. Once established it can spread to neighboring populations and become established in them, given that they also are small enough to allow the stochastic deviations from the 1:1 sex ratio.

Another interesting aspect of this analysis is that although the population consists of two different strategies producing 1:1 or variable sex ratios, the average sex ratio in the population in the long run is, in statistical terms, 1:1. This occurs because the mutant strategy is continuously working to respond to random changes in the population sex ratio. This feature is consistent with an increasing amount of empirical evidence showing that – despite the fact that animals are capable of adjusting their progeny sex ratio in an adaptive fashion (Cassinello and Gomedino 1996; Ellegren *et al.* 1996; Bereczkei and Dunbar 1997; Komdeur *et al.* 1997; Lummaa *et al.* 1998) – the population sex ratio is still usually 1:1. Hence, there is no apparent conflict with Fisher's (1930) theory predicting a 1:1 sex ratio and the theories (e.g., Trivers and Williard 1973) predicting individual sex ratio adjustments. There is, however, a subtle snag with this reasoning. As pointed out by Frank (1990), the Fisherian theory assumes random matings in large, panmictic populations, and that both sexes gain equal fitness returns per unit investment. However, if the fitness returns, for example, are sex-specific or nonlinear, or if the population is inbred, then other mechanisms come into play, e.g., local mate competition.

# Life histories

### Iteroparity versus semelparity

One of the central issues in evolution of life histories is to understand the fitness trade-off between reproducing only once (semelparity) or reproducing repeated times (iteroparity) during a lifetime. The research on this topic has been keen since Cole's (1954) influential treatise (Murphy 1968; Charnov and Schaffer 1973; Goodman 1974; Stearns 1976, 1992; Horn 1978; Emlen 1984; Bulmer 1985, 1994; Roff 1992; Charlesworth 1994). Charnov and Schaffer (1973) made progress in addressing the question of whether to reproduce once or repeatedly per lifetime by proposing the following model to account for the temporal dynamics of semelparous, S, and iteroparous, I, reproducers

$$N_S(t+1) = P_J b_S N_S(t)$$
$$N_I(t+1) = (P_J b_I + P_A) N_I(t).$$

(12.2)

Here $N_\bullet$ refers to population size in different years (the symbol $\bullet$ refers in turn to semelparity and iteroparity), and $b_\bullet$ is the offspring number. Juvenile survival is $P_J$, while $P_A$ is the adult survival rate in iteroparous breeders (for the geometric rate of increase we have $\lambda_S = P_J b_S$ and $\lambda_I = P_J b_I + P_A$). If the populations are to increase at the same geometric rate, $\lambda_S = \lambda_I$, one has

$$b_S = b_I + \frac{P_A}{P_J}.$$

(12.3)

If $P_J = P_A$, resulting in $b_S = b_I + 1$, this is known as Cole's paradox: "[for a semelparous breeder] a clutch of 101 at age 1 would serve the same purpose as [for an iteroparous breeder] having a clutch of size 100 every year forever" (Charnov and Schaffer 1973, p. 791). That is, by just having one offspring more, the semelparous breeder would beat iteroparity. Yet, iteroparity is a widespread strategy. The paradox dissolves when realizing that, usually, $P_J < P_A$. Bulmer (1985, 1994) pointed out that the Charnov–Schaffer model, eq. 12.2, ignores density dependence affecting the dynamics of the populations. We shall here adopt the way Bulmer (1994) introduced density dependence in the model

$$P_J = p_j \exp\{-\alpha[b_S N_S(k) + b_I N_I(t)]\}.$$

(12.4)

The term $p_j$ is the maximum juvenile survival in the absence of competition from other juveniles. The parameter $\alpha$ is a scaling coefficient affecting the maximum values of $N_\bullet$. When eq. 12.4 is substituted into eq. 12.2, the desired effect of density dependence is achieved (eq. 12.4 is one version of the Ricker equation).

We can now pose two questions: is it possible for iteroparous breeders to invade a population of semelparous breeders? Is it possible for semelparous breeders to invade a population of iteroparous breeders? Answers will reveal whether one of the two lifetime reproductive strategies is an ESS. This calls for an invasion analysis (Box 12.2). Bulmer (1985, 1994) was able to show that an iteroparous breeding strategy – when initially

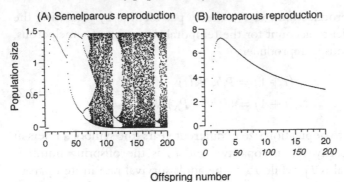

*Fig. 12.7.* Bifurcation diagrams for semelparous (A) and iteroparous (B) breeding strategies when alone. The bifurcation parameter is the number of offspring produced ($b_S$ and $b_I$). For both reproductive strategies, population dynamics obey eq. 12.1 where juvenile mortality ($P_J$) is replaced by eq. 12.4. The other parameter values are: $p_j = 0.2$, $P_A = 0.8$, $b_I = \nu b_S$ with $\nu = 0.1\alpha = 0.05$ (for semelparous breeders the $x$ axis labels in italics indicate $b_I$).

rare – will be unable to invade a population of semelparous breeders when the following inequalities hold (Bulmer 1994)

$$b_S > \frac{b_I}{1 - P_A} \text{ or } \frac{b_I}{b_S} + P_A < 1. \tag{12.5}$$

The same inequality gives the answer when a semelparous mutant strategy is capable of invading a population of iteroparous reproducers. From eq. 12.5 one can see that the inequality is independent of the juvenile survival rate, $P_J$. Bulmer (1994, p. 74, equating semelparity and iteroparity with annual and perennial breeding systems of plants) concludes, "Thus the annual form is at an unconditional advantage, and we should expect to find only annual plants under these circumstances. When the opposite inequality holds, the reverse is true and we should expect to find only perennials." For the Bulmer's inequality to hold, the only condition needed is that eq. 12.2, enhanced with the density-dependent juvenile mortality, leads to a stable equilibrium population size.

## Nonlinear dynamics, iteroparity versus semelparity

The invasion analysis here follows Ranta *et al.* (2000d). The first step is to describe the dynamics of populations obeying the two breeding strategies without the influence of the other. Figure 12.7 shows the stability properties of a system with semelparous breeders or iteroparous breeders only.

The dynamics are first stable (up to $b_S \approx 40$ after which they turn to more complex dynamics), while the dynamics for iteroparous breeders are stable for the entire range of $b_I$ used. Invasion of the iteroparous breeding system is not possible under stable dynamics (when $b_S < 40$), while invasion of the semelparous breeding system is possible. It is now of interest to look at the invasion success of a given reproductive strategy into a well-established resident population with the opposite reproductive strategy. For these analyses the parameter values were selected so that the inequality (12.5) holds (i.e., under stable dynamics, invasion by iteroparous breeders is not possible, while invasion by semelparous breeders is). Examples of the bifurcation graphs of the invasion analysis are displayed in fig. 12.8.

Successful invasion and long-term persistence by iteroparously breeding individuals is possible in the entire nonlinear (periodic and chaotic) region of the population dynamics of semelparous breeders examined here (fig. 12.8(A)). Second, the invaders often display complex population dynamics (this is much dependent on the parameter values used). In addition, the dynamics of the resident semelparous population may change, even drastically, as comparison of figs. 12.7 and 12.8 indicates. This is in line with the earlier finding of co-existence in temporally heterogeneous environments.

With an extensive analysis, Ranta *et al.* (2000d) could conclude that under complex (i.e., unstable) population dynamics the parameter range (offspring numbers, survival rates) for co-existence of iteroparous and semelparous life histories is much wider than in Bulmer's (1994) analysis under stable dynamics.

### Space–modulated dynamics, iteroparity versus semelparity

Our argument throughout this book is that spatial extension appears to give a better understanding of a wide range of ecological and evolutionary phenomena. This we will demonstrate here when addressing the question of whether a mutant life history can invade a population with a resident life history. We are especially interested in the region of stable population dynamics, i.e., the conditions in which invasion by the rare strategy is governed by the inequality in eq. 12.5, and when invasion otherwise is improbable or impossible.

In this system, we shall assume the environment consists of $n$ dispersal-coupled habitable patches of equal quality (Ranta *et al.* 2000e). As earlier, we assume that a constant fraction $m_{\bullet}$ ($0 \leq m_{\bullet} \leq 1$) of any given

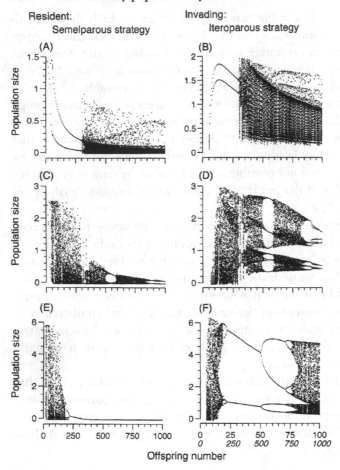

Resident: Semelparous strategy

Invading: Iteroparous strategy

*Fig. 12.8.* Examples of the invasion analysis. For these bifurcation graphs, semelparous breeders have been the resident reproductive strategy for 500 generations. Then, after a short randomly selected period, $1 \times 10^{-8}$ iteroparous breeders are introduced. Semelparous and iteroparous breeders are left to reproduce 600 generations, and their population sizes are displayed for the final 100 generations against the strategy-specific offspring numbers, $b$. The parameter values used are as follows: (A), (B) $p_j = 0.2$, $P_A = 0.8$; (C), (D) $p_j = 0.35$, $P_A = 0.35$; (E), (F) $p_j = 0.8$, $P_A = 0.2$. Here, always, $v = 0.1$ (for semelparous breeders, the $x$ axis labels in italics indicate $b_I$). The protocol of the invasion analysis (Ranta *et al.* 2000c) is: first, either the semelparous or the iteroparous breeding strategy was the resident strategy. The selected values of $b_\bullet$, $p_j$ and $P_A$ were initiated with $N_\bullet(1) = 1$ and left to renew for 500 generations. Then a period drawn from uniform random distribution [0, 30] was applied, after which the opposite reproductive strategy was introduced with a frequency of $1 \times 10^{-8}$. The populations were left to renew for 600 generations, and the final 100 generations were taken as a sample of the achieved dynamics to illustrate bifurcation against $b_S$ and to score whether the invasion was successful or not.

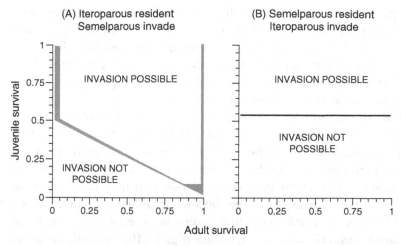

*Fig. 12.9.* Parameter space (juvenile and adult survival rates, $p_j$ and $P_A$, respectively), where invasion by a rare mutant is possible. (A) For the invading semelparous breeders, values above the lower rim of the gray border indicate parameter combinations calling for spatially explicit systems; the borderline indicates juvenile and adult survival rates enabling invasions. The upper rim of the gray border approximates parameter combinations where the spatially implicit structure allows successful invasion (note that close to adult survival values the divider zone goes up steeply). In the white area surrounded by the gray border, both spatially explicit and spatially implicit population structures enable invasion. (B) Invasion of resident populations of semelparous breeders are possible above the graph dividing line (close to $P_A = 0.51$). Above the demarcation line, both spatially implicit and explicit systems are possible.

population disperses annually. Thus, the spatially linked dynamics have the familiar form

$$N_{\bullet,i}(t+1) = (1 - m_\bullet)F[N_{\bullet,i}(t)] + \Sigma_{s,s\neq i}M_{\bullet,si}(t) \qquad (12.6)$$

where $N_{\bullet,i}(t)$ is the population size of semelparous or iteroparous breeders in patch $i$ at time $t$. The term $M_{\bullet,si}(t)$ refers to the number of immigrants arriving in patch $i$ from patch $s$ as a consequence of dispersal after dispersal kernel I (p. 53). An exhaustive parameter search ($m_\bullet$, $c_\bullet$, $p_j$, $P_A$, $b_S$, and $b_I$) was done to uncover the possibilities of co-existence of the two strategies (Ranta *et al.* 2000d). In their explorations the parameters $v$, $b_S$, and $P_A$ were selected so that under stable dynamics invasion of the rare strategy is impossible in spatially nonstructured populations.

The results are clear-cut: invasion of the mutant strategy is possible for both reproductive strategies being the invaders (fig. 12.9). However, there are clear differences in the parameter space enabling the invasion.

Perhaps the most pronounced is that, for invading iteroparous breeders, the spatial structure of the population system can be both implicit and explicit (fig. 12.9(B)). For invading semelparous breeders, there is a narrow range of parameters in which invasion is possible in the implicit spatial structure alone, whereas most of the parameter space (fig. 12.9(A)) also allows invasion in spatially explicit systems.

For demonstration purposes only, we shall redo a few of the simulations reported by Ranta et al. (2000d). These simulations were run to illustrate some features of the rich behavior of population dynamics emerging after introducing the two life histories into a spatially variable environment. This exploration of population dynamics of semelparous and iteroparous breeders in the invasion-enabling parameter space (fig. 12.9) reveals a few interesting observations. First, the co-existing populations assume nonlinear dynamics (fig. 12.10). Second, the non-linearity emerges after a transient period of stable dynamics ruled by the established strategy. During the transient phase, the invader population builds a foothold in terms of $N_\bullet$. Third, the transient period is sensitive to many of the parameter values selected, as well as to the spatial configuration in the spatially explicit systems. Fourth, the locally co-existing populations often fluctuate in opposite phase. In the invasion-allowing parameter space, the percentage of negative cross correlation coefficients was 96% for the cases of invading semelparous breeders, and 100% for the cases of invading iteroparous breeders. This again underscores the fact that co-existence is promoted by heteorogeneity, whether it is spatial or temporal (e.g., Fryxell and Lundberg 1997; Ranta et al. 2001). Further details of the semelpary versus iteropary story are given in Ranta et al. (2001), Kaitala et al. (2002) and Tesar et al. (2002).

## Seed bank in annuals

Plants may also delay their reproduction by allocating a part of their offspring into a dormant state, known as the seed bank (e.g., Baskin and Baskin 1998). Delaying reproduction is an apparent maladaptation since the mean annual number of offspring is reduced. Seed bank gives a "storage effect" (Chesson 1986), which enables the renewal of the banker morph even after years of complete reproductive failure. Germination from the seed bank reduces variance in the yearly number of recruits. Allocation of seeds into bank may thus function as an evolutionary bet-hedging strategy (Cohen 1966; Bulmer 1984; Silvertown 1988; Rees and Long 1992; Rees 1994; Baskin and Baskin 1998; Clauss and Venable

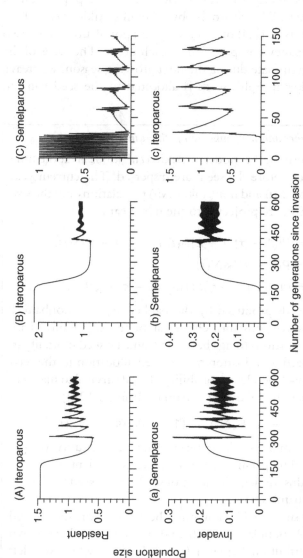

*Fig. 12.10.* Examples of population dynamics of the resident and the invading strategy. The resident population has been extant for 1000 generations, and an invading strategy enters with $N_{(1001),Invader} = 1 \times 10^{-8}$ into a randomly chosen subunit ($n = 25$ randomly allocated subunits in a $25 \times 25$ distance-unit space). The dynamics of the resident (upper row of panels) and the invader (lower row of panels) is then graphed against time for a population subunit the furthest away from the locus of invasion (note the differences both in $x$ and $y$ axis scales). The parameters in these examples were: $m_S = m_I = 0.1$; $c_S = c_I = 0.75$; $v = 0.2$, $p_j = 0.09$, $b = 75$; in (A), (a), (B), (b) $P_A = 0.92$, whereas $P_A = 0.8$ in (C), (c). See text for more details.

2000; Easterling and Ellner 2000). Environmental unpredictability favors allocation to the seed bank, especially when the environmental variation affects the success of seedling establishment (Cohen 1966; Rees 1994). A notable effect is also that part of the banker morph population is able to escape the density-dependent effects of competition, which the nonbanker morph is unable to do. Allocating seeds into seed bank as a life history strategy should be evaluated in a competitive situation, where its viability is determined by its ability to invade (Box 12.2) the population of conspecific nonbankers. We will do this by following Aikio *et al.* (2002).

Consider a model (Box 12.4) of an annual plant population where two morphs of seed dormancy compete against each other. The seeds of the nonbanker morph germinate during the next growing season, whereas a fraction of the banker morph's seeds is allocated to the seed bank to

---

**Box 12.4** · *Banker and nonbanker competition in annuals*

The population census in the model by Aikio *et al.* (2002) is at the time of seed ripening but before the seeds are dispersed. The current year density of the banker $B(t)$ and nonbanker $N(t)$ populations and the size of the seed bank, $S(t)$, are projected to the next year as

$$B(t + 1) = P[(1 - \sigma)(1 - \pi)b_B B(t) + \gamma(1 - \pi)S(t)]$$
$$N(t + 1) = P(1 - \pi)b_N N(t)$$
$$S(t + 1) = (1 - \pi)(1 - \gamma)S(t) + (1 - \pi)\sigma b_B B(t).$$

The numbers of seeds produced by the banker, $b_B$, and nonbanker morph, $b_N$, are positive constants. Other parameters of the model are probabilities or proportions (range between 0 and 1): seed mortality in the seed bank $\pi$, seed germination $\gamma$, and seed allocation to the seed bank $\sigma$. The density-dependent probability of seed survival to maturity $P$ is assumed to obey the following dynamics (Bulmer 1994)

$$P = p \exp\{-\nu(1 - \pi)[b_N N(t) + (1 - \sigma)b_B B(t) + \gamma S(t)]\}.$$

Here $p$ is the maximum seedling survival rate to maturity (ranges between 0 and 1). The parameter $\nu$ is a scaling coefficient, set to the value $\nu = 0.01$. In this model, allocation of seeds into a seed bank is set in a population dynamics frame.

The system was simulated by initiating the nonbanking morph and the banking morph populations with i.i.d. random numbers drawn from uniform distribution between 0 and 1. The system was left running for 10 000 generation, of which the final 500 were used to score the long-term average population size for the two strategists. If the average population density of either of the morphs was $<1 \times 10^{-4}$ it was scored to lose in the competition against the other morph. The stability properties of the system were evaluated from the value of the Lyapunov exponent (p. 109) $\lambda$ of the system. These were computed after von Bremen et al. (1997). Positive values of $\lambda$ suggest that the trajectory of population densities is sensitive to the initial conditions of simulation, which is an indication of complex population dynamics.

germinate one or several generations later. Seeds in the soil are subject to density-independent mortality. Regardless of the morph, population renewal follows the Ricker model (e.g., Bulmer 1994; Box 12.4), with the density dependence affecting directly the maturation of seedlings to adults. Germination from seed bank increases population density and thus brings an annual extra to the density dependence of seedling mortality. Therefore, not all of the seeds that survive to spring will produce viable seedlings to mature. Now, one can ask: when is it beneficial for an annual plant to allocate a fraction of seeds into the seed bank and when should a plant produce seeds that germinate without delay? Aikio et al. (2002) explored conditions that enable either of the morphs to become an ESS, or allow the co-existence of the two morphs. The outcome is conditional on the number of seeds produced, seedling survival to seed set, the fraction of seeds allocated to seed bank, the mortality of the seeds in the bank, and the proportion of the survivors being able to germinate.

The results of the competitive outcome split the parameter space into three regions: the banker morph wins (banker is ESS), the nonbanker morph wins (nonbanker is ESS) or the two morphs co-exist (figs. 12.11–12.13). All three competitive outcomes may show complex population dynamics. The banker morph wins if it has a higher seed production than the nonbanker one, regardless of seedling survival, seed mortality or germination rate (fig. 12.11). When both morphs have a low seed production rate, either one of the two morphs wins. Co-existence is possible when the nonbanker morph has a high seed production and the banker morph has a low seed production, if seedling survival is sufficiently high (fig. 12.11(A),(B),(D),(E)). When seedling survival is low, the parameter space where the morphs may co-exist is narrow or undetectable (fig. 12.11(C),(F)). At high levels of seed production, the banker morph

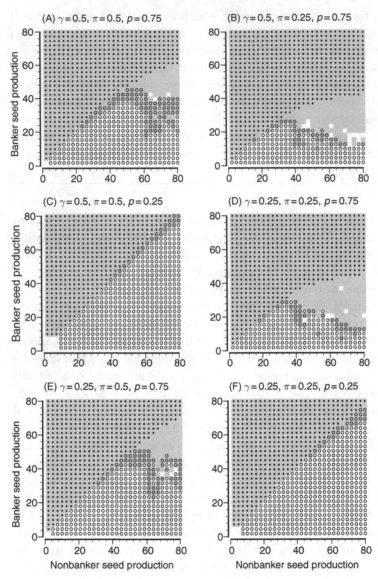

*Fig. 12.11.* The long-term outcome of the competition between banker and nonbanker morphs under different combinations of morph-specific seed production. The symbols indicate: ■ = the banker morphs wins (outcompetes the nonbanker), o = the nonbanker morph wins, no symbol = the morphs co-exist, gray background = complex population dynamics (indicated by a positive value of the Lyapunov exponent). Seed allocation rate was $\sigma = 0.1$, and other parameter values are as given above the corresponding panels. Modified from Aikio *et al.* (2002).

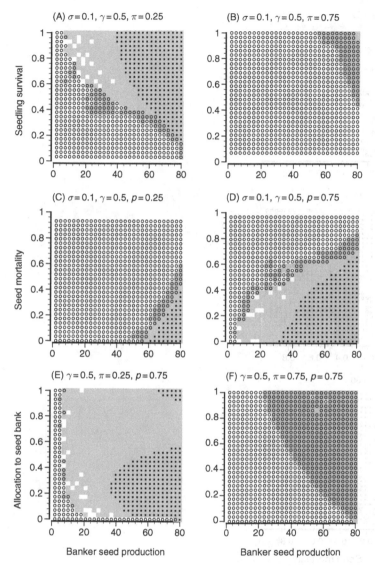

*Fig. 12.12.* The long-term outcome of competition between the banker and nonbanker morphs under different combinations of banker seed production and (A), (B) maximum seedling survival, (C), (D) = seed mortality and (E), (F) = allocation to seed bank. The seed production for the nonbanking morph was $b_N = 80$. The different panels indicate the effects of varying seed survival for the seedling survival (A), (B) and seedling survival for the seed survival (C), (D) inspections. Modified from Aikio *et al.* (2002).

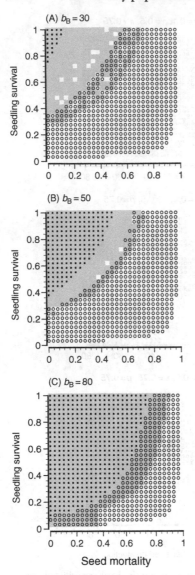

*Fig. 12.13.* The long-term outcome of competition between the banker and nonbanker morphs under different combinations of seed mortality and maximum seedling survival with three banker morph seed production rates. The seed production for the nonbanking morph was $b_N = 80$. The different panels indicate the effects of varying seed production by the banking morph, $b_B = 30$, $b_B = 50$, and $b_B = 80$. Other parameter values are: $\sigma = 0.1$, $\gamma = 0.5$. Modified from Aikio *et al.* (2002).

may win if it has a lower seed production than the nonbanker one. The parameter space of co-existence is wide when the nonbanker morph's seed production and seedling survival are high. The co-existence of the morphs is only slightly affected by seed mortality and germination rate (fig. 12.11(A),(B),(D),(E)).

The parameter space where the banker morph wins or the morphs co-exist increases with seedling survival rate (fig. 12.12 (A)) and decreases with increasing seed mortality rate (fig. 12.12(B)–(D)). Moderate allocation to the seed bank and high seed production by the banker morph benefit the banker morph, while high rates of allocation increase the possibility of co-existence by the two morphs (fig. 12.12(E)). A high seed mortality level disables banker morph persistence (fig. 12.12(F)). When the banker morph has a low seed production it wins only when seed mortality is low and seedling survival is high (fig. 12.12(A)). Low seedling survival and high seed mortality disables the persistence of both morphs (fig. 12.13). When seed production by the banker morph is close to that of the nonbanker morph, the banker morph wins at a larger parameter space, and the parameter space for co-existence becomes narrow (fig. 12.13). Again, neither of the morphs is viable under the combination of high seed mortality and low seedling survival.

The dynamics of the system were complex when the banker morph was the winning strategy. In addition, the co-existence of the morphs was strongly characterized by complex population dynamics. When the non-banker morph was ESS it showed complex population dynamics on those parts of the parameter space that are near to the parameter space for the co-existence of the two morphs. The long-term persistence of the morphs was increased by their own seed production and lowered by the seed production of the competing morph. Co-existence of the morphs was lowered by the banker morph seed production and increased by the nonbanker seed production. Seedling survival increased the long-term persistence of both morphs and their co-existence. High allocation to the seed bank and high seed mortality reduced the banker morph's persistence and increased the nonbanker morph's persistence and the co-existence of the morphs. Aikio et al. (2002) showed that all parameters of the model have nonlinear effects on the persistence of the morphs.

Complex population dynamics make the competitive pressure unpredictable for the two competing morphs. Results of the above analysis clearly tell us that even a simple question in evolutionary population ecology may yield a complex answer. The unpredictability of the environment favors allocation of seeds into the seed bank, which works as an evolutionary bet-hedging strategy (Cohen 1966; Bulmer 1984; Rees and Long 1992; Rees 1994; Baskin and Baskin 1998; Clauss and Venable 2000; Easterling and Ellner 2000). Seed bank appears to have a similar effect on competitive unpredictability as it has on environmental variation. Complex dynamics in plant population are most likely to be encountered in populations of annuals (Silvertown and Lovett Doust 1993; Gonzales-Andujar and Hughes 2000). These are also most dependent on the existence of a seed bank, since the lack of one could drive them to extinction in the event of a single reproductive failure. Unstable population dynamics may therefore have been closely involved in the evolution of a seed bank as a life history strategy.

Aikio *et al.* (2002) found that population dynamics were complex throughout the parameter space where the seed banking morph was an ESS. This seems to be opposite to the "storage effect" hypothesis, which predicts that seed bank stabilizes population dynamics by buffering population renewal against fluctuations in reproductive success (Chesson 1986). These results rather suggest that the instability of population dynamics is a requirement for seed banking to be ESS. Alternatively, strong environmental stochasticity could open for shifting relative advantages of the two morphs such that co-existence is promoted. Complex population dynamics do not, however, seem to make the banker morph a winner, since population dynamics are also complex in most of the parameter space where banker and nonbanker morphs co-exist. The complexity of population dynamics cannot be a sole consequence of the presence of the banker morph either, since the nonbanker morph may also have complex population dynamics when it is an ESS.

## Evolutionary genetics

Throughout this book, we have largely omitted genetics, not because we regard it as unimportant or irrelevant, but because we have had a strong focus on population and community dynamics in "ecological time." Clearly, there is no meaningful distinction between the time scales of ecological and evolutionary processes; they are inevitably intertwined. Fully aware of that, we have nevertheless had a somewhat biased approach. In this, and in some of the preceding chapters, we have, however, come closer to the genetics domain. Hence, a few words on evolutionary genetics.

Some of the conclusions we arrive at when analyzing population and community dynamics also have their counterparts in population genetics.

Hopefully, the similar results obtained by different approaches, e.g., population genetics and game theory (including both density- and frequency-dependent feedback) indicate that we have come closer to understanding the systems we study. Population genetics has been much concerned about the role of spatial heterogeneity for the evolution and maintenance of (co-existing alternative) adaptations. For example, the debate on the role of random factors (e.g., genetic drift) versus deterministic and directional forces (natural selection) originating with RA Fisher and Sewall Wright (e.g., Fisher 1930; Wright 1932, 1948) is directly related to the importance of spatial structure for creating the heterogeneity necessary for evolutionary change and the maintenance of, e.g., polymorphism. This debate is not settled (e.g., Coyne *et al.* 1997; Wade and Goodnight 1998).

The results shown in this chapter suggest that spatial and temporal variability may be equally important for the maintenance of alternative strategies. Genetics have that differently. If there is only temporal variability, it has to be strong and relatively "blue" (as in the chaotic dynamics promoting co-existence), otherwise alternative genotypes may have difficulties in coping with one another, as in more stable environments where one would be the winner (see also, e.g., Felsenstein 1976; Wade and Kalisz 1990; Svensson and Sinervo 2000 for the role of spatial heterogeneity).

As we have shown here, the density-dependent feedback is an important factor for the evolution of alternative life history strategies (cf. Charlesworth 1971; Roughgarden 1971). Recent work by Sinervo and Svensson and colleagues (Sinervo *et al.* 2000; Svensson *et al.* 2001a, 2001b) nicely show how a combination of frequency- and density-dependent selection may operate in nature for the maintenance of polymorphism.

The ecology of populations is indeed a rich branch of the scientific tree.

## Summary

The process of evolution can be abstracted as the course of strategy replacement, where "strategy" means a set of traits under selection. The evolution of life history traits (traits that determine age- (or stage-) specific survival and fecundity) is an important part of evolutionary population ecology. Theoretical explorations of life history evolution often conclude that there is only one optimal solution to a given life history problem and that the co-existence of two or more strategies is "impossible." In this chapter, we show that co-existence is indeed possible, and even likely, if we allow for heterogeneous environments, both spatially and temporally. We show, for example, that in a homogenous world, polygyny is always a better mating strategy (i.e., the ESS mating strategy) than monogamy (assuming monogamy to have the cost of finding mates), unless polygyny carries some costs that monogamy does not. However, in temporally varying environments (due to intrinsic variability – cycles or chaos – or environmental stochasticity) polygyny and monogamy might very well co-exist as alternative mating strategies. The same is often true for spatially heterogeneous environments. We show that the same conclusion holds for, e.g., iteroparity versus semelparity and in plants with or without seed banks. Spatial and temporal heterogeneity open alternating windows for the respective strategy to be more successful relative to the other. That is, at any given moment, the optimal solution may favor one over the other, but over longer time spans (or larger spatial scales), both strategies will thrive. We also show in this chapter that the evolutionary forces shaping life histories may have profound effects on the dynamics of populations, as well as the reverse: the dynamics of populations may have consequences for the possible evolutionary avenues.

# 13 · Epilogue

Twelve short chapters, some of them rather superficial, is, of course, not much for a book on the ecology of populations. In most chapters, we have caught glimpses of intriguing and sometimes unexpected population phenomena; in others, we have been able to reach more definitive and firm conclusions and somewhat deeper understanding. This book does not, however, primarily summarize and synthesize; rather, it illustrates a set of approaches and points of departure for studies and analyses yet to be done. This book is a manifestation of ecological and evolutionary significance of dispersal-linkage in spatially structured populations. If the book serves its purpose as a source of inspiration, we have performed well.

The power of modern computers has made it easy to simulate complicated population processes with various sources of environmental stochasticity, population structure, and spatial heterogeneity. That is not to say that we are therefore necessarily closer to a more robust understanding of the ecology of populations, but it helps. Real understanding can only be achieved if there is a theory to aid us in obtaining insights. Such a theory does arguably exist for the temporal structure of population abundance (Turchin 1999; Berryman and Turchin 2001). We have the data, means to analyze them, and the theory to interpret the results for single-population dynamics in uniform space. This is the "vector" approach to population processes, i.e., a vector $[x_1, x_2, x_3, \ldots x_k]$ of abundances $x$ from time step 1 to $k$ is under scrutiny (possibly together with vectors of relevant environmental variables). By analogy, the analysis is obviously extended by letting the system take a matrix format by structuring the system either by stage (of the focal population) or by adding other populations presumably interacting with the focal one. The latter is a multispecies community, the former a "community" of conspecifics. Likewise, by perfect analogy, a third dimension can be added, now representing spatial locations. This matrix version of population (and community) ecology hinges on the assumption that the measurable objects can be discretized. Although this may be an objectionable view of the natural world, it is nevertheless a very operational one – population and community *data* almost invariably come in discrete form. The correspondence between the theoretical constructs and what is observed is hence straightforward. As shown in Chapters 4 and 5, there are also indications that the identification of at least spatial location may be largely arbitrary without losing important information about spatial and temporal dynamics. We believe that the "matrix" view of population and community processes is indeed a fruitful one because it so transparently merges theory, statistics, and data.

## The feedback environment

The study of the distribution and abundance of organisms is still often regarded as different from the study of evolution. Of course, that is not really true. It is true that population ecologists often ask "how-questions" and not "why-questions." For example, the interest may be focused on describing and characterizing abundance and distribution patterns rather than asking which evolutionary forces have caused them. This putative dichotomy is neither valid nor true. As we show in the Introduction, understanding the (stage-specific) rates of births and deaths is the starting point for all ecology and evolutionary biology no matter how we want to label the respective approaches and brands of science. One could argue that there is a slight shift in focus between the two important biotic components of the feedback environment of an organism: density- and frequency-dependent processes. In this book, we have primarily studied the density-dependent feedback, although with some exceptions, especially in the last two chapters (Chapters 11 and 12). The feedback environment is becoming a key concept in ecology and evolutionary biology (e.g., Heino *et al.* 1997b; Meszéna and Metz 1999). The concept is not entirely unambiguously defined. When emphasizing the community level of organization, then the dimensionality of the feedback environment can be said to be equal to the number of discernible regulatory factors. In a two-species system (without population structure) the number of dimensions would be two (self + other species). With a stronger emphasis on genetics, there are potentially infinitely many dimensions, each one representing every possible mutant (genotype or strategy). Since we have largely omitted genetics in this treatise, we have regarded the dimensionality equal to the number of relevant population stages, and the number of interacting species (and their relevant stages), and finally other environmental factors (including abiotic ones) affecting the dynamics (and hence potentially the feedback structure). The conclusion that this reasoning implies "everything affects everything" is falsely drawn. A species-rich habitat is not necessarily more multidimensional than a poor habitat from the point of view of the organism. The relevant feedback environment only consists of the factors affecting the rate of change (i.e., fitness) of a population (which is not necessarily the case if $A$ eats $B$ or vice versa). We refer to Meszéna and Metz (1999) for rigorous definitions of the dimensionality.

We raise the question of the feedback environment dimensionality for two reasons; for what we have, and what we have not, done. Throughout this book, we have shown that environmental variability plays a fundamental role as a feedback component. Environmental "noise" is therefore an unfortunate term in this respect. Environmental variability is rather the unknowns we tend to sweep under the carpet, but may contain crucial information about components of the environment that we have not identified (theoretically or empirically). The simple noise term in for example time series analysis is an illustrative example. Consider the model

$$R(t) = a_0 + a_1 X(t) + w(t), \tag{13.1}$$

where $R(t)$ is the rate of change of the population, $X(t)$ is the natural logarithm of population density, $a_i$ are parameters and $w(t)$ is noise (cf. Chapter 2 and Box 2.7). The entire feedback environment is encompassed in $a_1$ and the noise term; although $a_1$ is usually interpreted as the strength of density dependence, being a metaphor for that feedback environment. The problem here is that it is impossible to know what aspects of it are captured by $a_1$ and the noise term. The last term encompasses, as a scalar, the entire feedback environment except self-regulation by population $X$.

What we have not done, at least not in any significant depth, is to consider simultaneously the ecological and evolutionary consequences of a serious feedback environment analysis. The tools of that trade go under the name of adaptive dynamics. Although adaptive dynamics are likely to become one of the most fundamental concepts in evolutionary biology during the years to come, we have felt that widening the scope of this book into that realm would simply be too much.

The feedback environment dimensionality problem is related to the two versions of the visibility problem we have outlined here (Chapter 2). If the feedback environment is ill-defined or poorly understood, then model estimation and interpretation can be very problematic (Jonzén et al. 2002a). Royama's problem (Chapter 2, p. 38) is an illustrative example of that challenge. The other side of the coin is the correlation between the environmental variability and population dynamics. As shown in Chapter 2, such correlations tend to be very risky to interpret, which raises caution against using biological time series as "bio-assays" of environmental change (cf. also Chapter 9, p. 219).

At the core of the feedback problem lie the issues brought up in Chapter 1 and the whole of Chapter 12. In that chapter, we show that the static view of life history solutions is restricted and that nonequilibrium situations are more promising and realistic scenarios for life history (phenotypic) evolution. We also argue that the trade-offs imposed when selection acts at whole organisms and whole-organism traits directly related to fitness must necessarily go beyond and extend the analysis of evolutionary genetics. This was also succinctly pointed out by Stearns (2000) in a recent review of the significance of population dynamics in the study of life history evolution.

## Generalizations

Throughout this book, we have attempted generalizations. However, as indicated at the very beginning, we are aware of the obvious bias we have had towards animal (especially vertebrate) populations. There is no question that this is a rather serious bias and we fully realize that there are a number of traits and life history properties that are unique to this restricted group of organisms. It is, however, also quite clear that most of the theory and statistical analysis of populations, i.e., the analytical tools we attempt to acquire, are indeed general. Chapters 2 and 3 briefly review the major building blocks of such general analyses of populations, although, for example, the modal life-style of most plants is not included. Chapter 4 is rich in examples of spatially synchronous dynamics in birds and mammals, but the principles and statistical analyses are perfectly general. Chapter 5 deals with more abstract pattern formation and is perhaps the chapter least related to vertebrate populations. Some of the most convincing examples of spatial pattern formation and self-organization come from invertebrates and infectious diseases (Grenfell et al. 2001). The brief exposé of structured populations in Chapter 6 is not very taxonomically biased, but is restricted in other ways. Again, we have assumed here that the population state can be discretized in unambiguous stages (e.g., age) although that may indeed be a questionable assumption (e.g., Metz et al. 1992; Tuljapurkar and Caswell 1997). It is likely, for example, that many organisms are better modeled by continuous time models than the discrete ones we have restricted ourselves to here. We still believe, however, that the suggested way of representing and analyzing maternal and cohort effects, as well as ontogenetic niche shifts has sufficient generality to merit closer attention.

Although single-population dynamics may be rather accurately dealt with by a very general approach, the problem expands quickly when studying communities. In Chapter 8, we made a limited number of excursions into that rich field, but without, we hope, too much loss of generality. The major problem in multispecies analyses is to accurately define the interaction terms. Model communities are relatively easily put together, but the question is to what extent such rather tightly connected systems can actually be mapped to real ones.

The generality of our treatment of habitat loss and landscape fragmentation (Chapter 8) remains an open question and must do so until we have reached a more fundamental agreement on basic definitions and concepts. The field of landscape ecology has by now an explicit history of several decades. Yet, the basic notion of, e.g., fragmentation remains open to debate. In Chapter 9 we become more explicitly applied, hence in a sense by definition reducing the degree of generality. The models reviewed are, however, primarily aimed at generalizations rather than management tools. The three last chapters address very fundamental concepts in evolutionary ecology but are, to the extent that it is possible to find examples and empirical support, primarily exemplified by a limited set of organisms. We feel, however, that this drawback is outweighed by the general aim. This book is neither a comprehensive catalog of animal and plant population dynamics, nor a handbook for the field biologist. Our aim has been to highlight a set of population and community ecology problems that are in need of firmer establishment in the ecological scientific community, or need to be picked up and developed further and in more breadth and depth. Hence, although there are examples found throughout, and although we strongly feel that there is a not fully used possibility to tie theory and empirical evolutionary ecology closer together (for example, by recent developments in model selection procedures; Burnham and Anderson 1998), this book has become "theoretical," i.e., general. Obviously, many of the models used for that theory building are not "realistic" in the sense that they mimic the details of real systems. No theory should do that. But we strongly believe that the combination of broad theoretical generalizations (preferably in analytically tractable mathematical format) together with numerical explorations of larger systems (simulations), and model selection procedures by data confrontation is the most fruitful way to expand the science of evolutionary ecology.

# References

Abrahams, M. V. 1986. Patch choice under perceptual constraints: a cause for departures from the IFD. *Behavioral Ecology and Sociobiology* **10**:409–415. Chp 10★

Abrams, P. A. 2001. Modelling the adaptive dynamics of traits involved in inter- and intraspecific interactions: an assessment of three methods. *Ecology Letters* **4**:166–175. Chp 1

Acosta, C. A. 2002. Spatially explicit dispersal dynamics and equilibrium population sizes in marine harvest refuges. *ICES Journal of Marine Science* **59**:458–468. Chp 9

Aikio, S., Ranta, E., Kaitala, V. and Lundberg, P. 2002. Seed bank in annuals: competition between banker and non-banker morphs. *Journal of Theoretical Biology* **217**:341–349. Chp 12

Allen, J. C., Schaffer, W. M. and Rosko, D. 1993. Chaos reduces species extinction by amplifying local population noise. *Nature* **364**:229–232. Chp 3, Chp 8

Alonso, J. C., Martin, E., Alonso, J. A. and Morales, M. B. 1998. Proximate and ultimate causes of natal dispersal in the great bustard *Otis tarda*. *Behavioral Ecology* **9**:243–252. Chp 12

Anderson, H. M., Hutson, V. and Law, R. 1992. On the conditions for permanence of species in ecological communities. *American Naturalist* **139**:663–668. Chp 7

Andrewartha, H. G. and Birch, L. C. 1954. *The Distribution and Abundance of Animals*. Chicago, Iu: University of Chicago Press. Chp 1

Anholt, B. R. 1997. Sexual size dimorphism and sex-specific survival in adults of the damselfly *Lestes disjunctus*. *Ecological Entomology* **22**:127–132. Chp 12

Arendt, J. D. and Wilson, D. S. 1997. Optimistic growth: competition and an ontogenetic niche-shift select for rapid growth in pumpkinseed sunfish (*Lepomis gibbosus*). *Evolution* **51**:1946–1954. Chp 6

Axelrod, R. 1984. *The Evolution of Co-operation*. London: Penguin Books. Chp 11

Axelrod, R. and Hamilton, W. D. 1981. The evolution of cooperation. *Science* **211**:1390–1396. Chp 11

Bascompte, J. and Solé, R. V. eds. 1997. *Modeling Spatiotemporal Dynamics in Ecology*. Berlin: Springer-Verlag. Chp 3, Chp 8

Bascompte, J., Solé, R. V. and Martinez, N. 1997. Population cycles and spatial patterns in snowshoe hares: an individual-oriented simulation. *Journal of Theoretical Biology* **187**:213–222. Chp 5

★ Refers to the chapter where the reference appears

Baskin, C. C. and Baskin, J. M. 1998. *Seeds – Ecology, Biogeography, and Evolution of Dormancy and Germination*. London: Academic Press. Chp 12

Batzli, G. O. 1992. Dynamics of small mammal populations: a review. In McCullough, D. R. and Barrett, R. H. eds. *Wildlife 2001: Populations*. London: Elsevier, pp. 831–850. Chp 8

Beckerman, A., Benton, T., Ranta, E., Kaitala, V. and Lundberg, P. 2001. Population dynamic consequences of delayed life history effects. *Trends in Ecology and Evolution* **70**:590–599. Chp 2, Chp 6

Beddington, J. R. and May, R. M. 1977. Harvesting natural populations in a randomly fluctuating environment. *Science* **197**:463–465. Chp 9

Begon, M., Harper, J. L. and Townsend, C. R. 1996. *Ecology*. Oxford: Blackwell Science. Chp 2

Benado, M. 1997. Second-order density-dependence in a Drosophilid community in La Florida, Santiago, Chile. *Revista Chilena de Historia Natural* **70**:415–420. Chp 6

Benton, T. G., Ranta, E., Kaitala, V. and Beckerman, A. P. 2001. Maternal effects and the stability of population dynamics in noisy environments. *Journal of Animal Ecology* **70**:590–599. Chp 6

Bereczkei, T. and Dunbar, R. I. M. 1997. Female-biased reproductive strategies in a Hungarian Gypsy population. *Proceedings of the Royal Society of London B* **264**:17–22. Chp 12

Berger, J. and Cunningham, C. 1995. Predation, sensitivity and sex: why female black rhinoceroses outlive males. *Behavioral Ecology* **6**:57–64. Chp 12

Bernardo, J. 1996. The particular maternal effect of propagule size, especially egg size: patterns, models, quality of evidence and interpretations. *American Zoologist* **36**:216–236. Chp 6

Berryman, A. A. 1996. What causes population cycles of forest Lepidoptera? *Trends in Ecology and Evolution* **11**:28–32. Chp 6

Berryman, A. A. 2002. Population: a central concept for ecology? *Oikos* **97**:439–442. Chp 1, Chp 2

Berryman, A. A. and Turchin, P. 1997. Detection of delayed density dependence: comment. *Ecology* **78**:318–320. Chp 8

Berryman, A. A. and Turchin, P. 2001. Identifying the density-dependent structure underlying ecological time series. *Oikos* **92**:265–270. Chp 2, Chp 13

Beverton, R. J. H. and Holt, S. J. 1957. On the dynamics of exploited fish populations. *Fisheries Investigation Series 2* **19**:1–533. Chp 2, Chp 9

Bisonette, J. A. ed. 1997. *Wildlife and Landscape Ecology: Effects of Pattern and Scale*. New York: Springer. Chp 9

Bjørnstad, O. N. and Bascompte, J. 2001. Synchrony and second-order correlation in host-parasitoid systems. *Journal of Animal Ecology* **70**:924–933. Chp 5

Bjørnstad, O. N., Falck, W. and Stenseth, N. C. 1995. Geographic gradient in small rodent density-fluctuations – a statistical modelling approach. *Proceedings of the Royal Society of London B* **262**:127–133. Chp 6

Bjørnstad, O. N., Begon, M., Stenseth, N. C., Falck, W., Sait, S. M. and Thompson, D. J. 1998. Population dynamics of the Indian meal moth: demographic stochasticity and delayed regulatory mechanisms. *Journal of Animal Ecology* **67**:110–126. Chp 6

Bjørnstad, O. N., Ims, R. A. and Lambin, X. 1999a. Spatial population dynamics: analysing patterns and processes of population synchrony. *Trends in Ecology and Evolution* **14**:427–432. Chp 4, Chp 5

Bjørnstad, O. N., Stenseth, N. C. and Saitoh, T. 1999b. Synchrony and scaling in dynamics of voles and mice in northern Japan. *Ecology* **80**:622–637. Chp 4

Bjørnstad, O. N., Sait, S. M., Stenseth, N. C., Thompson, D. J. and Begon, M. 2001. The impact of specialized enemies on the dimensionality of host dynamics. *Nature* **409**:1001–1006. Chp 2, Chp 6

Bjørnstad, O. N., Finkenstaedt, B. F. and Grenfell, B. T. 2002a. Dynamics of measles epidemics: estimating scaling of transmission rates using a time series SIR model. *Ecological Monographs* **72**:169–184. Chp 5

Bjørnstad, O. N., Peltonen, M., Liebhold, A. M. and Baltensweiler, W. 2002b. Waves of larch budmoth outbreaks in the European Alps. *Science* **298**:1020–1023. Chp 5

Blasius, B. and Stone, L. 2000. Nonlinearity and the Moran effect. *Nature* **406**:846–847. Chp 4

Blumstein, D. and Armitage, K. 1999. Cooperative breeding in marmots. *Oikos* **84**:369–382. Chp 6

den Boer, P. J. and Reddingius, J. 1996. *Regulation and Stabilization Paradigms in Ecology*. London: Chapman and Hall. Chp 2

Bolker, B. and Pacala S. W. 1999. Spatial moment equations for plant competition: understanding spatial strategies and the advantage of short dispersal. *American Naturalist* **153**:572–602. Chp 5

Boonstra, R., Krebs, C. J. and Stenseth, N. C. 1998. Population cycles in small mammals: the problem of explaining the low phase. *Ecology* **79**:1479–1488. Chp 6

Borrvall, C., Ebenman, B. and Johnson, T. 2000. Biodiversity lessens the risk of cascading extinction in model food webs. *Ecology Letters* **3**:131–136. Chp 7

Botsford, L. W., Method, R. D. and Johnston, W. E. 1983. Effort dynamics of the northern California Dungeness crab (*Cancer magister*) fishery. *Canadian Journal of Fisheries and Aquatic Sciences* **40**:337–346. Chp 9

Botsford, L. W., Castilla, J. C. and Peterson, C. H. 1997. The management of fisheries and marine ecosystems. *Science* **277**:509–515. Chp 9

Box, G. E. P., Jenkins, G. M. and Reinsel, G. C. 1994. *Time Series Analysis: Forecasting and Control*. Englewood Cliffs, N. J.: Prentice Hall. Chp 2, Chp 4, Chp 5, Chp 10

Boyce, M. S., Sinclair, A. R. E. and White, G. A. 2000. Seasonal compensation of predation and harvesting. *Oikos* **87**:419–426. Chp 9

Bradbury, R. B. and Blakey, J. K. 1998. Diet, maternal condition, and offspring sex ratio in the zebra finch, *Poephila guttata*. *Proceedings of the Royal Society of London B* **265**:895–899. Chp 12

von Bremen, H. F., Udwadia, F. E. and Proskurowski, W. 1997. An efficient QR based method for the computation of Lyapunov exponents. *Physica D* **101**:1–16. Chp 5, Chp 12

Brommer, J. E., Kokko, H. and Pietiäinen, H. 2000. Reproductive effort and reproductive value in periodic environments. *American Naturalist* **155**:454–472. Chp 6

Brown, J. H. 1984. On the relationship between abundance and distribution of species. *American Naturalist* **124**:255–279. Chp 7

Brown, J. L. 1987. *Helping and Communal Breeding in Birds: Ecology and Evolution.* Princeton, N. J.: Princeton University Press. Chp 6

Brown, J. S. and Vincent, T. L. 1992. Organization of predator-prey communities as an evolutionary game. *Evolution* **46**:1269–1283. Chp 7

Bulmer, M. G. 1974. A statistical analysis of the 10-year cycle in Canada. *Journal of Animal Ecology* **43**:701–718. Chp 5

Bulmer, M. G. 1984. Delayed germination of seeds: Cohen's model revisited. *Theoretical Population Biology* **26**:367–377. Chp 12

Bulmer, M. G. 1985. Selection of iteroparity in a variable environment. *American Naturalist* **126**:63–71. Chp 12

Bulmer, M. G. 1994. *Theoretical Evolutionary Ecology.* Sunderland, Mass.: Sinauer Associates. Chp 12

Buonaccorsi, J. P., Elkinton, J., Evans, S. R. and Liebhold, A. M. 2001. Measuring and testing for spatial synchrony. *Ecology* **82**:1668–1679. Chp 4

Burgman, M. A., Freson, S. and Akcakaya, H. R. 1993. *Risk Assessment in Conservation Biology.* London: Chapman & Hall. Chp 2, Chp 8, Chp 9

Burnham, K. P. and Andersson, D. R. 1984. Tests of compensatory vs. additive hypothesis of mortality in mallards. *Ecology* **65**:105–112. Chp 9

Burnham, K. P. and Anderson, D. R. 1998. *Model Selection and Inference: A Practical Information- theoretic Approach.* New York: Springer-Verlag. Chp 1, Chp 2, Chp 9, Chp 11, Chp 13

Burroughs, W. J. 1992. *Weather Cycles: Real or Imaginary?* Cambridge: Cambridge University Press. Chp 4, Chp 8

Butler, L. 1953. The nature of cycles in populations of Canadian mammals. *Canadian Journal of Zoology* **31**:242–262. Chp 4, Chp 5

Byholm, P., Ranta, E., Kaitala, V., Lindén, H., Saurola, P. and Wikman, M. 2002. Resource availability and goshawk offspring sex ratio: a large-scale ecological phenomenon. *Journal of Animal Ecology* **71**:994–1001. Chp 12

Cassinello, J. and Gomedino, M. 1996. Adaptive variation in litter size and sex ratio at birth in a sexually dimorphic ungulate. *Proceedings of the Royal Society of London B* **264**:1461–1466. Chp 12

Castillo-Chavez, C. and Huang, W. 1995. the logistic equation revisited: the two-sex case. *Mathematical Bioscience* **128**:299–316. Chp 12

Caswell, H. 2001. *Matrix Population Models*, 2nd edn. Sunderland, Mass.: Sinauer. Chp 1, Chp 2, Chp 3, Chp 6, Chp 12

Caswell, H. and Weeks, D. E. 1986. Two-sex models: chaos, extinction, and other dynamic consequences of sex. *American Naturalist* **128**:707–735. Chp 12

Cattadori, I. M. and Hudson, P. J. 1999. Temporal dynamics of grouse populations at the southern edge of their distribution: the Dolomitic Alps. *Ecography* **22**:374–3874. Chp 4

Cattadori, I. M., Merler, S. and Hudson, P. J. 2000. Searching for mechanisms of synchrony in spatially structured gamebird populations. *Journal of Animal Ecology* **69**:620–638. Chp 4

Caughley, G. 1994. Directions in conservation biology. *Journal of Animal Ecology* **63**:215–244. Chp 9

Charlesworth, B. 1971. Selection in density-regulated populations. *Ecology* **52**:469–474. Chp 12

Charlesworth, B. 1994. *Evolution in Age-structured Populations*. Cambridge: Cambridge University Press. Chp 12

Charnov, E. L. and Schaffer, W. M. 1973. Life-history consequences of natural selection: Cole's result revisited. *American Naturalist* **107**:791–793. Chp 12

Chatfield, C. 1999. *The Analysis of Time Series*, 5th edn. London: Chapman & Hall. Chp 1, Chp 2, Chp 4

Chesson, P. L. 1986. Environmental variation and the coexistence of species. In Diamond, J. and Case, T. J. eds. *Community Ecology*. New York: Harper and Row, pp. 240–256. Chp 12

Clark, C. J. 1990. *Mathematical Bioeconomics*, 2nd edn. New York: Wiley. Chp 9

Clauss, M. J. and Venable, D. L. 2000. Seed germination in desert annuals: an empirical test of adaptive bet hedging. *American Naturalist* **155**:168–186. Chp 12

Clinton, W. L. and LeBoeuf, B. J. 1993. Sexual selection's effects on male life-history and the pattern of male mortality. *Ecology* **74**:1884–1892. Chp 12

Clutton-Brock, T. H. 1991. *The Evolution of Parental Care*. Princeton, N. J.: Princeton University Press. Chp 6, Chp 12

Clutton-Brock, T. H. 2002. Behavioral ecology: breeding together. Kin selection and mutualism in cooperative vertebrates. *Science* **296**:69–72. Chp 11

Clutton-Brock, T. and Parker, G. A. 1995. Punishment in animal societies. *Nature* **373**:209–216. Chp 11

Clutton-Brock, T. H., Illius, A. W., Wilson, K., Grenfell, B. T., MacColl, A. D. C. and Albon, S. D. 1997. Stability and instability in ungulate populations: an empirical analysis. *American Naturalist* **149**:195–219. Chp 9

Cockburn, A., Scott, M. P. and Scotts, D. J. 1985. Inbreeding avoidance and male-biased dispersal in Antechinus spp (Marsupialia, Dasyuridae). *Animal Behavior* **33**:908–915. Chp 12

Cody, M. L. and Diamond, J. M. 1975. *Ecology and Evolution of Communities*. Cambridge, Mass.: Harvard University Press. Chp 7

Cohen, D. 1966. Optimizing reproduction in a randomly varying environment. *Journal of Theoretical Biology* **12**:110–126. Chp 12

Cohen, J. E. 1995. Unexpected dominance of high frequencies in chaotic nonlinear population models. *Nature* **378**:610–612. Chp 2

Cohen, Y., Vincent, T. L. and Brown, J. S. 1999. A G-function approach to fitness minima, fitness maxima, evolutionarily stable strategies, and adaptive landscapes. *Evolutionary Ecology Research* **1**:923–942. Chp 1

Cole, L. C. 1954. The population consequences of life history phenomena. *Quarterly Reviews of Biology* **29**:103–137. Chp 12

Comins, H. N., Hassell, M. P. and May, R. M. 1992. The spatial dynamics of host-parasitoid system. *Journal of Animal Ecology* **61**:735–748. Chp 5

Connell, J. H. 1980. Diversity and the coevolution of competitors, or the ghost of competition past. *Oikos* **35**:131–138. Chp 7

Connor, R. C. 1995. Altruism among non-relatives – alternatives to prisoners dilemma. *Trends in Ecology and Evolution* **10**:84–86. Chp 11

Cooch, E. G., Lank, D. B. and Cooke, F. 1996. Intraseasonal variation in the development of sexual size dimorphism in a precocial bird: evidence from the lesser snow goose. *Journal of Animal Ecology* **65**:439–450. Chp 12

Cooper, A. B. and Mangel, M. S. 1999. The dangers of undetected metapopulation structure for the conservation of salmonids. *Fishery Bulletin* **97**:213–226. Chp 9

Costantino, R. F., Cushing, J. M., Dennis, B. and Desharnais, R. A. 1995. Experimentally induced transitions in the dynamic behaviour of insect populations. *Nature* **375**:227–230. Chp 6

Costantino, R. F., Desharnais, R. A., Cushing, J. M. and Dennis, B. 1997. Chaotic dynamics in an insect population. *Science* **275**:389–391. Chp 6

Costantino, R. F., Cushing, J. M., Dennis, B., Desharnais, R. A. and Henson, S. M. 1998. Resonant population cycles in temporally fluctuating habitats. *Bulletin of Mathematical Biology* **60**:247–273. Chp 6

Coyne, J. A., Barton, N. H. and Turelli, M. 1997. Perspective: a critique of Sewall Wright's shifting balance theory of evolution. *Evolution* **51**:643–671. Chp 12

Creel, S. R. and Macdonald, D. W. 1995. Sociality, group size and reproductive suppression among carnivores. *Advances in Studies of Behavior* **24**:203–257. Chp 6

Crone, E. E. 1997. Parental environmental effects and cyclical dynamics in plant populations. *American Naturalist* **150**:708–729. Chp 6

Crone, E. E. and Taylor, D. R. 1996. Complex dynamics in experimental populations of an annual plant, *Cardamine pensylvanica*. *Ecology* **77**:289–299. Chp 6

Curry, R. 1989. Geographic variation in social organization of Galapagos mockingbirds: ecological correlates of group territoriality and cooperative breeding. *Behavioral Ecology and Sociobiology* **25**:147–160. Chp 6

Cushing, J. M., Costantino, R. F., Dennis, B., Desharnais, R. A. and Henson, S. M. 1998a. Non-linear population dynamics: models, experiments, and data. *Journal of Theoretical Biology* **194**:1–9. Chp 6

Cushing, J. M., Dennis, B., Desharnais, R. A. and Costantino R. F. 1998b. Moving toward an unstable equilibrium: saddle nodes in population systems. *Journal of Animal Ecology* **67**:298–306. Chp 6

Das Gupta, P. 1972. On two-sex models leading to stable populations. *Theoretical Population Biology* **3**:358–375. Chp 12

Dash, A. T. and Cressman, R. 1988. Polygamy in human and animal species. *Mathematical Bioscience* **88**:49–66. Chp 12

Davies, N. B. 1991. Mating systems. In Krebs, J. R. and Davies, N. B. eds. *Behavioural Ecology: An Evolutionary Approach*. Oxford: Blackwell, pp. 263–294. Chp 12

Dawkins, R. 1976. *The Selfish Gene*. Oxford: Oxford University Press. Chp 11

DeAngelis, D. and Gross, L. eds. 1992. *Individual-based Models and Approaches in Ecology*. New York: Chapman and Hall. Chp 6

Dennis, B., Desharnais, R. A., Cushing, J. M. and Costantino R. F. 1997. Transitions in population dynamics: equilibria to periodic cycles to aperiodic cycles. *Journal of Animal Ecology* **66**:704–729. Chp 6

Dennis, B., Desharnais, R. A., Cushing, J. M., Henson, S. M. and Costantino, R. F. 2001. Estimating chaos and complex dynamics in an insect population. *Ecological Monographs* **71**:277–303. Chp 6

Diamond, J. M. 1975. Assembly of species communities. In Cody, M. L. and Diamond, J. M. eds. *Ecology and Evolution of Communities*. Cambridge, Mass.: Harvard University Press, pp. 460–490. Chp 7

Diamond, J. M. and Case, T. J. 1986. *Community Ecology*. New York: Harper and Row. Chp 7

Dieckmann, U. 1997. Can adaptive dynamics invade? *Trends in Ecology and Evolution* **12**:128–131. Chp 1

Dieckmann, U. and Law, R. 1996. The dynamical theory of coevolution: a derivation from stochastic ecological processes. *Journal of Mathematical Biology* **34**:579–612. Chp 1

Diffendorfer, J. E. 1998. Testing models of source-sink dynamics and balanced dispersal. *Oikos* **81**:417–433. Chp 9

Doebeli, M. 1995. Dispersal and dynamics. *Theoretical Population Biology* **47**:82–106. Chp 3

Doebeli, M. 1996. Quantitative genetics and population dynamics. *Evolution* **50**:532–546. Chp 12

Doebeli, M. 1997. Genetic variation and persistence of predator-prey interactions in the Nicholson-Bailey model. *Journal of Theoretical Biology* **188**:109–120. Chp 12

Doebeli, M. and Koella, J. C. 1994. Sex and population dynamics. *Proceedings of the Royal Society of London B* **257**:17–23. Chp 12

Doncaster, C. P., Pound, G. E. and Cox, S. J. 2000. The ecological cost of sex. *Nature* **404**:281–285. Chp 12

Drake, J. A. 1991. Community-assembly mechanics and the structure of an experimental species ensemble. *American Naturalist* **137**:1–26. Chp 7

Dugatkin, L. A. 1997. *Cooperation Among Animals: an Evolutionary Perspective.* Oxford: Oxford University Press. Chp 11

Dugatkin, L. A. 1998. Game theory and cooperation. In Dukatkin, L. A. and Reeve, H. K. eds. *Game Theory and Animal Behaviour.* Oxford: Oxford University Press, pp. 38–63. Chp 11

Dugatkin, L. A. and Reeve, H. K. 1998. *Game Theory and Animal Behaviour.* Oxford: Oxford University Press. Chp 11

Earn, D. J. D., Rohani, P., Bolker, B. M. and Grenfell, B. T. 2000. A simple model for complex dynamical transitions in epidemics. *Science* **287**:667–670. Chp 5

Easterling, M. R. and Ellner, S. P. 2000. Dormancy strategies in a random environment: comparing structured and unstructured models. *Evolutionary Ecology Research* **2**:387–407. Chp 12

Edelstein-Keshet, L. 1988. *Mathematical Models in Biology.* New York: Random House. Chp 1, Chp 2

Efron, B. and Tbshirani, R. J. 1983. *An Introduction to Bootstrap.* London: Chapman & Hall. Chp 4

Ellegren, H., Gustafsson, L. and Sheldon, B. C. 1996. Sex ratio adjustment in relation to paternal attractiveness in a wild bird population. *Proceedings of the National Academy of Science USA* **93**:11 723–11 728. Chp 12

Elton, C. S. 1924. Periodic fluctuations in the numbers of animals: their causes and effects. *British Journal of Experimental Biology* **2**:119–163. Chp 2, Chp 4, Chp 8

Elton, C. S. and Nicholson, M. 1942. The ten-year cycle in numbers of the lynx in Canada. *Journal of Animal Ecology* **11**:215–244. Chp 4, Chp 5, Chp 8

Emlen, J. M. 1984. *Population Biology. The Coevolution of Population Dynamics and Behavior.* New York: Macmillan Publishing Company. Chp 2, Chp 7, Chp 12

Engen, S., Bakke, Ø. and Islam, A. 1998. Demographic and environmental stochasticity – concepts and definitions. *Biometrics* **54**:840–846. Chp 9

Erb, J., Boyce, M. S. and Stenseth, N. C. 2001. Population dynamics of large and small mammals. *Oikos* **92**:3–12. Chp 6

Fagen, R. 1987. A generalized habitat matching rule. *Evolutionary Ecology* **1**:5–10. Chp 10

Farnsworth, K. D. and Beecham, J. A. 1997. Beyond the ideal free distribution: more general models of predator distribution. *Journal of Theoretical Biology* **1187**:389–396. Chp 10

Felsenstein, J. 1976. The theoretical population genetics of variable selection and migration. *Annual Review of Genetics* **10**:253–280. Chp 12

Fisher, R. A. 1930. *The Genetical Theory of Natural Selection.* Oxford: Oxford University Press. Chp 12

Fowler, J. and Cohen, L. 1990. *Practical Statistics for Field Biology.* Chichester: Wiley. Chp 10

Frank, S. A. 1990. Sex allocation theory for birds and mammals. *Annual Review of Ecology and Systematics* **21**:13–55. Chp 12

Frean, M. and Abraham, E. R. 2001. Rock-scissors-paper and the survival of the weakest. *Proceedings of the Royal Society, London,* B **268**:1323–1327. Chp 11

Fretwell, S. D. 1972. *Populations in a Seasonal Environment.* Princeton, N. J.: Princeton University Press. Chp 10

Fretwell, S. D. and Lucas, H. L. 1970. On territorial behavior and other factors influencing habitat distribution in birds. I. Theoretical development. *Acta Biotheoretica* **19**:16–36. Chp 9, Chp 10

Fritts, H. C. 1976. *Three Rings and Climate.* London: Academic Press. Chp 4

Fromentin, J. M., Myers, R. A., Bjørnstad, O. N., Stenseth, N. C., Gjosaeter, J. and Christie, H. 2001. Effects of density-dependent and stochastic processes on the regulation of cod populations. *Ecology* **82**:567–579. Chp 6

Fryxell, J. M. and Lundberg, P. 1993. Optimal patch use and metapopulation dynamics. *Evolutionary Ecology* **7**:379–393. Chp 3, Chp 8

Fryxell, J. M. and Lundberg, P. 1997. *Individual Behavior and Community Dynamics.* London: Chapman & Hall. Chp 6, Chp 8, Chp 12

Gabriel, W. and Bürger, R. 1992. Survival of small populations under demographic stochasticity. *Theoretical Population Biology* **41**:44–71. Chp 2

Gaillard, J.-M., Boutin, J.-M., Delorme, D., Van Laere, G., Duncan, P. and Lebreton, J.-D. 1997. Early survival in roe deer: causes and consequences of cohort variation in two contrasted populations. *Oecologia* **112**:502–513. Chp 12

Gaston, K. 1996. The multiple forms of the interspecific abundance-distribution relationship. *Oikos* **76**:211–220. Chp 7

Gaston, K. J., Blackburn, T. M. and Lawton, J. H. 1997. Interspecific abundance-range size relationships: an appraisal of mechanisms. *Journal of Animal Ecology* **66**:579–601. Chp 7

Gauch, H. G. 1982. *Multivariate Analysis in Community Ecology.* Cambridge: Cambridge University Press. Chp 7

Gause, G. F. 1934. *The Struggle for Existence.* Baltimore, Md.: Williams and Wilkins. Chp 7

Gelman, A., Carlin, J. B., Stern, H. S. and Rubin, D. B. 1995. *Bayesian Data Analysis.* London: Chapman & Hall. Chp 9

Getz, W. M. and Kaitala, V. 1993. Ecogenetic analysis and evolutionary stable strategies in harvested populations. In Stokes, T. K., McGlade, J. M. and Law R. eds. *The Exploitation of Evolving Resources, Lecture Notes in Biomathematics*, vol. 99, pp. 187–203. Berlin: Springer-Verlag. Chp 9

Gillman, M. P. and Dodd, M. 2000. Detection of delayed density dependence in an orchid population 11. *Journal of Ecology* **88**:204–212. Chp 6

Gilpin, M. and Hanski, I. eds. 1991. *Metapopulation Dynamics: Empirical and Theoretical Investigations*. London: Academic Press. Chp 8

Ginzburg, L. R. and Taneyhill, D. E. 1994. Population cycles of forest Lepidoptera – a maternal effect hypothesis. *Journal of Animal Ecology* **63**:79–92. Chp 6

Gonzalez-Andujar, J. L. 1998. Effect of immigration on a chaotic insect population. *Ecological Research* **13**:259–261. Chp 3

Gonzales-Andujar, J. L. and Hughes, G. 2000. Complex dynamics in weed populations. *Functional Ecology* **14**:524–526. Chp 12

Goodman, D. 1974. Natural selection and a cost ceiling on reproductive effort. *American Naturalist* **108**:247–268. Chp 12

Gotelli, N. J. 1995. *A Primer of Ecology*. Sunderland, Mass.: Sinauer Associates. Chp 2

Grebogi, C., Ott, E. and Yorke, J. A. 1987. Chaos, strange attractors, and fractal basin boundaries in nonlinear dynamics. *Science* **238**:632–638. Chp 3

Greenman, J. V. and Benton, T. G. 2001. The impact of stochasticity on the behaviour of nonlinearpopulation models: synchrony and the Moran effect. *Oikos* **93**:343–351. Chp 2, Chp 4

Grenfell, B. T. and Bolker, B. M. 1998. Cities and villages: infection hierarchies in a measles metapopulation. *Ecology Letters* **1**:63–70. Chp 5

Grenfell, B. T. and Harwood, J. 1997. (Meta)population dynamics of infectious diseases. *Trends in Ecology and Evolution* **12**:395–399. Chp 5

Grenfell, B. T., Wilson, K., Finkenstädt, B. F., Coulson, T. N., Murray, S., Albon, S. D., Pemberton, J. M., Clutton-Brock, T. H. and Crawley, M. J. 1998. Noise and determinism in synchronized sheep dynamics. *Nature* **394**:674–677. Chp 4

Grenfell, B. T., Finkenstädt, B. F., Wilson, K. T., Coulson, T. N. and Crawley, M. J. 2000. Reply: nonlinearity and the Moran effect. *Nature* **406**:847. Chp 4

Grenfell, B. T., Bjørnstad, O. N. and Kappey, J. 2001. Travelling waves and spatial hierarchies in measles epidemics. *Nature* **414**:716–723. Chp 5, Chp 13

Grenfell, B. T., Bjørnstad, O. N. and Finkenstaedt, B. F. 2002. Dynamics of measles epidemics: scaling noise, determinism, and predictability with the TSIR model. *Ecological Monographs* **72**:185–202. Chp 5

Gurney, W. S. C. and Nisbet, R. M. 1998. *Ecological Dynamics*. New York: Oxford University Press. Chp 2

Hall, A. J., McConnell, B. J. and Barker, R. J. 2001. Factors affecting first-year survival in grey seals and their implications for life history strategy. *Journal of Animal Ecology* **70**:138–149. Chp 6

Halley, J. M. 1995. Ecology, evolution and 1/f noise. *Trends in Ecolology and Evolution* **11**:33–38. Chp 2, Chp 11

Halley, J. M. and Iwasa, Y. 1998. Extinction rate of a population under both demographic and environmental stochasticity. *Theoretical Population Biology* **53**:1–15. Chp 2

Halme, E. and Niemelä, J. 1993. Carabid beetles in fragments of coniferous forests. *Annales Zoologici Fennici* **30**:17–30. Chp 8

Hamilton, W. D. 1964a. The genetical evolution of social behaviour. I. *Journal of Theoretical Biology* **7**:1–16. Chp 10, Chp 11

Hamilton, W. D. 1964b. The genetical evolution of social behaviour. II. *Journal of Theoretical Biology* **7**:17–32. Chp 10, Chp 11

Hamilton, W. D. 1967. Extraordinary sex ratios. *Science* **156**:477–488. Chp 12

Hanski, I. 1999a. *Metapopulation Ecology. Oxford Series in Ecology and Evolution.* Oxford: Oxford University Press. Chp 3, Chp 9

Hanski, I. 1999b. Habitat connectivity, habitat continuity, and metapopulations in dynamic landscapes. *Oikos* **87**:209–219. Chp 8

Hanski, I. and Gilpin, M. 1991. Metapopulation dynamics: brief history and conceptual domain. *Biological Journal of the Linnean Society* **42**:3–16. Chp 3

Hanski, I. and Woiwod, I. P. 1993. Spatial synchrony in the dynamics of moth and aphid populations. *Journal of Animal Ecology* **62**:656–668. Chp 4

Hanski, I., Hansson, L. and Henttonen, H. 1991. Specialist predators, generalist predators, and the microtine rodent cycles. *Journal of Animal Ecology* **60**:353–367. Chp 8

Hanski, I., Turchin, P., Korpimäki, E. and Henttonen, H. 1993. Population oscillations of boreal rodents: regulation by mustelid predators leads to chaos. *Nature* **364**:232–235. Chp 8

Hansson, L. 1999. Interspecific variation in dynamics: small rodents between food and predation in changing landscapes. *Oikos* **86**:159–169. Chp 8

Hansson, L. and Henttonen, H. 1985. Gradients in density variations of small rodents: the importance of latitude and snow cover. *Oecologia* **67**:394–402. Chp 5, Chp 8

Hansson, L. and Henttonen, H. 1988. Rodent dynamics as community processes. *Trends in Ecology and Evolution* **3**:195–200. Chp 8

Hassell, M. P. 2000. Host-parasitoid population dynamics. *Journal of Animal Ecology* **69**:543–566. Chp 5

Hassell, M. P. and May, R. M. 1973. Stability in insect host-parasite models. *Journal of Animal Ecology* **42**:693–726. Chp 5

Hassell, M. P., Comins, H. N. and May, R. M. 1991. Spatial structure and chaos in insect population dynamics. *Nature* **353**:255–258. Chp 5

Hassell, M. P., Comins, H. N. and May, R. M. 1994. Species coexistence and self-organizing spatial dynamics. *Nature* **370**:290–292. Chp 5

Hastings, A. 1990. Spatial heterogeneity and ecological models. *Ecology* **71**:426–428. Chp 3

Hastings, A. 1993. Complex interaction between dispersal and dynamics: lessons from coupled logistic equations. *Ecology* **74**:1362–1372. Chp 3

Hastings, A. 1997. *Population Biology.* New York: Springer-Verlag. Chp 2

Hastings, A. and Higgins, K. 1994. Persistence of transients in spatially structured ecological models. *Science* **263**:1133–1136. Chp 3

Hawkins, B. A. and Holyoak, M. 1998. Transcontinental crashes of insect populations? *American Naturalist* **152**:480–484. Chp 4

Heino, M. 1998. Management of evolving fish stocks. *Canadian Journal of Fisheries and Aquatic Sciences* **55**:1971–1982. Chp 9

Heino, M., Kaitala, V., Ranta, E. and Lindström, J. 1997a. Synchronous dynamics and rates of extinction in spatially structured populations. *Proceedings of the Royal Society, London, B* **264**:481–486. Chp 4, Chp 7, Chp 8

Heino, M., Metz, J. A. J. and Kaitala, V. 1997b. Evolution of mixed maturation strategies in semelparous life histories: the crucial role of dimensionality of the feedback environment. *Philosophical Transactions of the Royal Society, London, B* **352**:1647–1655. Chp 13

Heino, M., Metz, J. A. J. and Kaitala, V. 1998. The enigma of frequence-dependent selection. *Trends in Ecology and Evolution* **13**:367–370. Chp 1

Heino, M., Ripa, J. and Kaitala, V. 2000. Extinction risk under coloured environmental noise. *Ecography* **23**:177–184. Chp 2

Heino, R. 1994. *Climate in Finland During the Period of Meteorological Observations.* Finnish Meteorological Institute Contributions No. 12. Chp 4

Henttonen, H. 1985. Predation causing extended low densities in microtine cycles: further evidence from shrew dynamics. *Oikos* **45**:156–157. Chp 4, Chp 8

Hilborn, R. and Mangel, M. 1997. *The Ecological Detective.* Princeton, N. J.: Princeton University Press. Chp 1, Chp 2, Chp 9

Hilborn, R. and Walters, C. J. 1992. *Quantitative Fisheries Stock Assessment.* London: Chapman & Hall. Chp 2, Chp 6, Chp 9

Hofbauer, J. and Sigmund, K. 1998. *Evolutionary Games and Population Dynamics.* Cambridge: Cambridge University Press. Chp 11

Holt, R. D. 1983a. Immigration and the dynamics of peripheral populations. In Miyata, K. and Rhodin, A. eds. *Advances in Herpetology and Evolutionary Biology.* Harvard University, Mass.: Museum of Comparative Zoology, pp. 680–694. Chp 3

Holt, R. D. 1983b. Models for peripheral populations: the role of immigration. In Freeman, H. I. and Strobeck, C. eds. *Lecture Notes in Biomathematics*, pp. 25–32. Berlin: Springer-Verlag. Chp 3

Holt, R. D. 1997. On the evolutionary stability of sink populations. *Evolutionary Ecology* **11**:723–731. Chp 9

Holt, R. D. 2002. Food webs in space: on the interplay of dynamic instability and spatial processes. *Ecological Research* **17**:261–273. Chp 3

Horn, H. S. 1978. Optimal tactics of reproduction and life-history. In Krebs, J. R. and Davies, N. B. eds. *Behavioural Ecology: An Evolutionary Approach.* Oxford: Blackwell Scientific Publications, pp. 411–429. Chp 12

Hörnfeldt, B. 1994. Delayed density dependence a determinant of vole cycles. *Ecology* **75**:791–806. Chp 8

Hubbell, S. P. 2001. *The Unified Neutral Theory of Biodiversity and Biogeography.* Princeton, N. J.: Princeton University Press. Chp 7

Hudson, P. J. 1992. *Grouse in Space and Time.* Fordingbridge: Game Conservancy Trust. Chp 4

Hudson, P. J. and Cattadori, I. M. 1999. The Moran effect: a cause of population synchrony. *Trends in Ecology and Evolution.* **14**:1–2. Chp 4

Hudson, P. J., Dobson, A. P. and Newbor, D. 1998. Prevention of population cycles by parasite removal. *Science* **282**:2256–2258. Chp 5

Hughes, N. F. 1992. Ranking of feeding positions by drift-feeding Arctic grayling (*Thymallus arcticus*) in dominance hierarchies. *Canadian Journal of Fisheries and Aquatic Sciences* **49**:1994–1998. Chp 10

Huntingford, F. A. and Turner, A. K. 1987. *Animal Conflict*. London: Chapman & Hall. Chp 10

Hutchings, J. A. 2000. Collapse and recovery of marine fishes. *Nature* **406**:882–885. Chp 9

Hutchinson, G. E. 1961. The paradox of the plankton. *American Naturalist* **95**:137–145. Chp 7

ICES. 1999. *Cooperative Research Report. No. 236*. Report of the Advisory Committee on Fisheries Management. Chp 7

Ims, R. A. and Andreassen, H. P. 2000. Spatial synchronization of vole population dynamics by predatory birds. *Nature* **408**:194–196. Chp 4

Ims, R. A. and Steen, H. 1990. Geographical synchrony in microtine population cycles: a theoretical evaluation of the role of the nomadic avian predators. *Oikos* **57**:381–387. Chp 4

Inchausti, P. and Ginzburg, L. R. 1998. Small mammals cycles in northern Europe: patterns and evidence for a maternal effect hypothesis. *Journal of Animal Ecology* **67**:180–194. Chp 6

Ives, A. R., Gross, K. and Jansen, V. A. A. 2000. Periodic mortality events in predator-prey systems. *Ecology* **81**:3330–3340. Chp 3

Jackson, A. L., Ranta, E., Lundberg, P., Kaitala, V. and Ruxton, G. D. 2004. Consumer-resource matching in a food chain when both predators and prey are free to move. *Oikos* **106**:445–450. Chp 10

Jansen, V. A. A. 2001. The dynamics of two diffusively coupled predator-prey populations. *Theoretical Population Biology* **59**:119–131. Chp 3

Jeffrey, H. J. 1992. Chaos game visualization of sequences. *Computer Graphics* **16**:25–33. Chp 5

Jenkins, D. G. and Buikema, A. L. 1998. Do similar communities develop in similar sites? A test with zooplankton structure and function. *Ecological Monographs* **68**:421–443. Chp 7

Johnson, C. N. 1998. Species extinction and the relationship between distribution and abundance. *Nature* **394**:272–274. Chp 7

Johnson, S. D. 1994. Sex ratio and population stability. *Oikos* **69**:172–176. Chp 12

Jonzén, N. and Lundberg, P. 1999. Temporally structured density-dependence and population management. *Annales Zoologici Fennici* **36**:39–44. Chp 9

Jonzén, N., Lundberg, P., Ranta, E. and Kaitala, V. 2002a. The irreducible uncertainty of the demography-environment interaction in ecology. *Proceedings of the Royal Society of London B* **269**:221–226. Chp 2, Chp 4, Chp 13

Jonzén, N., Ripa, J. and Lundberg, P. 2002b. A theory of stochastic harvesting in stochastic environments. *American Naturalist* **159**:427–437. Chp 9

Jonzén, N., Ranta, E., Lundberg, P., Kaitala, V. and Lindén, H. 2003. Harvesting induced population fluctuations? *Wildlife Biology* **9**:59–65. Chp 9

Jordan, D. W. and Smith, P. 1997. *Mathematical Techniques*, 2nd edn. Oxford: Oxford University Press. Chp 2

Jorgenson, J. T., Festa-Bianchet, M., Gaillard, J. M. and Wishart, W. D. 1997. Effects of age, sex, disease, and density on survival of bighorn sheep. *Ecology* **78**:1019–1032. Chp 12

Joshi, N. V. and Gadgil, M. 1991. On the role of refugia in promoting prudent use of biological resources. *Theoretical Population Biology* **40**:211–229. Chp 9

Kaitala, V. 2002. Travelling waves in spatial population dynamics. *Annales Zoologici Fennici* **39**:161–171. Chp 5

Kaitala, V. and Getz, W. M. 1995. Population dynamics and harvesting of semelparous species with phenotypic and genotypic variability in reproductive age, *Journal of Mathematical Biology* **33**:521–556. Chp 9

Kaitala, V. and Heino, M. 1996. Complex nonuniqueness in ecological interactions. *Proceedings of the Royal Society of London B* **263**:1011–1013. Chp 3

Kaitala, V. and Ranta, E. 1996. Red/blue chaotic power spectra. *Nature* **381**:198–199. Chp 2

Kaitala, V. and Ranta, E. 1998. Travelling wave dynamics and self-organization in spatio-temporally structured populations. *Ecology Letters* **1**:186–192. Chp 5

Kaitala, V. and Ranta, E. 2001. Is the impact of environmental noise visible in the dynamics of age-structured populations? *Proceedings of the Royal Society of London B* **268**:1769–1774. Chp 2, Chp 4

Kaitala, V., Ranta, E. and Lindström, J. 1996a. Cyclic population dynamics and random perturbations. *Journal of Animal Ecology* **65**:249–251. Chp 3, Chp 9

Kaitala, V., Ranta, E. and Lindström, J. 1996b. External perturbations and cyclic dynamics in stable populations. *Annales Zoologici Fennici* **33**:275–282. Chp 9

Kaitala, V., Lundberg, P., Ripa, J. and Ylikarjula, J. 1997a. Red, blue and green – dyeing population dynamics. *Annales Zoologici Fennici* **34**:217–228. Chp 2

Kaitala, V., Ylikarjula, J., Ranta, E. and Lundberg, P. 1997b. Population dynamics and the colour of environmental noise. *Proceedings of the Royal Society of London B* **264**:943–948. Chp 2

Kaitala, V., Ylikarjula, J. and Heino, M. 1999. Dynamic complexities in host-parasitoid interaction. *Journal of theoretical Biology* **197**:331–341. Chp 3

Kaitala, V., Ylikarjula, J. and Heino, M. 2000. Non-unique population dynamics: basic patterns. *Ecological Modelling* **135**:127–134. Chp 3

Kaitala, V., Ranta, E. and Lundberg, P. 2001a. Self-organised dynamics in spatially structured populations. *Proceedings of the Royal Society of London B* **268**:1655–1660. Chp 5, Chp 7, Chp 11

Kaitala, V., Alaja, S. and Ranta, E. 2001b. Temporal self-similarity created by spatial individual-based population dynamics. *Oikos* **94**:273–278. Chp 11

Kaitala, V., Tesar, D. and Ranta, E. 2002. Semelparity vs. iteroparity and the number of age groups, *Evolutionary Ecology Research* **4**:169–179. Chp 12

Kaitala, V., Jonzén, N. and Enberg, K. 2003. Harvesting strategies in a fish stock dominated by low-frequency variability: the Norwegian spring spawning herring (*Clupea harengus*). *Marine Resourse Economics* **18**:263–274. Chp 9

Kaitala, V., Enberg, K. and Ranta, E. 2004. Harvesting, conservation reserves, and distribution of individuals over space. *Biological Letters* **41**:3–10. Chp 9

Kalela, O. 1962. On the fluctuations in the numbers of arctic and boreal small rodents as a problem of population biology. *Annales Academiae Scientiarum Fennicae, Series A IV* **66**:1–38. Chp 8

Kareiva, P. 1990. Population dynamics in spatially complex environments: theory and data. *Philosophical Transactions of the Royal Society of London, B* **330**:175–190. Chp 3

Karlin, S. and Lessard, S. 1986. *Theoretical Studies on Sex Ratio Evolution*. Princeton, N.J.: Princeton University Press. Chp 12

Kauffman, S. A. 1993. *The Origins of Order*. Oxford: Oxford University Press. Chp 11

Kelly, D. 1994. The evolutionary ecology of mast seeding. *Trends in Ecology and Evolution* **9**:465–470. Chp 4

Kendall, B. E. and Fox, G. A. 1998. Spatial structure, environmental heterogeneity, and population dynamics: analysis of the coupled logistic map. *Theoretical Population Biology* **54**:11–37. Chp 3, Chp 10

Kendall, B. E., Prendergast, J. and Bjørnstad, O. N. 1998. The macroecology of population dynamics: taxonomic and biogeographic patterns in population cycles. *Ecology Letters* **1**:160–164. Chp 9

Kendall, B. E., Bjørnstad, O. N., Bascompte, J., Keitt, T. H. and Fagan, W. F. 2000. Dispersal, environmental correlation, and spatial synchrony in population dynamics. *American Naturalist* **155**:628–636. Chp 4

Kennedy, M. and Gray, R. D. 1993. Can ecological theory predict the distribution of foraging animals? A critical analysis of experiments on the Ideal Free Distribution. *Oikos* **68**:158–166. Chp 10

Killingback, T. and Doebeli, M. 1998. Self-organized criticality in spatial evolutionary game theory. *Journal of Theoretical Biology* **199**:135–140. Chp 11

Koenig, W. D. and Knops, J. M. 1998. Testing for spatial autocorrelation in ecological studies. *Ecography* **21**:423–429. Chp 4

Koenig, W. D. and Knops, J. M. 2000a. Scale of mast-seeding and three-ring growth. *Nature* **396**:225–226. Chp 4

Koenig, W. D. and Knops, J. M. 2000b. Patterns of annual seed production by northern hemisphere trees: a global perspective. *American Naturalist* **155**:59–69. Chp 4

Kokko, H. and Eberhardt, T. 1996. Measuring the strength of demographic stochasticity. *Journal of Theoretical Biology* **183**:169–178. Chp 9

Kokko, H. and Lindström, J. 1998. Seasonal density dependence, timing of mortality and sustainable harvesting. *Ecological Modelling* **110**:293–304. Chp 9

Kokko, H., Helle, E., Lindström, J., Ranta, E., Sipilä, T. and Courchamp, F. 1999. Backcasting population sizes of ringed and grey seals in the Baltic and Lake Saimaa during the 20th century. *Annales Zoologici Fennici* **36**:65–73. Chp 9

Kokko, H., Lindström, J. and Ranta, E. 2001. Life histories and sustainable harvesting. In Reynolds, J. D., Mace, G. M., Redford, K. H. and Robinson, J. G. eds. *Conservation of Exploited Species*. Cambridge: Cambridge University Press, pp. 301–322. Chp 9

Kokko, H., Ranta, E., Ruxton, G. and Lundberg, P. 2002. Sexually transmitted disease and the evolution of mating systems. *Evolution* **56**:1091–1100. Chp 12

Komdeur, J. and Hatchwell, B. J. 1999. Kin recognition: function and mechanism in avian societies. *Trends in Ecology and Evolution* **14**:237–241. Chp 1

Komdeur, J., Daan, S. and Tinbergen, J. 1997. Extreme adaptive modification in sex ratio of the Seychelles warbler's eggs. *Nature* **358**:522–525. Chp 12

Korpimäki, E. 1987. Clutch size, breeding success and brood size experiments in Tengmalm's owl *Aegolius funereus*: a test of hypotheses. *Ornis Scandinavica* **18**:277–284. Chp 6

Korpimäki, E. and Hakkarainen H. 1991. Fluctuating food supply affects the clutch size of Tengmalm's owl independent of laying date. *Oecologia* **85**:543–552. Chp 6

Korpimäki, E. and Norrdahl, K. 1991. Do breeding nomadic avian predators dampen population fluctuations of small mammals? *Oikos* **62**:195–208. Chp 4

Korpimäki, E. and Norrdahl, K. 1998. Experimental reduction of predators reverses the crash phase of small-rodent cycles. *Ecology* **79**:2448–2455. Chp 4, Chp 8

Kot, M. 1992. Discrete-time travelling waves – ecological examples. *Journal of Mathematical Biology* **30**:413–436. Chp 5

Krebs, C. J. 1972. *Ecology. The Experimental Analysis of Distribution and Abundance.* New York: Harper and Row. Chp 10

Krebs, C. J., Boutin, S., Boonstra, R., Sinclair, A. R. E., Smith, J. N. M., Dale, M. R. T. and Turkington, R. 1995. Impact of food and predation on the snowshoe hare cycle. *Science* **269**:1112–1115. Chp 8

Krebs, C. J., Boonstra, R., Boutin, S. and Sinclair, A. R. E. 2001. What drives the 10-year cycle of snowshoe hares? *Bioscience* **51**:25–35. Chp 4

Kruuk, L. E. B., Clutton-Brock, T. H., Albon, S. D., Pemberton, J. M. and Guinness, F. E. 1999. Population density affects sex ratio variation in red deer. *Nature* **399**:459–461. Chp 12

Kuikka, S., Hildén, M., Gislason, H., Hansson, S., Sparholt, H. and Varis, O. 1999. Modelling environmentally driven uncertainties in Baltic cod (*Gadus morhua*) management by Bayesian influence diagrams. *Canadian Journal of Fisheries and Aquatic Sciences* **56**:629–641. Chp 7

Kvarnemo, C., Moore, G. I., Jones, A. G., Nelson, W. S. and Avise, J. C. 2000. Monogamous pair bonds and mate switching in the Western Australian seahorse *Hippocampus subelongatus*. *Journal of Evolutionary Biology* **13**:882–888. Chp 12

Laakso, J., Kaitala, V. and Ranta, E. 2001. How does environmental variation translate into biological processes? *Oikos* **92**:119–122. Chp 2, Chp 4

Lack, D. 1968. *Ecological Adaptations for Breeding in Birds.* London: Chapman and Hall. Chp 12

Lambin, X., Elston, D. A., Petty, S. J. and MacKinnon, J. L. 1998. Spatial asynchrony and periodic travelling waves in cyclic populations of field voles. *Proceedings of the Royal Society of London B* **265**:1491–1496. Chp 5

Lande, R. 1987. Extinction thresholds in demographic models of territorial populations. *American Naturalist* **130**:624–635. Chp 9

Lande, R. 1993. Risk of population extinction from demographic and environmental stochasticity, and random catastrophes. *American Naturalist* **142**:911–927. Chp 2, Chp 9

Lande, R. 1998. Demographic stochasticity and Allee effect on a scale with isotropic noise. *Oikos* **83**:353–358. Chp 9

Lande, R., Engen, S. and Sæther, B.-E. 1995. Optimal harvesting of fluctuating populations with a risk of extinction. *American Naturalist* **145**:728–745. Chp 9

Lande, R., Sæther, B.-E. and Engen, S. 1997. Threshold harvesting for sustainability of fluctuating resources. *Ecology* **78**:1341–1350. Chp 9

Lande, R., Engen, S. and Sæther, B.-E. 1999. Spatial scale of population synchrony: environmental correlation versus dispersal and density regulation. *American Naturalist* **154**:271–281. Chp 4

Lande, R., Sæther, B.-E. and Engen, S. 2001. Sustainable exploitation of fluctuating populations. In Reynolds, J. D., Mace, G. M., Redford, K. H. and Robinson, J. G. eds. *Conservation of Exploited Species.* Cambridge: Cambridge University Press, pp. 67–86. Chp 9

Lauck, T., Clark, C. W., Mangel, M. and Munro, G. R. 1998. Implementing the precautionary principle in fisheries management through marine reserves. *Ecological Applications* **8**:72–78. Chp 9

Law, R. 1999. Theoretical aspects of community assembly. In McGlade, J. ed. *Advanced Ecological Theory*. Oxford: Blackwell Science, pp. 143–171. Chp 7

Law, R. 2000. Fishing, selection, and phenotypic evolution. *ICES Journal of Marine Science* **57**:659–669. Chp 9

Law, R. 2001. Phenotypic and genetic changes due to selective exploitation. In Reynolds, J. D., Mace, G. M., Redford, K. H. and Robinson, J. G. eds. *Conservation of Exploited Species*. Cambridge: Cambridge University Press, pp. 323–342. Chp 9

Law, R. and Grey, D. R. 1989. Evolution of yields from populations with age-specific cropping. *Evolutionary Ecology* **3**:343–359. Chp 9

Law, R. and Morton, R. D. 1996. Permanence and the assembly of ecological communities. *Ecology* **77**:762–775. Chp 7

Law, R., Weatherby, A. J. and Warren, P. H. 2000. On the invasibility of persistent protist communities. *Oikos* **88**:319–326. Chp 7

Lawton, J. H., Nee, S., Letcher, A. J. and Harvey, P. H. 1994. In Edwards, P. J., May, R. M. and Webb, N. R. eds. *Large-scale Ecology and Conservation Biology*. London: Blackwell Scientific, pp. 41–58. Chp 7

Lefkovitch, L. P. 1965. The study of population growth in organisms grouped by stages. *Biometrics* **21**:1–18. Chp 2

Legendre, P. and Fortin M.-J. 1989. Spatial pattern and ecological analysis. *Vegetatio* **80**:107–138. Chp 5

Leimar, O. and Hammerstein, P. 2001. Evolution of cooperation through indirect reciprocity. *Proccedings of the Royal Society*, London, B **268**:745–753. Chp 11

Leopold, A. 1933. *Game Management*. New York: Charles Scribner's Sons. Chp 9

Leslie, P. H. 1945. On the use of matrices in certain population mathematics. *Biometrika* **33**:183–212. Chp 2, Chp 6

Leslie, P. H. 1948. Some further notes on the use of matrices in population mathematics. *Biometrika* **35**:213–245. Chp 2, Chp 6

Leslie, P. H. 1959. The properties of a certain lag type of population growth and the influence of an external random factor on a number of such populations. *Physiological Zoology* **32**:151–159. Chp 4, Chp 6, Chp 9

Leslie, P. H. and Gower, J. C. 1958. The properties of a stochastic model for two competing species. *Biometrika* **45**:316–330. Chp 8

Leslie, P. H. and Gower, J. C. 1960. The properties of a stochastic model for the predator-prey type of interaction between two species. *Biometrika* **47**:219–234. Chp 4, Chp 8, Chp 9

Levins, R. 1968. *Evolution in Changing Environments*. Princeton, N.J.: Princeton University Press. Chp 7, Chp 10, Chp 12

Levins, R. 1969. Some demographic and genetic consequences of environmental heterogeneity for biological control. *Bulletin Entomological Society of America* **15**:237–240. Chp 3

Liebholt, A., Koenig, W. D. and Bjørnstad, O. N. 2004. Spatial synchrony in population dynamics. *Annual Review of Ecology, Evolution, and Systematics* **35**:467–490. Chp 4

Lima, S. L. and Zollner, P. A. 1996. Towards a behavioural ecology of ecological landscapes. *Trends in Ecology and Evolution* **11**:131–135. Chp 6

Lindén, H. 1981. Changes in Finnish tetraonid populations and some factors influencing mortality. *Finnish Game Research* **30**:3–11. Chp 6, Chp 8

Lindén, H. 1988. Latitudinal gradients in predator-prey interactions, cyclicity and synchronism in voles and small game populations in Finland. *Oikos* **52**:341–349. Chp 4

Lindén, H. 1989. Characteristics of tetraonid cycles in Finland. *Finnish Game Research* **46**:34–42. Chp 4

Lindén, H. 1991. Patterns of grouse shooting in Finland. *Ornis Scandinavica* **22**:241–244. Chp 9

Lindén, H. and Rajala, P. 1981. Fluctuations in long-term trends in the relative densities of tetraonid populations in Finland, 1964–77. *Finnish Game Research* **39**:13–34. Chp 4, Chp 6

Lindström, J. 1994. Tetraonid population studies – state of the art. *Annales Zoologici Fennici* **31**:347–364. Chp 8

Lindström, J. 1996. *Modelling grouse population dynamics.* Ph.D.thesis. Finland: University of Helsinki. Chp 9

Lindström, J. 1999. Early development and fitness in birds and mammals. *Trends in Ecology and Evolution* **14**:343–347. Chp 6

Lindström, J. and Kokko, H. 1998. Sexual reproduction and population dynamics: the role of polygyny and demographic sex differences. *Proceedings of the Royal Society of London B* **265**:483–488. Chp 12

Lindström, J., Ranta, E., Kaitala, V. and Lindén, H. 1995. The clockwork of Finnish tetraonid population dynamics. *Oikos* **74**:185–194. Chp 4, Chp 6, Chp 8

Lindström, J., Ranta, E. and Lindén, H. 1996. Large-scale synchrony in the dynamics of Capercaillie, Black Grouse and Hazel Grouse populations in Finland. *Oikos* **76**:221–227. Chp 4

Lindström, J., Kokko, H. and Ranta, E. 1997a. Detecting periodicity in short and noisy time series data. *Oikos* **78**:406–410. Chp 4

Lindström, J., Ranta, E., Lindén, M. and Lindén, H. 1997b. Reproductive output, population structure and cyclic dynamics in Capercaillie, Black Grouse and Hazel Grouse. *Journal of Avian Biology* **28**:1–8. Chp 6

Lindström, J., Kokko, H., Ranta, E. and Lindén, H. 1999. Density dependence and the response surface methodology. *Oikos* **85**:40–52. Chp 4, Chp 6, Chp 8

Lindström, J., Ranta, E., Kokko, H., Lundberg, P. and Kaitala, V. 2001. From arctic lemmings to adaptive dynamics: Charles Elton's legacy in population ecology. *Biological Reviews* **76**:129–158. Chp 4, Chp 5, Chp 8

Lloyd, A. L. 1995. The coupled logistic map – a simple model for the effects of population dynamics. *Journal of Theoretical Biology* **173**:217–230. Chp 3

Lockwood, D. R., Hastings, A. and Botsford, L. W. 2002. The effects of dispersal patterns on marine reserves: does the tail wag the dog? *Theoretical Population Biology* **61**:297–309. Chp 9

Ludwig, D. 1996. The distribution of population survival times. *American Naturalist* **147**:506–526. Chp 2, Chp 9

Ludwig, D. 1998. Management of stocks that may collapse. *Oikos* **83**:397–402. Chp 9

Lummaa, V., Merilä, J. and Kause, A. 1998. Adaptive sex ratio varation in pre-industrial human (*Homo sapiens*) populations? *Proceedings of the Royal Society of London B* **265**:563–568. Chp 12

Lundberg, P. and Jonzén, N. 1999a. Optimal population harvesting in a source-sink environment. *Evolutionary Ecology Research* **1**:719–729. Chp 9

Lundberg, P. and Jonzén, N. 1999b. Spatial population dynamics and the design of marine reserves. *Ecology Letters* **2**:129–134. Chp 9

Lundberg, P., Ranta, E., Ripa, J. and Kaitala, V. 2000a. Population variability in space and time. *Trends in Ecology and Evolution* **15**:460–464. Chp 3

Lundberg, P., Ranta, E., Kaitala, V. and Jonzén, N. 2000b. Coexistence and resource competition. *Nature* **407**:694. Chp 7

Lundberg, P., Ranta, E. and Kaitala, V. 2000c. Species loss leads to community closure. *Ecology Letters* **3**:465–468. Chp 7

Lundberg, P., Ripa, J., Kaitala, V. and Ranta, E. 2001. Visibility of demography-modulating noise in population dynamics. *Oikos* **96**:379–382. Chp 2, Chp 4

MacArthur, R. H. 1972. *Geographical Ecology*. Princeton, N. J.: Princeton University Press. Chp 7

MacCall, A. D. 1990. *Dynamic Geography of Marine Fish Populations*. University of Washington Press, Seattle: Washington Sea Grant Program, pp. 153. Chp 9

Mackenzie, J. M. D. 1952. Fluctuations in the numbers of British tetraonids. *Journal of Animal Ecology* **21**:128–153. Chp 4

Mackinnon, J. L., Petty, S. J., Elston, D. A., Thomas, C. J., Sherratt, T. N. and Lambin, X. 2001. Scale invariant spatio-temporal patterns of field vole density. *Journal of Animal Ecology* **70**:101–111. Chp 5

Mangel, M. 1994. Spatial patterning in resource exploitation and conservation. *Philosophical Transactions of the Royal Society, London B* **343**:93–98. Chp 9

Mangel, M. 1998. No-take areas for sustainability of harvested species and a conservation invariant for marine reserves. *Ecology Letters* **1**:87–90. Chp 9

Mangel, M. 2000a. On the fraction of habitat allocated to marine reserves. *Ecology Letters* **3**:15–22. Chp 9

Mangel, M. 2000b. Irreducible uncertainties, sustainable fisheries and marine reserves. *Evolutionary Ecology Research* **2**:547–557. Chp 9

Marcström, V., Höglund, N. and Krebs, C. 1990. Periodic fluctuations in small mammals at Boda, Sweden from 1961 to 1988. *Journal of Animal Ecology* **59**:753–761. Chp 4

Matter, S. F. 1996. Interpatch movement of the red milkweed beetle, *Tetraopes tetraophthalmus* – individual responses to patch size and isolation. *Oecologia* **105**:447–453. Chp 12

May, R. M. 1974. Biological populations with non-overlapping generations: stable points, stable cycles, and chaos. *Science* **186**:645–747. Chp 2, Chp 7

May, R. M. 1975. *Stability and Complexity in Model Ecosystems*, 2nd edn. Princeton, N. J.: Princeton University Press. Chp 2, Chp 7

May, R. M. 1976. Simple mathematical models with very complicated dynamics. *Nature* **261**:459–467. Chp 2

May, R. M. 1981. The evolution of cooperation. *Nature* **292**:291–292. Chp 11

May, R. M. 1987. More evolution of cooperation. *Nature* **327**:15–16. Chp 11

356 · References

May, R. M., Beddington, J. R., Horwood, J. W. and Shepherd, J. G. 1978. Exploiting natural populations in an uncertain world. *Mathematical Biosciences* **42**:219–252. Chp 9

Maynard Smith, J. M. 1974. *Models in Ecology*. Cambridge: Cambridge University Press. Chp 4

Maynard Smith, J. M. 1982. *Evolution and the Theory of Games*. Cambridge: Cambridge University Press. Chp 10, Chp 11, Chp 12

Maynard Smith, J. and Price, G. R. 1973. The logic of animal conflict. *Nature* **246**:15–18. Chp 11

McCallum, H. I. 1992. Effects of immigration on chaotic population dynamics. *Journal of Theoretical Biology* **154**:177–184. Chp 3

McCullough, D. R. 1996. Spatially structured populations and harvest theory. *Journal of Wildlife Management* **60**:1–9. Chp 9

Meszéna, G. and Metz, J. A. J. 1999. *Species diversity and population regulation: the importance of environmental feedback dimensionality*. IIASA Report IR–99–045. Laxenburg, Austria: IAASA. Chp 13

Metcalfe, N. B. and Monaghan, P. 2001. Compensation of a bad start: grow now, pay later? *Trends in Ecology and Evolution* **16**:254–260. Chp 6

Metz, J. A. J. and Diekmann, O. 1986. *The Dynamics of Physiologically Structured Populations*. New York: Springer-Verlag. Chp 6

Metz, J. A. J., Nisbet, R. M. and Gertiz, S. A. H. 1992. How should we define "fitness" for general ecological scenarios? *Trends in Ecology and Evolution* **7**:198–202. Chp 12, Chp 13

Milinski, M. 1979. Evolutionarily stable feeding strategy in sticklebacks. *Zeitschrift für Tierpsychologie* **51**:36–40. Chp 10

Milinski, M. 1994. Ideal free theory predicts more than only input matching – critique of Kennefy and Gray's review. *Oikos* **71**:163–166. Chp 10

Milinski, M. and Parker, G. A. 1991. Competition for resources. In Krebs, J. R. and Davies, N. B. eds. *Behavioural Ecology*. Oxford: Blackwell Scientific Publications, pp. 137–168. Chp 10

Milinski, M., Seemann, D., Bakker, T. C. M. and Krambeck, H.-J. 2001. Cooperation through indirect reciprocy: image scoring or standing strategy? *Proceedings from the Royal Society London B* **268**:2495–2501. Chp 11

Moran, P. A. P. 1950a. Some remarks on animal population dynamics. *Biometrika* **6**:250–258. Chp 2

Moran, P. A. P. 1950b. Notes on continuous stochastic phenomena. *Biometrika* **37**:17–23. Chp 5

Moran, P. A. P. 1953a. The statistical analysis of the Canadian lynx cycle. I. Structure and prediction. *Australian Journal of Zoology* **1**:163–173. Chp 4

Moran, P. A. P. 1953b. The statistical analysis of the Canadian lynx cycle. II. Synchronization and meteorology. *Australian Journal of Zoology* **1**:291–298. Chp 4, Chp 5, Chp 10

Morris, D. W. 1988. Habitat-dependent population regulation and community structure. *Evolutionary Ecology* **2**:253–269. Chp 10

Morris, D. W. 1994. Habitat matching: alternatives and implications to populations and communities. *Evolutionary Ecology* **8**:387–406. Chp 10

Morris, D. W., Lundberg, P. and Ripa, J. 2001. Hamilton's rule confronts ideal free habitat selection. *Proceedings of the Royal Society London B* **268**:921–924. Chp 10

Morton, R. D., Law, R., Pimm, S. L. and Drake, J. A. 1996. On models for assembling ecological communities. *Oikos* **75**:493–499. Chp 7

Moss, R., Elston, D. A. and Watson, A. 2000. Spatial asynchrony and demographic travelling waves during Red Grouse population cycles. *Ecology* **81**:981–989. Chp 5

Mousseau, T. A. and Fox, C. W. 1998. The adaptive significance of maternal effect. *Trends in Ecology and Evolution* **13**:403. Chp 6

Murphy, G. I. 1968. Pattern in life history and the environment. *American Naturalist* **102**:391–403. Chp 12

Murray, B. G. Jr. 1999. Can the population regulation controversy be buried and forgotten? *Oikos* **84**:148–152. Chp 2

Murray, B. G. Jr. 2001. Are ecological and evolutionary theories scientific? *Biological Reviews* **76**:255–289. Chp 1

Myers, J. H. 1998. Synchrony in outbreaks of forest lepidoptera: a possible example of the Moran effect. *Ecology* **79**:1111–1117. Chp 4

Myers, R. A. and Hoenig, J. M. 1997. Direct estimates of gear selectivity from multiple tagging experiments. *Canadian Journal of Fisheries and Aquatic Sciences* **54**:1–9. Chp 9

Myers, R. A., Mertz, G. and Birdson, J. 1997a. Spatial scales of interannual recruitmen variations of marine, anadromous, and freshwater fish. *Canadian Journal of Fisheries and Aquatic Sciences* **54**:1400–1407. Chp 4, Chp 6

Myers, R. A., Bradford, M. J., Bridson, J. M. and Mertz, G. 1997b. Estimating delayed density-dependent mortality in sockeye salmon (*Oncorhynchus nerka*): a meta-analytic approach. *Canadian Journal of Fisheries and Aquatic Sciences* **54**:2449–2462. Chp 6

Myers, R. A., Hutchings, J. A. and Barrowman, N. J. 1997c. Why do fish stocks collapse? *Ecological Applications* **7**:91–106. Chp 7

Nakano, S. 1995. Individual differences in resource use, growth and emigration under the influence of a dominance hierarchy in fluvial red-spotted masu salmon in a natural habitat. *Journal of Animal Ecology* **64**:75–84. Chp 10

Nason, G. P. and von Sachs, R. 1999. Wavelets in time series analysis. *Philosophical Transactions of the Royal Society of London A* **357**:2511–2526. Chp 5

Nicholson, A. J. and Bailey, V. A. 1935. The balance of animal populations. *Proceedings of the Zoological Society of London* **3**:551–598. Chp 5

Nisbet, R. M. and Gurney, W. S. C. 1982. *Modelling Fluctuating Populations*. Chichester: John Wiley and Sons. Chp 9

Nowak, M. A. and May, R. M. 1992. Evolutionary games and spatial chaos. *Nature* **359**:826–829. Chp 11

Nowak, M. A. and Sigmund, K. 1998. Evolution of indirect reciprocity by image scoring. *Nature* **393**:573–577. Chp 11

Olsen, E. M., Heino, M., Lilly, G. R., Morgan, M. J., Brattey, J., Ernande, B. and Dieckmann, U. 2004. Maturation trends indicative of rapid evolution preceded the collapse of northern cod. *Nature* **428**:32–935. Chp 9

O'Riain, M., Bennett, N., Brotherton, P., McIlrath, G. and Clutton-Brock, T. H. 2000. Reproductive suppression and inbreeding avoidance in wild populations of

358 · References

co-operatively breeding meerkats (*Suricata suricatta*). *Behavioral Ecology and Sociobiology* **48**:471–477. Chp 6

Palmer, A. R. 2000. Quasi-replication and the contract of error: lessons from sex ratios, heritabilities, and fluctuating asymmetry. *Annual Review of Ecology and Systematics* **31**:441–480. Chp 12

Palmqvist, E., Lundberg, P. and Jonzén, N. 2000. Linking resource matching and dispersal. *Evolutionary Ecology* **14**:1–12. Chp 3, Chp 10

Paradis, E. 1997. Metapopulations and chaos: on the stabilizing influence of dispersal. *Journal of Theoretical Biology* **186**:261–266. Chp 3

Paradis, E., Baillie, S., Sutherland, W. and Gregory, R. D. 2000. Spatial synchrony in populations of birds: effects of habitat, population trend, and spatial scale. *Ecology* **81**:2112–2125. Chp 4

Park, T. 1948. Experimental studies of interspecific competition. 1. Competition between populations of the flour beetle *Tribolium confusum* Duval and *Tribolium castaneoum* Herbst. *Ecological Monographs* **18**:267–307. Chp 6

Parker, G. A. 1974. The reproductive behaviour and the nature of sexual selection in *Scatophaga stercoraria* L. (Diptera: Scatophagia). IX. Spatial distribution of fertilization rates and evolution of male search strategy within the reproductive area. *Evolution* **28**:93–108. Chp 10

Parker, G. A. and Sutherland, W. J. 1986. Ideal free distributions when individuals differ in competitive ability: phenotype-limited ideal free models. *Animal Behaviour* **34**:1222–1242. Chp 10

Patterson, K. R. 1999. Evaluating uncertainty in harvest control law catches using Bayesian Markov chain Monte Carlo virtual population analysis with adaptive rejection resampling and including structural uncertainty. *Canadian Journal of Fisheries and Aquatic Sciences* **56**:208–221. Chp 9

Petersen, C. W. and Levitan, D. R. 2001. The Allee effect: a barrier to recovery by exploited species. In Reynolds, J. D., Mace, G. M., Redford, K. H. and Robinson, J. G. eds. *Conservation of Exploited Species.* Cambridge: Cambridge University Press, pp. 281–300. Chp 9

Peterson, R. 1997. *Ecological studies of wolves on Isle Royale. Annual Report 1996–1997.* Houghton, Mich.: Isle Royale Natural History Association. Chp 9

Pielou, E. C. 1974. *Population and Community Ecology.* New York: Gordon and Breach Science Publishers. Chp 7

Pimm, S. L. 1980. Food web design and the effect of species deletion. *Oikos* **35**:139–149. Chp 7

Pimm, S. L. 1982. *Food Webs.* London: Chapman & Hall. Chp 7

Pimm, S. L. 1991. *The Balance of Nature?* Chicago, Ill.: The University of Chicago Press. Chp 7

Pimm, S. L. and Redfearn, A. 1988. The variability of animal populations. *Nature* **334**:616–614. Chp 2

Plaistow, S. and Siva-Jothy, M. T. 1999. The ontogenetic switch between odonate life history stages: effects on fitness when time and food are limited. *Animal Behaviour* **58**:659–667. Chp 6

Poethke, H. J. and Hovestadt, T. 2002. Evolution of density- and patch-size-dependent dispersal rates. *Proceedings of the Royal Society of London B* **269**:637–645. Chp 3

Pollard, E. 1991. Synchrony in population fluctuations: the dominant influence of widespread factors on local butterfly populations. *Oikos* **60**:7–10. Chp 4

Poole, R. W. 1974. *An Introduction to Quantitative Ecology.* London: McGraw-Hill. Chp 9

Post, E. and Stenseth, N. C. 1999. Climatic variability, plant phenology and northern ungulates. *Ecology* **80**:1322–1339. Chp 4

Post, E., Forchhammer, M. C., Stenseth, N. C. and Callaghan, T. V. 2001. The timing of life-history events in changing climate. *Proceedings of the Royal Society London B* **268**:15–23. Chp 4

Post, W. M. and Pimm, S. L. 1983. Community assembly and food web stability. *Mathematical Biosciences* **64**:169–192. Chp 7

Potts, G. R., Tapper, S. C. and Hudson, P. J. 1984. Population fluctuations in red grouse: analysis of bag records and a simulation model. *Journal of Animal Ecology* **53**:21–36. Chp 9

Powell, T. M. and Steel, J. H. eds. 1995. *Ecological Time Series.* New York: Chapman and Hall. Chp 2

Priestly, M. B. 1981. *Spectral Analysis and Time Series.* London: Academic Press. Chp 2

Pulliam, H. R. 1988. Sources, sinks and population regulation. *American Naturalist* **132**:652–661. Chp 3

Pulliam, H. R. and Caraco, T. 1984. Living in groups: is there an optimal group size? In Krebs, J. R. and Davies, N. B. eds. *Behavioural Ecology*, 2nd edn. Oxford: Blackwell Scientific Publications, pp. 122–147. Chp 10

Quinn, T. J. II and Deriso, R. B. 1999. *Quantitative Fish Dynamics.* Oxford: Oxford University Press. Chp 2, Chp 9

Ranta, E. and Kaitala, V. 1997. Travelling waves in vole population dynamics. *Nature* **390**:456. Chp 5

Ranta, E. and Kaitala, V. 1999. Punishment of polygyny. *Proceedings of the Royal Society of London B* **266**:2337–2341. Chp 12

Ranta, E. and Kaitala, V. 2000. Resource matching and population dynamics in a two-patch system. *Oikos* **91**:507–511. Chp 4, Chp 10

Ranta, E., Kaitala, V., Lindström, J. and Lindén, H. 1995a. Synchrony in population dynamics. *Proceedings of the Royal Society London B* **262**:113–118. Chp 3, Chp 4, Chp 5, Chp 8, Chp 10

Ranta, E., Lindström, J. and Lindén, H. 1995b. Synchrony in tetraonid population dynamics. *Journal of Animal Ecology* **64**:767–776. Chp 4

Ranta, E., Kaitala, V. and Lundberg, P. 1997a. The spatial dimension of population fluctuations. *Science* **278**:1621–1623. Chp 3, Chp 4, Chp 5, Chp 8

Ranta, E., Kaitala, V. and Lindström, J. 1997b. Dynamics of Canadian lynx populations in space and time. *Ecography* **20**:454–460. Chp 3, Chp 4, Chp 8

Ranta, E., Kaitala, V. and Lindström, J. 1997c. Spatial dynamics of populations. In Bascompte, J. and Solé, R. V. eds. *Modelling Spatiotemporal Dynamics in Ecology.* New York: Springer-Verlag, pp. 45–60. Chp 3, Chp 4, Chp 8, Chp 10

Ranta, E., Lindström, J., Kaitala, V., Kokko, H., Lindén, H. and Helle, E. 1997d. Solar activity and hare dynamics – a cross-continental comparison. *American Naturalist* **149**:765–775. Chp 4

Ranta, E., Kaitala, V., Lindström, J. and Helle, E. 1997e. Moran effect and synchrony in population dynamics. *Oikos* **78**:136–142. Chp 4, Chp 10

360 · **References**

Ranta, E., Kaitala, V. and Lindström, J. 1997f. External disturbances and population dynamics. *Annales Zoologici Fennici* **34**:127–132. Chp 4

Ranta, E., Kaitala, V. and Lundberg, P. 1998. Population variability in space and time: the dynamics of synchronous population fluctuations. *Oikos* **83**:376–382. Chp 4, Chp 8

Ranta, E., Kaitala, V. and Lindström, J. 1999a. Spatially autocorrelated disturbances and patterns in population synchrony. *Proceedings of the Royal Society London B* **266**:1851–1956. Chp 4, Chp , Chp 8, Chp 10

Ranta, E., Kaitala, V. and Lindström, J. 1999b. Synchronicity in population dynamics: cause and consequence mixed. *Trends in Evolution and Ecology* **14**:400–401. Chp 5, Chp 8

Ranta, E., Lundberg, P. and Kaitala, V. 1999c. Resource matching with limited knowledge. *Oikos* **86**:383–385. Chp 10

Ranta, E., Kaitala, V. and Lindström, J. 1999d. Sex in space: population dynamic consequences. *Proceedings of the Royal Society of London B* **266**:1155–1160. Chp 12

Ranta, E., Lundberg, P., Kaitala, V. and Laakso, J. 2000a. Visibility of the environmental noise modulating population dynamics. *Proceedings of the Royal Society of London B* **267**:1851–1856. Chp 2, Chp 4

Ranta, E., Lundberg, P. and Kaitala, V. 2000b. Size of environmental grain and resource matching. *Oikos* **89**:573–576. Chp 10

Ranta, E., Lummaa, V., Kaitala, V. and Merilä, J. 2000c. Spatial dynamics of adaptive sex ratios. *Ecology Letters* **3**:30–34. Chp 12

Ranta, E., Kaitala, V., Tesar, D. and Alaja, S. 2000d. Complex dynamics and evolution of semelparous and iteroparous reproductive strategies. *American Naturalist* **155**:294–300. Chp 12

Ranta, E., Tesar, D., Alaja, S. and Kaitala, V. 2000e. Does evolution of iteroparous and semelparous reproduction call for spatially structured systems? *Evolution* **54**:145–150. Chp 12

Ranta, E., Tesar, D. and Kaitala, V. 2001. Local extinctions promote coexistence of semelparous and iteroparous life histories. *Evolutionary Ecology Research* **3**:759–766. Chp 12

Ratner, S. and Lande, R. 2001. Demographic and evolutionary responses to selective harvesting in populations with discrete generations. *Ecology* **82**:3093–3104. Chp 9

Rees, M. 1994. Delayed germination of seeds: a look at the effects of adult longevity, the timing of reproduction. and population age/stage structure. *American Naturalist* **144**:43–64. Chp 12

Rees, M. and Long, M. J. 1992. Germination biology and the ecology of annual plants. *American Naturalist* **139**:484–508. Chp 12

Restrepo, V. R., Hoenig, J. M., Powers, J. E., Baird, J. W. and Turner, S. C. 1992. A simple simulation approach to risk and cost analysis, with applications to swordfish and cod fisheries. *Fishery Bulletin* **90**:736–748. Chp 9

Reynolds, J. D., Mace, G. M., Redford, K. H. and Robinson, J. G. eds. 2001. *Conservation of Exploited Species*. Cambridge: Cambridge University Press. Chp 9

Ricker, W. E. 1954. Stock and recruitment. *Journal of the Fisheries Board of Canada* **2**:559–623 Chp 2, Chp 3

Rijnsdorp, A. D. 1993. Fisheries as a large-scale experiment on life-history evolution: disentangling phenotypic and genetic effects in changes in maturation and reproduction of North Sea plaice, *Pleuronectes platessa* L. *Oecologia* **96**:391–401. Chp 9

Riolo, R. L., Cohen, M. D. and Axelrod, R. 2001. Evolution of cooperation without reciprocity. *Nature* **414**:441–443. Chp 11

Ripa, J. 2000. Analysing the Moran effect and dispersal: their significance and interaction in synchronous population dynamics. *Oikos* **89**:175–187. Chp 4

Ripa, J. and Heino, M. 1999. Linear analysis solves two puzzles in population dynamics: the route to extinction and extinction in coloured environments. *Ecology Letters* **2**:219–222. Chp 2, Chp 9

Ripa, J. and Lundberg, P. 1996. Noise colour and the risk of population extinctions. *Proceedings of the Royal Society of London B* **263**:1751–1753 Chp 2, Chp 6

Ripa, J. and Lundberg, P. 2000. The route to extinction in variable environments. *Oikos* **90**:89–96. Chp 2, Chp 9

Ripa, J., Lundberg, P. and Kaitala, V. 1998. A general theory of environmental noise in ecological food webs. *American Naturalist* **151**:256–263. Chp 2

Roff, D. A. 1992. *The Evolution of Life Histories.* New York: Chapman & Hall. Chp 2, Chp 6, Chp 11, Chp 12

Rohani, P. and Miramontes, O. 1995a. Immigration and the persistence of chaos in population models. *Journal of Theoretical Biology* **175**:203–206. Chp 3

Rohani, P. and Miramontes, O. 1995b. Host-parasitoid metapopulations – the consequences of parasitoid aggregation on spatial dynamics and searching efficiency. *Proceedings of the Royal Society of London B* **260**:335–342. Chp 5

Rohner, C. 1996. The numerical response of great horned owls during a peak and decline of the snowshoe hare cycle. *Journal of Animal Ecology* **65**:359–370. Chp 6

de Roos, A. M., Diekmann, O. and Metz, J. A. J. 1992. Studying the dynamics of structured population models: a versatile technique and its applications to Daphnia. *American Naturalist* **139**:123–147. Chp 6

Rose, M. R. and Bradley, T. J. 1998. Evolutionary physiology of the cost of reproduction. *Oikos* **83**:443–451. Chp 6

Roughgarden, J. 1971. Density-dependent natural selection. *Ecology* **52**:453–468. Chp 12

Roughgarden, J. 1998. *Primer of Theoretical Ecology.* Princeton, N. J.: Prentice-Hall. Chp 1, Chp 2

Royama, T. 1981. Fundamental concepts and methodology for the analysis of animal population dynamics, with particular reference to univoltine species. *Ecological Monographs* **51**:473–493. Chp 2

Royama, T. 1992. *Analytical Population Dynamics.* New York: Chapman & Hall. Chp 1, Chp 2, Chp 4, Chp 5, Chp 9

Rusak, J. A., Norman, D. Y., Somers, K. M. and McQueen, D. J. 1999. The temporal coherence of zooplankton population abundances in neighboring north-temperate lakes. *American Naturalist* **153**:46–58. Chp 4

Ruxton, G. D. 1994. Low levels of immigration between chaotic populations can reduce system extinction by reducing asynchronous regular cycles. *Proceedings of the Royal Society of London B* **256**:189–193. Chp 3

Ruxton, G. D. 1995a. The effect of emigration and immigration on the dynamics of a discrete-generation population. *Journal of Bioscience* **20**:397–407. Chp 3

Ruxton, G. D. 1995b. Population models with sexual reproduction show a reduced propensity to exhibit chaos. *Journal of Theoretical Biology* **175**:595–601. Chp 12

Ruxton, G. D. 1996a. Density-dependent migration and stability in a system of linked populations. *Bulletin of Mathematical Biology* **58**:643–660. Chp 3

Ruxton, G. D. 1996b. Synchronisation between individuals and the dynamics of linked populations. *Journal of Theoretical Biology* **183**:47–54. Chp 10

Ruxton, G. D. and Doebeli, M. 1996. Spatial self-organization and persistence of transients in a metapopulation model. *Proceedings of the Royal Society of London B* **263**:1153–1158. Chp 3

Ruxton, G. D. and Humphries, S. 1999. Multiple ideal free distributions of unequal competitors. *Evolutionary Ecology Research* **1**:635–640. Chp 10

Ruxton, G. D. and Rohani, P. 1998. Population floors and the persistence of chaos in ecological models. *Theoretical Population Biology* **53**:175–183. Chp 3

Ruxton, G. D., Gonzales-Andujar, J. L. and Perry, J. N. 1997. Mortality during dispersal and the stability of a metapopulation. *Journal of Theoretical Biology* **186**:389–396. Chp 10

Ruxton, G. D., Armstrong, J. D. and Humphries, S. 1999. Modelling territorial behaviour of animals in variable environments. *Animal Behaviour* **58**:113–120. Chp 10

Sæther, B.-E., Engen, S. and Lande, R. 1996. Density-dependence and optimal harvesting of fluctuating populations. *Oikos* **76**:40–46. Chp 9

Sæther, B.-E., Ringsby, T. H., Bakke, O. and Solberg, E.-J. 1999. Spatial and temporal variation in demography of a house sparrow metapopulation. *Journal of Animal Ecology* **68**:628–637. Chp 9

Satake, A. and Iwasa, Y. 2000. Pollen coupling of forest trees: forming synchronized and periodic reproduction out of chaos. *Journal of Theoretical Biology* **203**:63–84. Chp 4

Savill, N. J. and Hogeweg, P. 1999. Competition and dispersal in predator-prey waves. *Journal of Theoretical Biology* **56**:243–263. Chp 5

Schmitz, O. J., Beckerman, A. P. and O'Brien, K. M. 1997. Behaviourally mediated trophic cascades: effects of predation risk on food web interactions. *Ecology* **78**:1388–1399. Chp 6

Schoen, R. 1983. Relationships in a simple harmonic mean two-sex fertility model. *Journal of Mathematical Biology* **18**:201–211. Chp 12

Scott, F. A. M. & Grant, A. 2004 Visibility of the impact of environmental noise: a response to Kaitala and Ranta. *Proceedings of the Royal Society of London B* **271**:1119–1124. Chp 2

Sherratt, J. A. 2001. Periodic travelling waves in cyclic predator-prey system. *Ecology Letters* **4**:30–37. Chp 5

Sherratt, J. A., Lambin, X., Thomas, C. J. and Sherratt, T. N. 2002. Generation of periodic waves by landscape features in cyclic predator-prey system. *Proceedings of the Royal Society of London B* **269**:327–334. Chp 5

Sherratt, T. N., Lambin, X., Petty S. J., Mackinnon, J. L., Coles, C. F. and Thomas, C. J. 2000. Use of coupled oscillator models to understand synchrony and travelling waves in populations of the field vole *Microtus agrestis* in northern England. *Journal of Applied Ecology* **37**:148–158. Chp 5

Shigesada, N., Kawasaki, K. and Teramoto, E. 1986. Travelling periodic waves in heterogeneous environments. *Theoretical Population Biology* **30**:143–160. Chp 5

Shikesada, N. and Kawasaki, K. 1997. *Biological Invasions: Theory and Practice.* Oxford: Oxford University Press. Chp 5

Sigmund, K. 1993. *Games of Life: Explorations in Ecology, Evolution and Behaviour.* Oxford: Oxford University Press. Chp 11

Silvertown, J. 1988. The demographic and evolutionary consequences of seed dormancy. In Davy, A. J., Hutchings, M. J. and Watkinson, A. R. eds. *Plant Population Ecology.* Oxford: Blackwell Scientific Publication, pp. 205–219. Chp 12

Silvertown, J. and Lovett Doust, J. 1993. *Introduction to Plant Population Biology.* Oxford: Blackwell Scientific Publication. Chp 12

Sinervo, B. and Lively, C. M. 1996. The rock-paper-scissors game and the evolution of alternative male strategies. *Nature* **380**:240–243. Chp 11

Sinervo, B., Svensson, E. and Comendant, T. 2000. Density cycles and an offspring quality and quantity game driven by natural selection. *Nature* **406**:985–988. Chp 12

Sinha, S. and Parthasarathy, S. 1996. Unusual dynamics of extinction in an unusual ecological model. *Proceedings of the National Academy of Sciences* **93**:1504–1508. Chp 3

Small, R. J., Marcström, V. and Willebrand, T. 1993. Synchronous and non-synchronous population fluctuations of some predators and their prey in central Sweden. *Ecography* **16**:360–364. Chp 4

Smith, C. H. 1983. Spatial trends in Canadian snowshoe hare, *Lepus americanus*, population cycles. *Canadian Field-Naturalist* **97**:151–160. Chp 4, Chp 5, Chp 8

Sokal, R. R. and Oden, F. M. L. S. 1978. Spatial autocorrelation in biology. 1. Methodology. *Biological Journal of the Linnean Society* **10**:199–228. Chp 5

Solé, R. V. and Bascompte, J. 1993. Chaotic Turing structures. *Physics Letters A* **179**:325–331. Chp 5

Solé, R. V. and Gamarra, J. G. P. 1998. Chaos, dispersal and extinction in coupled ecosystems. *Journal of Theoretical Biology* **193**:539–541. Chp 3

Solé, R. V. and Valls, J. 1991. Order and chaos in a 2D Lotka-Volterra coupled map lattice. *Physics Letters A* **153**:330–336. Chp 5

Solé, R. V., Bascompte, J. and Valls, J. 1992a. Stability and complexity of extended 2-species competition. *Journal of Theoretical Biology* **159**:469–480. Chp 5

Solé, R. V., Valls, J. and Bascompte, J. 1992b. Spiral waves, chaos and multiple attractors in lattice models of interacting populations. *Physics Letters A* **166**:123–128. Chp 5

Stearns, S. C. 1976. Life history tactics: a review of the ideas. *Quarterly Review of Biology* **51**:3–47. Chp 12

Stearns, S. 1992. *The Evolution of Life Histories.* Oxford: Oxford University Press. Chp 6, Chp 11, Chp 12

Stearns, S. C. 2000. Life history evolution: successes, limitations, and prospects. *Naturwissenshaften* **87**:476–486. Chp 13

Steel, J. H. 1985. A comparison of terrestrial and marine ecological systems. *Nature* **313**:355–358. Chp 2

Steen, H., Yoccoz, N. G. and Ims, R. A. 1990. Predators and small rodent cycles: an analysis of 79-year time series of small rodent population fluctuations. *Oikos* **59**:115–120. Chp 4

Stenseth, N. C. 1977. Evolutionary aspects of demographic cycles: the relevance of some models of cycles for microtine fluctuations. *Oikos* **29**:525–538. Chp 8

Stenseth, N. C. 1999. Population cycles in voles and lemmings: density dependence and phase dependence in a stochastic world. *Oikos* **87**:427–461. Chp 2, Chp 4, Chp 5, Chp 8, Chp 9

Stenseth, N. C., Chan, K.-S., Framstad, E. and Tong, H. 1998a. Phase- and density-dependent population dynamics in Norwegian lemmings: interaction between deterministic and stochastic processes. *Proceedings from the Royal Society of London B* **265**:1957–1968. Chp 2, Chp 8

Stenseth, N. C., Falck, W., Chan, K. S., Bjørnstad, O. N., O'Donoghue, M., Tong, H., Boonstra, R., Boutin, S., Krebs, C. J. and Yoccoz, N. G. 1998b. From patterns to processes: phase and density dependencies in the Canadian lynx cycle. *Proceedings of the National Academy of Sciences of the USA* **95**:15430–15435. Chp 6, Chp 8

Stenseth, N. C., Bjørnstad, O. N. and Saitoh, T. 1998c. Seasonal forcing on the dynamics of *Clethrionomys rufocanus*: modeling geographic gradients in population dynamics. *Research on Population Ecology* **40**:85–95. Chp 8

Stenseth, N. C., Chan, K.-S., Tong, H., Boonstra, R., Boutin, S., Krebs, C. J., Post, E., O'Donoghue, M., Yoccoz, N. G., Forchhammer, M. C. and Hurrell, J. W. 1999. Common dynamic structure of Canada lynx populations within three climatic regions. *Science* **285**:1071–1073. Chp 5, Chp 8

Stephens, P. A., Sutherland, W. J. and Freckleton, R. P. 1999. What is the Allee effect? *Oikos* **87**:185–190. Chp 9

Stokes, T. K., McGlade, J. M. and Law, R. eds. 1993. *The Exploitation of Evolving Resources. Lecture Notes in Biomathematics 99*. Berlin: Springer-Verlag. Chp 9

Stone, L. and Hart, D. 1999. Effects of immigration on the dynamics of simple population models. *Theoretical Population Biology* **55**:227–234. Chp 3

Strong, D. R., Simberloff, D., Abele, L. G. and Thistle, A. B. 1984. *Ecological Communities*. Princeton, N. J.: Princeton University Press. Chp 7

Sutcliffe, O. L., Thomas, C. D. and Moss, D. 1996. Spatial synchrony and asynchrony in butterfly population dynamics. *Journal of Animal Ecology* **65**:85–95. Chp 4

Sutherland, W. J. 1983. Aggregation and the "ideal free" distribution. *Journal of Animal Ecology* **52**:821–828. Chp 10

Sutherland, W. J. 1996. *From Individual Behaviour to Population Ecology*. Oxford: Oxford University Press. Chp 6, Chp 9, Chp 10

Sutherland, W. J. and Parker, G. A. 1985. Distribution of unequal competitors. In Sibly, R. M. and Smith, R. H. eds. *Behavioural Ecology: Ecological Consequences of Adaptive Behaviour*. Oxford: Blackwell Scientific Publications. pp. 255–274. Chp 10

Supriatna, A. K. and Possingham, H. P. 1998. Optimal harvesting for a predator-prey metapopulation. *Bulletin of Mathematical Biology* **60**:49–65. Chp 9

Svensson, E. and Nilsson, J.-Å. 1996. Mate quality affects offspring sex ratio in blue tits. *Proceedings of the Royal Society of London B* **263**:357–361. Chp 12

Svensson, E. and Sinervo, B. 2000. Experimental excursions on adaptive landscapes: density-dependent selection on egg-size. *Evolution* **54**:1396–1403. Chp 12

Svensson, E., Sinervo, B. and Comendant, T. 2001a. Condition, genotype-by-environment interaction, and correlational selection in lizard life-history morphs. *Evolution* **55**:2053–2069. Chp 12

Svensson, E., Sinervo, B. and Comendant, T. 2001b. Density-dependent competition and selection on immune function in genetic lizard morphs. *Proceedings from the National Academy of Science* **98**:12561–12565. Chp 12

Templeton, A. R. and Lawlor, L. R. 1981. The fallacy of averages in ecological optimization theory. *American Naturalist* **117**:390–393. Chp 3

Tesar, D., Kaitala, V. and Ranta, E. 2002. Stochasticity and spatial coexistence of semelparity and iteroparity as life histories. *Evolutionary Ecology* **15**:193–204. Chp 12

Thomas, C. D. 1991. Spatial and temporal variability in a butterfly population. *Oecologia* **87**:577–580. Chp 4

Thomas, C. D. and Kunin, W. E. 1999. The spatial structure of populations. *Journal of Animal Ecology* **68**:647–657. Chp 3

Tilman, D. 1999. The ecological consequences of changes in biodiversity: a search for general principles. *Ecology* **80**:1455–1474. Chp 7

Tilman, D. and Kareiva, P. eds. 1997. *Spatial Ecology. The Role of Space in Population Dynamics and Interspecific Interactions.* Princeton, N. J.: Princeton University Press. Chp 3, Chp 8

Tokeshi, M. 1999. *Species Coexistence.* Oxford: Blackwell Science. Chp 7

Tregenza, T. 1995. Building on the ideal free distribution. *Advances in Ecological Research* **26**:253–307. Chp 10

Trivers, R. 1971. The evolution of reciprocal altruism. *Quaterly Review in Biology* **46**:35–57. Chp 11

Trivers, R. L. and Williard, D. E. 1973. Natural selection of parental ability to vary the sex ratio of offspring. *Science* **191**:249–263. Chp 12

Tuck, G. N. and Possingham, H. P. 1994. Optimal harvesting strategies for a meta-population. *Bulletin of Mathematical Biology* **56**:107–127. Chp 9

Tuck, G. N. and Possingham, H. P. 2000. Marine protected areas for spatially structured exploited stocks. *Marine Ecology Progress Series* **192**:89–101. Chp 9

Tuljapurkar, S. and Caswell, H. eds. 1997. *Structured Population Models in Marine, Terrestrial, and Freshwater Systems.* London: Chapman & Hall. Chp 6, Ch 13

Turchin, P. 1990. Rarity of density dependence or population regulation with lags? *Nature* **344**:660–663. Chp 3, Chp 5, Chp 6

Turchin, P. 1998. *Quantitative Analysis of Movement. Measuring and Modeling Population Redistribution in Animals and Plants.* Sunderland, Mass: Sinauer Associates. Chp 8

Turchin, P. 1999. Population regulation: a synthetic view. *Oikos* **84**:153–159. Chp 2, Chp 13

Turchin, P. 2001. Does population ecology have general laws? *Oikos* **94**:17–26. Chp 6

Turchin, P. and Batzli, G. O. 2001. Availability of food and population dynamics of arvicoline rodents. *Ecology* **82**:1521–1534. Chp 2

Turchin, P. and Hanski, I. 1997. An empirically based model for latitudinal gradient in vole population dynamics. *American Naturalist* **149**:842–874. Chp 8

Turchin, P., Taylor, A. D. and Reeve, J. D. 1999. Dynamical role of predators in population cycles of forest insects: an experimental test. *Science* **285**:1068–1071. Chp 2, Chp 6

Veiga, J. P. 1990. Sexual conflict in the house sparrow: interference between poly-gynously mated females versus asymmetric male investment. *Behavioral Ecology and Sociobiology* **27**:345–350. Chp 12

Wade, M. J. and Goodnight, C. J. 1998. Perspective. The theories of Fisher and Wright in the context of metapopulations: when nature does many small experiments. *Evolution* **52**:1537–1553. Chp 12

Wade, M. J. and Kalisz, S. 1990. The causes of natural selection. *Evolution* **44**:1947–1955. Chp 12

Wade, P. R. 2001. The conservation of species in an uncertain world: novel methods and the failure of traditional techniques. In Reynolds, J. D., Mace, G. M., Redford, K. H. and Robinson, J. G. eds. *Conservation of Exploited Species*. Cambridge: Cambridge University Press, pp. 110–143. Chp 9

Walters, C. J. and Parma, A. M. 1996. Fixed exploitation rate strategies for coping with effects of climate change. *Canadian Journal of Fisheries and Aquatic Sciences* **53**:148–158. Chp 9

Waser, P. M., Elliot, L. F., Creel, N. M. and Creel, S. R. 1995. Habitat variation and mongoose demography. In Sinclair, A. R. E. and Arcese, P. eds. *Serengeti II: Dynamics, Management and Conservation of an Ecosystem*. Chicago, Ill.: University of Chicago Press, pp. 421–447. Chp 6

Watkinson, A. R. and Sutherland, W. J. 1995. Sources, sinks and pseudosinks. *Journal of Animal Ecology* **64**:126–130. Chp 9

Watson, A., Moss, R. and Rae, S. 1998. Population dynamics of Scottish Rock Ptarmigan cycles. *Ecology* **79**:1174–1192. Chp 6

Watson, A., Moss, R. and Rothery, P. 2000. Weather and synchrony in 10-year population cycles of Rock Ptarmigan and Red Grouse in Scotland. *Ecology* **81**:2126–2136. Chp 6

Watt, K. E. F. 1968. *Ecology and Resource Management*. Toronto: McCraw-Hill Book Company. Chp 4

Weiher, E. and Keddy, P. A. 1999. *Ecological Assembly Rules*. Cambridge: Cambridge University Press. Chp 7

Wellborn, G. A., Skelly, D. K. and Werner, E. E. 1996. Mechanisms creating community structure across a freshwater habitat gradient. *Annual Review of Ecology and Systematics* **27**:337–363. Chp 6

Werner, E. E. and Anholt, B. R. 1993. Ecological consequences of the trade-off between growth and mortality rates mediated by foraging activity. *American Naturalist* **142**:242–272. Chp 6

Werner, E. E. and Gilliam, J. F. 1984. The ontogenetic niche and species interactions in size structured populations. *Annual Review of Ecology and Systematics* **15**:393–425. Chp 6

White, T. C. R. 2001. Opposing paradigms: regulation and limitation of populations. *Oikos* **93**:148–152. Chp 2

Wilbur, H. M. 1980. Complex life cycles. *Annual Review of Ecology and Systematics* **11**:67–93. Chp 6

Wilbur, H. M. 1996. Multistage life cycles. In Rhodes, O. J., Chesser, R. K. and Smith M. H. eds. *Population Dynamics in Ecological Space and Time*. Chicago, Ill.: University of Chicago Press, pp. 754–108. Chp 3

Williams, G. C. 1966. *Adaptation and Natural Selection: A Critique of Some Current Evolutionary Thought*. Princeton, N. J.: Princeton University Press. Chp 11

Williams, D. W. and Liebhold, A. M. 2000. Spatial synchrony of spruce budworm outbreaks in Eastern North America. *Ecology* **81**:2753–2766. Chp 4

Wilson, D. S. 1975. The adequacy of body size as a niche difference. *American Naturalist* **109**:769–784. Chp 6

Wilson, D. S. 1980. *The Natural Selection of Populations and Communities.* Menlo Park: Benjamin Cummings. Chp 11

Wilson, D. S. 1983. The group selection controversy: history and current status. *Annual Review of Ecology and Systematics* **14**:159–187. Chp 11

Wilson, W. G., Lundberg, P., Vázquez, D. P., Shurin, J. P., Smith, M., Langford, W., Gross, K. L. and Mittelbach, G. G. 2003. Biodiversity and species interactions: extending Lotka-Volterra community theory. *Ecology Letters* **6**:944–952. Chp 7.

Wood, S. N. 1997. Inverse problems and structured-population dynamics. In Tuljapurkar, S. and Caswell, H. eds. *Structured Population Models in Marine, Terrestrial, and Freshwater Systems.* London: Chapman & Hall, pp. 555–586 Chp 2

Wood, S. N. and Thomas, M. B. 1996. Space, time and persistence of virulent pathogens. *Proceedings of the Royal Society of London B* **263**:673–680. Chp 5

Wright, S. 1932. The roles of mutation, inbreeding, crossbreeding and selection in evolution. *Proceedings from the 6th International Congress of Genetics* **1**:356–366. Chp 12

Wright, S. 1948. On the roles of directed and random changes in gene frequency in the genetics of populations. *Evolution* **2**:279–294. Chp 12

Wynne-Edwards, V. C. 1962. *Animal Dispersion in Relation to Social Behaviour.* Edinburgh: Oliver and Boyd, Chp 11

Ydenberg, R. C. 1987. Nomadic predators and geographical synchrony in microtine population cycles. *Oikos* **50**:270–272. Chp 4

Ylikarjula, J., Alaja, S., Laakso, J. and Tesar, D. 2000. Effects of patch number and dispersal patterns on population dynamics and synchrony. *Journal of Theoretical Biology* **207**:377–387. Chp 3, Chp 10

Yodzis, P. 1989. *Introduction to Theoretical Ecology.* New York: Harper and Row. Chp 2, Chp 6

# Index